Current Advances in Anaerobic Digestion Technology

Current Advances in Anaerobic Digestion Technology

Editors

Marcell Nikolausz
Jörg Kretzschmar

MDPI • Basel • Beijing • Wuhan • Barcelona • Belgrade • Manchester • Tokyo • Cluj • Tianjin

Editors

Marcell Nikolausz
Helmholtz Centre for
Environmental Research -UFZ
Germany

Jörg Kretzschmar
DBFZ Deutsches Biomasseforschungszentrum gemeinnützige
GmbH (German Biomass Research Centre)
Germany

Editorial Office
MDPI
St. Alban-Anlage 66
4052 Basel, Switzerland

This is a reprint of articles from the Special Issue published online in the open access journal *Bioengineering* (ISSN 2306-5354) (available at: https://www.mdpi.com/journal/bioengineering/special_issues/Anaerobic_Digest).

For citation purposes, cite each article independently as indicated on the article page online and as indicated below:

LastName, A.A.; LastName, B.B.; LastName, C.C. Article Title. *Journal Name* **Year**, *Volume Number*, Page Range.

ISBN 978-3-0365-0222-9 (Hbk)
ISBN 978-3-0365-0223-6 (PDF)

Contents

About the Editors . . . . **vii**

Marcell Nikolausz and Jörg Kretzschmar
Anaerobic Digestion in the 21st Century
Reprinted from: *Bioengineering* **2020**, *7*, 157, doi:10.3390/bioengineering7040157 . . . . **1**

Matia Mainardis, Marco Buttazzoni and Daniele Goi
Up-Flow Anaerobic Sludge Blanket (UASB) Technology for Energy Recovery: A Review on
State-of-the-Art and Recent Technological Advances
Reprinted from: *Bioengineering* **2020**, *7*, 43, doi:10.3390/bioengineering7020043 . . . . **5**

**Amir Izzuddin Adnan, Mei Yin Ong, Saifuddin Nomanbhay, Kit Wayne Chew
and Pau Loke Show**
Technologies for Biogas Upgrading to Biomethane: A Review
Reprinted from: *Bioengineering* **2019**, *6*, 92, doi:10.3390/bioengineering6040092 . . . . **35**

Arpit H. Bhatt and Ling Tao
Economic Perspectives of Biogas Production via Anaerobic Digestion
Reprinted from: *Bioengineering* **2020**, *7*, 74, doi:10.3390/bioengineering7030074 . . . . **59**

**Harald Wedwitschka, Daniela Gallegos Ibanez, Franziska Schäfer, Earl Jenson
and Michael Nelles**
Material Characterization and Substrate Suitability Assessment of Chicken Manure for Dry
Batch Anaerobic Digestion Processes
Reprinted from: *Bioengineering* **2020**, *7*, 106, doi:10.3390/bioengineering7030106 . . . . **79**

Prativa Mahato, Bernard Goyette, Md. Saifur Rahaman and Rajinikanth Rajagopal
Processing High-Solid and High-Ammonia Rich Manures in a Two-Stage (Liquid-Solid)
Low-Temperature Anaerobic Digestion Process: *Start-Up and Operating Strategies*
Reprinted from: *Bioengineering* **2020**, *7*, 80, doi:10.3390/bioengineering7030080 . . . . **95**

Liane Müller, Nils Engler, Kay Rostalsky, Ulf Müller, Christian Krebs and Sandra Hinz
Influence of Enzyme Additives on the Rheological Properties of Digester Slurry and on
Biomethane Yield
Reprinted from: *Bioengineering* **2020**, *7*, 51, doi:10.3390/bioengineering7020051 . . . . **111**

**Britt Schumacher, Timo Zerback, Harald Wedwitschka, Sören Weinrich, Josephine Hofmann
and Michael Nelles**
The Influence of Pressure-Swing Conditioning Pre-Treatment of Cattle Manure on
Methane Production
Reprinted from: *Bioengineering* **2020**, *7*, 6, doi:10.3390/bioengineering7010006 . . . . **125**

Florian Monlau, Cecilia Sambusiti and Abdellatif Barakat
Comparison of Dry Versus Wet Milling to Improve Bioethanol or Methane Recovery from Solid
Anaerobic Digestate
Reprinted from: *Bioengineering* **2019**, *6*, 80, doi:10.3390/bioengineering6030080 . . . . **139**

Jan Moestedt, Maria Westerholm, Simon Isaksson and Anna Schnürer
Inoculum Source Determines Acetate and Lactate Production during Anaerobic Digestion of
Sewage Sludge and Food Waste
Reprinted from: *Bioengineering* **2020**, *7*, 3, doi:10.3390/bioengineering7010003 . . . . **153**

Ali Hosseini Taleghani, Teng-Teeh Lim, Chung-Ho Lin, Aaron C. Ericsson and Phuc H. Vo
Degradation of Veterinary Antibiotics in Swine Manure via Anaerobic Digestion
Reprinted from: *Bioengineering* **2020**, *7*, 123, doi:10.3390/bioengineering7040123 . . . . **173**

**Lauren Russell, Paul Whyte, Annetta Zintl, Stephen Gordon, Bryan Markey, Theo de Waal,
Enda Cummins, Stephen Nolan, Vincent O'Flaherty, Florence Abram, Karl Richards,
Owen Fenton and Declan Bolton**
A Small Study of Bacterial Contamination of Anaerobic Digestion Materials and Survival in
Different Feed Stocks
Reprinted from: *Bioengineering* **2020**, *7*, 116, doi:10.3390/bioengineering7030116 . . . . **201**

Alessandro Chiumenti, Giulio Fait, Sonia Limina and Francesco da Borso
Performances of Conventional and Hybrid Fixed Bed Anaerobic Reactors for the Treatment of
Aquaculture Sludge
Reprinted from: *Bioengineering* **2020**, *7*, 63, doi:10.3390/bioengineering7030063 . . . . **213**

About the Editors

Marcell Nikolausz is a senior scientist at the Department of Environmental Microbiology of the Helmholtz Centre for Environmental Research – UFZ, Germany. His current topics of research include utilization of lignocellulose-rich agricultural wastes by applying microorganisms from high-performance natural systems, assessment of the methanogenic pathways by using molecular biological techniques and stable isotope tools, energetic utilization of waste products of the Brazilian bioethanol industry, phytoremediation combined with biogas technology for the energetic utilization of contaminated landscapes, biomethanation of hydrogen from excess electricity with mixed cultures (power-to-gas), and anaerobic treatment of nitrogen-rich waste substrates. He has been involved in several international collaborative projects with Hungary, Brazil, Turkey, and China, and he has been frequently invited to give scientific talks at international conferences and workshops. He has supervised several M.Sc. and Ph.D. projects.

Jörg Kretzschmar is head of the working group "Process Monitoring and Simulation" at the Department of Biochemical Conversion at the DBFZ Deutsches Biomasseforschungszentrum gemeinnützige GmbH in Leipzig, Germany. He studied molecular biotechnology (B.Sc.) and environmental protection (Dipl. Ing.). In 2017, he received his Ph.D. in the field of applied microbial electrochemistry from the TU Dresden. He does research in anaerobic digestion, microbial electrochemistry, and bioprocess engineering. His key interests are (1) the development of technologies and processes at the interface of anaerobic digestion and microbial electrochemistry and (2) the monitoring and development of anaerobic bioprocesses.

 *bioengineering*

Editorial

Anaerobic Digestion in the 21st Century

Marcell Nikolausz [1],* and Jörg Kretzschmar [2]

1   Department of Environmental Microbiology, Helmholtz Centre for Environmental Research-UFZ, Permoserstrasse 15, 04318 Leipzig, Germany
2   Department of Biochemical Conversion, DBFZ Deutsches Biomasseforschungszentrum gemeinnützige GmbH, Torgauer Strasse 116, 04347 Leipzig, Germany; Joerg.Kretzschmar@dbfz.de
*   Correspondence: marcell.nikolausz@ufz.de

Received: 2 December 2020; Accepted: 2 December 2020; Published: 7 December 2020

Despite being a mature biotechnological process, anaerobic digestion is still attracting considerable research attention, mainly due to its versatility both in substrate and product spectra, as well as being a perfect test system for the microbial ecology of anaerobes. This Special Issue highlights some key topics of this research field.

Anaerobic digestion (AD) originally refers to biomass degradation under anoxic conditions in both natural and engineered systems. AD is one of the oldest biotechnologies used to produce an energy carrier, i.e., biogas, from organic waste. Therefore, it can be considered as one of the earliest approaches to a circular bioeconomy. Technological development was sparse until the beginning of the 20th century; however, with increasing industrial interest, research into anaerobic microbial processes was also intensified with the aim of identifying the important process parameters and to promote methane production from all kinds of residual organic matter, especially agricultural residues such as manure and slurry. Technological progress has been made, particularly with the development of the UASB reactor at the end of 1970s [1], which facilitated AD of substrates with a very low content of total solids such as municipal or industrial wastewater (reviewed in this Special Issue by Mainardis et al. [2]). From the industrial perspective of electrical power production, with 19 GW installed capacity worldwide and 6,586 TWh electrical power production in 2018, biogas plants are major players even if not reaching the podium of top three renewables. The majority of biogas plants are located in Asia (40%), Europe (20%) and North America (19%).

Extending the definition of AD, in addition to solid and liquid substrates, it can also convert gases rich in hydrogen and carbon dioxide into methane via hydrogenotrophic methanogenesis. This pathway can be used for biological upgrading of biogas, as reviewed in this Special Issue by Adnan and co-workers [3]. The resulting methane can substitute natural gas, which opens new opportunities by a direct link to traditional petrochemistry. Due to the wide substrate spectrum, AD is an ideal end-of-pipe technology for waste treatment and energy recovery in several (bio)technological applications. Furthermore, AD can be coupled with emerging biotechnological applications, such as microbial electrochemical technologies or the production of medium chain fatty acids by anaerobic fermentation. Ultimately, because of the wide range of applications, AD is still a very vital field in science. This is impressively shown by the number of scientific publications in 2019, which has been, with more than 2800 publications, the year with the most contributions in this field since the beginning of records in 1945. In the last five years, 12,529 papers on AD were published, accounting for 49.54 % of the total publications on that topic (Web of Science, www.webofknowledge.com, accessed 05.11.2020).

However, today, the AD sector faces new challenges, such as limited feedstock availability at increased price, the reduction of subsidies as well as the low competitiveness of the current products. Therefore, the techno-economical assessment of current and future technologies is important for investors in the waste management sector, which is addressed by Bhatt and Tao in this issue [4]. To avoid competition with food and feed production, the AD feedstock spectrum is currently being extended to waste products either rich in recalcitrant lignocellulose or containing inhibitory substances such as ammonia (see the studies of Wedwitschka et al. [5] and Mahato et al. [6] in this Special Issue). The development and evaluation of various pretreatment technologies for lignocellulosic biomass is a hot topic of AD research that several articles in this Special Issue deal with (see the studies of Müller et al. [7], Schumacher et al. [8], and Monlau et al. [9]). The effect of the inoculum on the microbial community structure and performance of the AD process is still an enigma. The study of Moestedt et al. [10] in this issue sheds some light on the microbiology of process inoculation and start-up, which was handled as a black box in the past. Although academic knowledge about the microbiome, the engine driving the AD process, has been accumulating, the use of this knowledge for the innovation of AD technologies is still scarce. With the rapid development of novel sequencing technologies, we also expect changes on that and the emergence of new reactor systems or technology concepts based on ecological knowledge in the future.

The fate of veterinary antibiotics, microorganisms resistant to antibiotics, resistance genes and pathogenic microorganisms in AD is a further important topic due to the massive application of antibiotics in livestock farming (see the studies of Hosseini Taleghani et al. [11] and Russel at al. [12]). We see AD plants more as treatment options than a threat, when the process parameters are properly adjusted to maximize attenuation. Aquacultures are also hotspots of direct antibiotics usage or indirect input from untreated manure and human wastes that are still applied in many Asian shrimp and tilapia farms. In general, the sludge from aquacultures is a very specific and problematic waste but AD technology can also contribute to its treatment (an example of reactor system development is presented in this issue by Chiumenti and co-workers [13]).

Germany is one of the European leaders in biogas technology, with regards to the number of large-scale plants and their installed capacity, partially due to the generous subsidy system of the German energy transition (Renewable Energy Sources Act). However, this support has gradually decreased in recent years. This situation, in addition to comparably high feedstock prices, enhances the competition with other renewables. Otherwise, AD plants are able to provide power on demand, thus balancing the fluctuations in power generation from wind turbines and photovoltaics. Therefore, the AD plants of the 21st century should be more flexible in terms of power generation, the substrate as well as the product spectra.

All of these examples highlight that there is still an enormous potential in AD to be an important engine of new biorefinery concepts and renewable power generation and to contribute substantially to greenhouse gas reduction as well as to a circular bioeconomy.

Funding: This research received no external funding.

Conflicts of Interest: The authors declare no conflict of interest.

References

1. Lettinga, G.; Van Velsen, A.F.M.; Hobma, S.W.; De Zeeuw, W.; Klapwijk, A. Use of the upflow sludge blanket (USB) reactor concept for biological wastewater treatment, especially for anaerobic treatment. *Biotechnol. Bioeng.* **1980**, *22*, 699–734. [CrossRef]
2. Mainardis, M.; Buttazzoni, M.; Goi, D. Up-Flow Anaerobic Sludge Blanket (UASB) Technology for Energy Recovery: A Review on State-of-the-Art and Recent Technological Advances. *Bioengineering* **2020**, *7*, 43. [CrossRef] [PubMed]
3. Adnan, A.I.; Ong, M.Y.; Nomanbhay, S.; Chew, K.W.; Show, P.L. Technologies for Biogas Upgrading to Biomethane: A Review. *Bioengineering* **2019**, *6*, 92. [CrossRef] [PubMed]
4. Bhatt, A.H.; Tao, L. Economic Perspectives of Biogas Production via Anaerobic Digestion. *Bioengineering* **2020**, *7*, 74. [CrossRef] [PubMed]
5. Wedwitschka, H.; Ibanez, D.G.; Schäfer, F.; Jenson, E.; Nelles, M. Material Characterization and Substrate Suitability Assessment of Chicken Manure for Dry Batch Anaerobic Digestion Processes. *Bioengineering* **2020**, *7*, 106. [CrossRef] [PubMed]

6. Mahato, P.; Goyette, B.; Rahaman, S.; Rajagopal, R. Processing High-Solid and High-Ammonia Rich Manures in a Two-Stage (Liquid-Solid) Low-Temperature Anaerobic Digestion Process: Start-Up and Operating Strategies. *Bioengineering* **2020**, *7*, 80. [CrossRef] [PubMed]

7. Müller, L.; Engler, N.; Rostalsky, K.; Müller, U.; Krebs, C.; Hinz, S.W.A. Influence of Enzyme Additives on the Rheological Properties of Digester Slurry and on Biomethane Yield. *Bioengineering* **2020**, *7*, 51. [CrossRef] [PubMed]

8. Schumacher, B.; Zerback, T.; Wedwitschka, H.; Weinrich, S.; Hofmann, J.; Nelles, M. The Influence of Pressure-Swing Conditioning Pre-Treatment of Cattle Manure on Methane Production. *Bioengineering* **2019**, *7*, 6. [CrossRef] [PubMed]

9. Monlau, F.; Sambusiti, C.; Barakat, A. Comparison of Dry Versus Wet Milling to Improve Bioethanol or Methane Recovery from Solid Anaerobic Digestate. *Bioengineering* **2019**, *6*, 80. [CrossRef] [PubMed]

10. Moestedt, J.; Sánchez-Laforga, A.M.; Isaksson, S.; Schnürer, A. Inoculum Source Determines Acetate and Lactate Production during Anaerobic Digestion of Sewage Sludge and Food Waste. *Bioengineering* **2019**, *7*, 3. [CrossRef] [PubMed]

11. Taleghani, A.H.H.; Lim, T.-T.; Lin, C.-H.; Ericsson, A.C.; Vo, P.H. Degradation of Veterinary Antibiotics in Swine Manure via Anaerobic Digestion. *Bioengineering* **2020**, *7*, 123. [CrossRef] [PubMed]

12. Russell, L.; Whyte, P.; Zintl, A.; Gordon, S.; Markey, B.; De Waal, T.; Cummins, E.; Nolan, S.; O'Flaherty, V.; Abram, F.; et al. A Small Study of Bacterial Contamination of Anaerobic Digestion Materials and Survival in Different Feed Stocks. *Bioengineering* **2020**, *7*, 116. [CrossRef] [PubMed]

13. Chiumenti, A.; Fait, G.; Limina, S.; Da Borso, F. Performances of Conventional and Hybrid Fixed Bed Anaerobic Reactors for the Treatment of Aquaculture Sludge. *Bioengineering* **2020**, *7*, 63. [CrossRef] [PubMed]

Publisher's Note: MDPI stays neutral with regard to jurisdictional claims in published maps and institutional affiliations.

Review

Up-Flow Anaerobic Sludge Blanket (UASB) Technology for Energy Recovery: A Review on State-of-the-Art and Recent Technological Advances

Matia Mainardis *, Marco Buttazzoni and Daniele Goi

Department Polytechnic of Engineering and Architecture (DPIA), University of Udine, Via del Cotonificio 108, 33100 Udine, Italy; buttazzoni.marco.1@spes.uniud.it (M.B.); daniele.goi@uniud.it (D.G.)
* Correspondence: matia.mainardis@uniud.it

Received: 23 March 2020; Accepted: 8 May 2020; Published: 10 May 2020

Abstract: Up-flow anaerobic sludge blanket (UASB) reactor belongs to high-rate systems, able to perform anaerobic reaction at reduced hydraulic retention time, if compared to traditional digesters. In this review, the most recent advances in UASB reactor applications are critically summarized and discussed, with outline on the most critical aspects for further possible future developments. Beside traditional anaerobic treatment of soluble and biodegradable substrates, research is actually focusing on the treatment of refractory and slowly degradable matrices, thanks to an improved understanding of microbial community composition and reactor hydrodynamics, together with utilization of powerful modeling tools. Innovative approaches include the use of UASB reactor for nitrogen removal, as well as for hydrogen and volatile fatty acid production. Co-digestion of complementary substrates available in the same territory is being extensively studied to increase biogas yield and provide smooth continuous operations in a circular economy perspective. Particular importance is being given to decentralized treatment, able to provide electricity and heat to local users with possible integration with other renewable energies. Proper pre-treatment application increases biogas yield, while a successive post-treatment is needed to meet required effluent standards, also from a toxicological perspective. An increased full-scale application of UASB technology is desirable to achieve circular economy and sustainability scopes, with efficient biogas exploitation, fulfilling renewable energy targets and green-house gases emission reduction, in particular in tropical countries, where limited reactor heating is required.

Keywords: UASB; co-digestion; biogas; high-rate anaerobic digestion; energy recovery; granular sludge; renewable energy; decentralized wastewater treatment; two-stage anaerobic digestion; Anammox

---

1. Introduction

Nowadays, the over dependence on fossil fuels poses global risks, such as resources depletion and increasing climate change, due to the net increase in CO_2 levels in the atmosphere [1]. Anaerobic digestion (AD) is one of the most promising technologies, breaking complex organic substrates into biogas [2] that is substantially composed of a mixture of methane and carbon dioxide. AD, being 100% renewable, is an effective and environmental-friendly waste and wastewater management technique and can be considered as one of the most important renewable energy sources, due to CH_4 generation during the digestion process [3]. However, biogas generation from different streams, together with utilization in energy applications, is still somewhat challenging due to complex waste physicochemical properties, affecting biomass metabolic pathways and methane yield [2].

AD requires less energy than other thermochemical methods, such as gasification and pyrolysis, due to the low operating temperature [4], and consequently AD application throughout the world has continuously increased in the last decades. Beside highly biodegradable streams, advances in research allowed to apply AD also to lignocellulosic substrates, characterized by slow hydrolysis kinetics, such as macroalgal biomass [5], switchgrass [6] and yard waste [7], widening the spectrum of suitable matrices for biogas production. Proper pre-treatment application before AD or creating a mixture of complementary substrates can significantly increase process efficiency and consequently biogas yield [4]. Apart from large-scale plants, AD can be applied also to small-to-medium enterprises (SMEs), contributing to local energy and environmental sustainability, if the produced organic substrates have a suitable methane potential (typically evaluated by means of Biochemical Methane Potential (BMP) tests) [8]. An increased biogas production helps to augment renewable energy production and penetration in the fossil fuel market, as sustained by European Union (EU) sustainable development programs [9].

High-rate anaerobic digesters, in particular, received great attention in recent years, due to their high loading capacity and low sludge production [10]. High-rate reactors, by uncoupling biomass retention (expressed as solid retention time, SRT) and liquid retention (hydraulic retention time, HRT), allow to significantly reduce the required reactor volume, if compared to traditional systems [11]. Among this wide category, up-flow anaerobic sludge blanket (UASB) reactor is the most widely applied system worldwide [10]. UASB reactor was developed in the 1970s in the Netherlands and its application rapidly increased, due to its excellent reported performances on different biodegradable wastewater streams [12]. The key feature of a UASB reactor is the granular sludge, that retains highly active biomass with excellent settling abilities in the reactor [11], showing a very low sludge volume index (SVI), consequently improving also sludge-effluent separation.

A simplified scheme of a UASB reactor is reported in Figure 1. Wastewater enters at the bottom of the reactor and flows upwards through a so-called "sludge blanket", consisting of a granular sludge bed [13]. UASB configuration enables an extremely efficient mixing between the biomass and the wastewater, leading to a rapid anaerobic decomposition [13]. The operation of a UASB reactor fundamentally revolves on its granular sludge bed, that gets expanded as the wastewater is made to flow vertically upwards through it [14]. The microflora attached to the sludge particles removes the pollutants contained in wastewater, thus biofilm quality and the intimacy of sludge-wastewater contact are among the key factors governing UASB reactor success [14]. The generated biogas facilitates the mixing and the contact between sludge and wastewater, and the three phase gas-liquid-solid separator, located in the upper part of the reactor, allows to extract biogas, separating it from liquid effluent and residual sludge particles [13]. Typical geometrical and operating characteristics of UASB reactor include a height to diameter ratio of 0.2–0.5 and an up-flow velocity of 0.5–1.0 m/h [12].

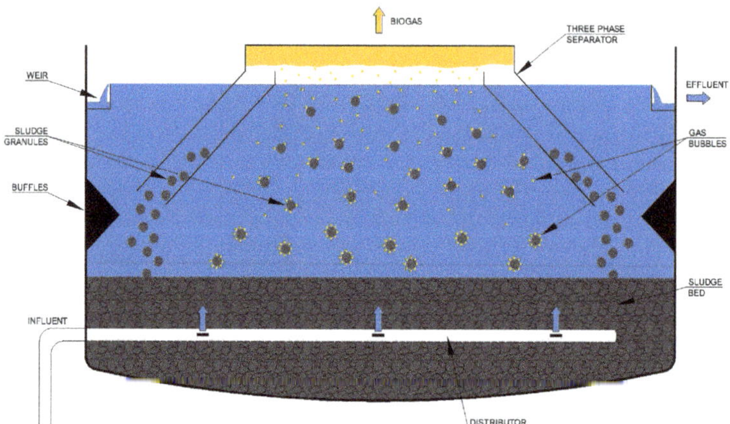

Figure 1. Up-flow anaerobic sludge blanket (UASB) reactor process scheme.

UASB treatment, if compared to aerobic stabilization, requires lower energy consumption, is efficient at higher loading rate and needs limited micro and macro-nutrients, producing a reduced amount of sludge, that is characterized by an improved dewaterability [12]. In fact, only about 5–10% of the organic matter in wastewater is transferred to the sludge fraction [12]. On the other hand, UASB treatment is known to have a limited effect on nutrients (nitrogen and phosphorous), as well as on micro-pollutants [15]. UASB treatment of high-strength industrial wastewater allows to significantly reduce energy expenses for aeration in wastewater treatment plants (WWTP), if UASB is applied as a pre-treatment before secondary biological process [16]. UASB reactor is able to efficiently treat various high-strength industrial wastewater (such as brewery wastewater [17], sugarcane vinasse [18], paper mill wastewater [16], dairy wastewater [16]), characterized by high chemical oxygen demand (COD) concentration and substantial biodegradability (high biochemical oxygen demand (BOD)/COD ratio). UASB treatment of high-loaded substrates allows to get high methane yields with a reduced energy requirement (if compared to aerobic stabilization) and a significantly lower excess sludge production [19]. Furthermore, recently UASB has proved to be efficient also on diluted streams, such as municipal wastewater [12], even at ambient operating temperature.

UASB treatment was compared to aerobic open lagoon in the work reported in [20] to treat wastewater from ethanol production, highlighting that the environmental cost of open lagoon is greater than UASB reactor. UASB reactor and anaerobic membrane bioreactor (AnMBR) are particularly indicated for treating chemical-industrial wastewater [21]. However, a further exploitation of anaerobic treatment is actually curbed by the ineffective mineralization of degradation-resistant organic substrates [21], thus a proper pre-treatment needs to be applied to enhance anaerobic degradability.

The most important aspects to be controlled when applying a UASB treatment are reactor start-up and granulation enhancement, coupling the anaerobic section with a post-treatment unit to efficiently abate organic matter, nutrients and pathogens [10]. A sufficient inoculation must be provided at the beginning of operations to reduce drawbacks such as process sensitivity, vulnerability, odor emission, long start-up period [12]. UASB operation start-up typically requires that 10–30% of the volume is inoculated with active granular biomass [12]. Given that the long start-up and the slow granulation are major constraints (in particular when treating complex and refractory wastewater), recently it was proved that granulation could be stimulated by chemical addition, such as calcium sulphate, enhancing granulation rate and improving methanogenic activity [22]. An increased granule formation (>0.25 mm) in the range of 7–40% was reported in a UASB reactor with $CaSO_4$ addition after 90 d in comparison with control, with an increased COD removal efficiency (3–9%) at moderate organic loading rate (OLR) (2.89 kg COD/m^3d) [22]. Granular biomass is able to tolerate higher hazardous and toxic compound concentration than traditional flocculent sludge: as an example, UASB was shown to tolerate higher organic loading rate (OLR) than anaerobic membrane bioreactor (AnMBR) when treating N, N-dimethylformamide [23].

In this review, the most recent advances in UASB anaerobic treatment are presented and discussed, with a focus on the most critical aspects for possible further developments. In Section 2, the most recent literature results regarding methane yields and operating parameters from a variety of substrates are presented, while in Section 3 the advances in UASB hydrodynamic understanding and microbial community composition are discussed. In Section 4, the most recent outcomes regarding two-stage UASB digestion are highlighted, while in Section 5 co-digestion applied to high-rate anaerobic treatment is presented. In Section 6, UASB application as Anammox process is introduced. Modified UASB systems are successively presented in Section 7. A particular focus on low-temperature decentralized UASB treatment of municipal streams is done in Section 8. Section 9 specifically deals with pre-treatments before AD and post-treatment of the treated anaerobic effluent with a plant-wide perspective. Section 10 analyzes toxicity aspects in UASB treatment, considering meaningful recent literature outcomes. The most critical aspects and future perspectives, with a focus on the energy and environmental aspects, are finally discussed in Section 11.

The main aim of the paper is to provide an up-to-date vision of the engineering aspects of UASB reactor application, together with forecasting possible further advancements. Most of the literature studies provide the results of laboratory and pilot-tests, while full-scale applications are still limited, in particular on slowly degradable substrates. An increased exploitation of biogas generation from liquid and solid substrates is strongly recommended to increase renewable energy share in the global market: high-rate anaerobic reactors can help to move towards an increased sustainable energy generation, in particular in tropical and decentralized areas.

2. UASB Reactor: Substrate Characteristics and Operating Conditions

In Section 2.1, the most recent literature outcomes regarding UASB treatment are critically summarized, while in Section 2.2 the influence of operating conditions is discussed; in Section 2.3, advanced high-rate reactors, developed from original UASB, are presented.

2.1. Substrate Characteristics

Substrate characteristics play a major role in UASB process efficiency: a high nitrogen concentration and a significant particulate matter content in the influent wastewater can lead to an excessive ammonia accumulation, which is notoriously toxic (after a certain threshold), and to a slow hydrolysis phase, reducing biogas production rate [24]. Some of the most recently reported literature outcomes regarding UASB treatment of different high-loaded substrates were summarized in Table 1. Besides traditional highly biodegradable substrates, research is actually focusing on refractory (chemical or industrial wastewater) and diluted (municipal wastewater) streams to extend the applicability of high-rate anaerobic reactors. Typically, complex wastewater requires a two-stage digestion (Section 4) or the selection of a proper pre-treatment (Section 9) to increase its biodegradability. UASB treatment of municipal wastewater, instead, is difficult to apply due to diluted stream characteristics, and is specifically described in Section 8. UASB reactor has shown to be particularly effective in degrading highly biodegradable substrates, where short HRTs (<24 h) can be applied, together with consistent OLRs (>20 kg COD/m^{3}d), obtaining high COD removal efficiency (>90%) (Table 1).

Table 1. Reported recent literature studies on UASB treatment of high-loaded substrates.

Substrate	Temperature (°C)	Influent Chemical Oxygen Demand (g/L)	Chemical Oxygen Demand Removal (%)	Hydraulic Retention Time (h)	Organic Loading Rate (g COD/L·d)	Methane Yield	Reference
Glutamate-rich wastewater	35	2.0	95.5–96.5	4.5–6	8.26–10.82	0.31 [1]	[25]
Monosodium glutamate	35	7.9	97	24	8	2.3 [2]	[26]
Sugarcane bagasse hydrolysate	20–30	1.82	86	18.4	2.4	0.27 [1]	[27]
Recycled paper mill wastewater	37	5.7	80.6	15.14	5.18	0.89 [2]	[28]
Vinasse	Ambient	120.2	91–93	40	72.1	0.46–0.53 [2]	[18]
Guar	37	1.1	79–84	10	2.78	0.15–0.16 [3]	[29]
Synthetic fiber wastewater	13.9–32.1	1.7–30.7	75.8	24	1.3–21.5	0.4–2 [2]	[30]
Pistachio wastewater	35	49.8	89.8	5.4 d	4.56	0.33 [1]	[31]
Perchlorate	30	-	84.7	2.2	9.96	-	[32]
Synthetic slaughterhouse wastewater	37	1.7	70	10	3.94–8.15	0.35 [2]	[33]
Chocolate wastewater	15–30	6.2	39–94	6	2–6	0.3–1.9 [4]	[34]
Pig slurry	36	21.5	-	1.5 d	14.3–16.4	0.25 [1]	[35]
Leachate from waste incineration	35	36.8	97.5–99.5	1.3–3 d	1.86–7.43	-	[36]

[1] L CH_4/g COD, [2] L CH_4/L·d, [3] L CH_4/g COD d, [4] L biogas/L·d.

Most of the studies regarding UASB treatment report mesophilic operations, which proved to be particularly effective in coupling a high biogas yield and a good process stability. A great variability in the applied operating conditions emerges from Table 1, both in terms of OLR and HRT, demonstrating that a substrate-specific approach is required to obtain a high COD abatement and a satisfactory methane yield. The highest abatements arise when treating highly biodegradable and soluble components, where the particulate fraction (difficult to hydrolyze) is limited. The treatment of extremely loaded substrates (COD > 100 g/L) leads to the necessity of adopting longer HRT, to have an efficient treatment, while streams characterized by lower COD concentration (<2 g/L) generally produce a lower methane yield, due to the reduced OLR. Moreover, it appears from Table 1 that standardization of methane yields from literature results is complex, due to the different adopted unity of measure.

Salinity is an important parameter in UASB anaerobic treatment when treating brackish streams: recent research proved that a high substrate removal can be achieved even under a salinity level of 10 g NaCl/L [29]. However, lower salt conditions stimulate the formation of larger granules and a faster degradation rate [29]. Moreover, it was seen that salinity does not substantially modify microbial community composition, even if methanogen abundance is reduced [29]. In a consistent way, a previous study on phenol UASB treatment demonstrated that the granular biomass is able to tolerate moderate salinity levels (up to 10 g Na^+/L), while higher salt levels (10–20 g Na^+/L) reduce reactor efficiency [37].

The treatment of substrates available in a specific territory is fundamental to achieve local circular economy and sustainability visions. Waste and wastewater can be valorized on-site to produce electric energy, fulfilling a high share of plant total need, and heat (that can be used for district heating). As an example, pistachio wastewater was tested as a possible feed for UASB reactor in the work in [31] and a potential to produce up to 28,200 MWh of electric energy from biogas was highlighted in Turkey by considering annual wastewater production (520,000 m^3/y) [31].

Wastewater containing high lipid concentration is particularly critical to be treated, given to a number of drawbacks such as clogging, sludge floatation, formation of foams and odor emission, biomass washout. However, lipid-rich wastewater has a higher methane potential (0.99 L CH_4/g) than proteins (0.63 L CH_4/g) and carbohydrates (0.42 L CH_4/g) [38]. Consistently, the treatment of slaughterhouse wastewater in a UASB reactor has to face with the inability to operate at high OLRs, as a result of suspended and colloidal impurities, including cellulose, proteins and fats, abundantly present in this stream [39]. In these situations, in order to achieve the desired efficiency, UASB reactor must be often coupled with an efficient post-treatment. In addition, substrate pre-treatment can be beneficial for increasing its biodegradability and the subsequent obtainable methane yield. The available pre- and post-treatment technologies are specifically described in Section 9. As an example, a semi-continuous process for slaughterhouse wastewater treatment was proposed in the work in [33], followed by a photoelectro-Fenton (PEF) treatment [33]. Wastewater from fish processing industry, instead, was studied in the work in [40] as another lipid-rich wastewater stream, suggesting a complex treatment scheme (baffled moving bed biofilm reactor followed by UASB reactor, fluidized immobilized cell carbon oxidation and chemoautotrophic activated carbon oxidation), with excellent COD, protein, lipid, oil and grease abatement [40].

Finally, particular attention has to be given to wastewater having significant sulfur concentration (such as sugar cane vinasse), considering that the inner part of the UASB reactor is exposed to a higher H_2S concentration than that measured in the treated effluent [41]. The influence of COD/SO_4^{2-} ratio on starch wastewater biodegradation in a UASB reactor was studied in the work in [42] with a progressive COD/SO_4^{2-} ratio decrease, highlighting a stable biogas production and a satisfactory COD and sulfate removal until COD/SO_4^{2-} ≥ 2 [42]. A further decrease in COD/SO_4^{2-} ratio suppressed methanogenesis through electron competition and sulfide inhibition [42]. Crude glycerol was again investigated in [43], where a sulfidogenic UASB reactor was proposed, showing maximum COD removal efficiency at COD/S-sulfate ratio of 8.5 g O_2/g S–SO_4^{2-}.

2.2. Influence of Operating Conditions

Beside substrate characteristics, the operational conditions play a major role in UASB process efficiency and stability. The main parameters that influence UASB performances are operating temperature (psychrophilic, mesophilic or thermophilic regime), pH, HRT, OLR, up-flow velocity. A stable pH, close to neutrality, is required to obtain a good-quality granular sludge, with sufficient alkalinity in the feeding substrate [14]. Up-flow velocity helps to maintain the mixing between sludge bed and wastewater, as well as to guarantee the desired HRT. The recommended up-flow velocity range in a UASB reactor is 0.5–1.5 m/h [11], even if values above 1 m/h in conventional UASB systems can lead to granule disintegration and biomass washout, due to shear stress that fragments the biomass [14]. A higher up-flow velocity is generally applied in the reactor start-up phase to select the biomass, removing smaller granules and maintaining the larger ones. As previously stated, the start-up of a UASB reactor is particularly critical and needs to be specifically controlled, with a progressive OLR increase.

Regarding temperature, most of the reported literature studies include mesophilic operations, which are widely accepted to be a good compromise between a sufficient biomass activity and reactor stability. The transition of operating conditions between different temperature regimes is another noteworthy aspect to be investigated, due to instability effects that can arise. As an example, the shift from mesophilic to thermophilic temperature regime was studied in the work in [44], with glucose and ethanol as feed: a better resistance to temperature variations was observed using ethanol as substrate, finding a significant correlation between granular sludge conductivity and COD removal rate, as well as between *Geobacter* abundance and COD abatement [44]. An enhanced sludge conductivity means a frequent electron transfer, subsequently increasing the efficiency of methane production, mainly through direct interspecies electron transfer (DIET) mechanism [44]. The influence of temperature and particle dimensions on UASB treatment of swine manure slurry was analyzed in [45], highlighting that temperature effect was more pronounced on high-particle content substrate. In fact, while at 25 °C the methane yield from raw and centrifuged manure was comparable, at 35 °C a significantly higher methane production was obtained from raw manure, showing that small particles and soluble components were mainly digested at the lower tested range [45].

Regarding pH effect, a stable pH close to neutrality is optimal for continuous operations. A sudden pH reduction leads to an imbalance in anaerobic trophic chain, with accumulation of undegraded volatile fatty acid (VFA) and a further tendency to pH decrease, particularly if a limited alkalinity is present. Sugar refinery wastewater was treated in a UASB reactor in the work in [46] by reducing pH to 5.0 to analyze bacterial dynamics, showing that COD and methane yield reduced of about 25%, with a significant change in acidogenic biomass, leading to propionate increase and accumulation, together with block of metabolic balance [46]. In another study, it was showed that intermediate compounds from glycerol degradation are mainly propionic and acetic acid and detrimental effects on system performances (with acetic acid accumulation) were highlighted when UASB reactor was subjected to pH shocks [43].

Finally, as for organic load, a low starting OLR is typically required in start-up and transitory periods, to allow biomass adaption, followed by a moderate progressive increase towards final target value. An increasing bacterial specialization and a progressive decrease in methanogenic population was highlighted by step-wise augmenting OLR in glycerol UASB treatment, due to a functional biomass organization, that corresponded to a proportional increase in methane yield [47].

2.3. Advanced High-Rate Reactors

Continuous research on UASB reactor performances in the treatment of different substrates led to the development of advanced high-rate reactors, starting from the basic UASB configuration (Figure 1). These advanced systems include expanded granular sludge bed (EGSB) reactor, which allows an improved contact between granules and wastewater and an enhanced sludge separation (due to the rapid up-flow velocity), and static granular bed reactor (SGBR), which acts as an anaerobic filter, with absence of mixing and opposite gas and wastewater flows [48]. Internal Circulation (IC) reactor is another advanced high-rate system, based on a series connection of two UASB reactors: IC has a higher height/diameter ratio and shows proved advantages over conventional UASB, such as increased operating OLR, powerful stress resistance, operational stability, economic space utilization [49]. Recently, a high-rate External Circulation Sludge Bed (ECSB) reactor was proved to be beneficial for high-rate anaerobic treatment of cheese industry wastewater at full-scale, capable also to tolerate high influent calcium loads [50].

The most recent advances in high-rate AD include Internal Circulation Experience (ICX) systems, having a two-stage separation system that enables an excellent biomass retention [51]. Higher OLRs (up to 20–35 kg COD/m^{3}d) can be applied in comparison to traditional UASB (10–15 kg COD/m^{3}d) and IC (20–30 kg COD/m^{3}d) reactors, maintaining at the same time a stable and high COD abatement [51]. The key feature of a ICX reactor is the two-stage phase separation, where the traditional three-phase separator is substituted by a top separator, that removes biogas from the reactor, and a bottom separator, that separates granular biomass from the treated effluent [51]. This system results in a higher biomass retention efficiency, if compared to UASB reactor [51]. More than 70 full-scale ICX reactors have been built since 2013, with a size ranging from 85 to 5000 m^3, showing a high and stable COD removal even under fluctuating operating conditions [51].

3. UASB Hydrodynamics and Microbial Community

The deepening of UASB hydrodynamics is fundamental to understand and model UASB reactor, providing useful insights for process optimization. Hydrodynamic UASB modeling is complex and should integrate AD process dynamics (for example by applying the widely known anaerobic digestion model number 1 (ADM1), developed by International Water Association (IWA) [52]), flow pattern, biofilm characteristics and sulfate reduction process [53]. A non-ideal flow was shown to better simulate UASB hydrodynamics by evaluating fundamental hydrodynamic parameters, such as Peclet number and dispersion coefficient [53]. UASB reactor hydraulics can be described with a dispersive model or a multi-continuously stirred tank reactor (CSTR) model [54]. It was highlighted that a dispersive model simulates in a better way experimental data (if compared to the multi-CSTR approach), and could be integrated with the ADM1 bioprocess model to include both biological and hydrodynamic aspects [54]. A three-dimensional numerical hydrodynamic study was proposed in [55] to simulate a UASB reactor with different inclinations of the gas deflector, showing a good fitting between modelled and experimental data regarding flowrate, influent and effluent organic matter, solid concentration at liquid-gas interface, pressure field [55].

The core aspect of UASB technology is undoubtedly the granular sludge bed, which is identified by an extreme compactness and a high capability to tolerate harsh environments. The granules in UASB reactors are characterized by a layered structure, where hydrolytic and acidifying bacteria are located in the outer granule shell, while methanogens are positioned in the inner stratus [12]. Cavities and holes in the granules allow transporting gases, substrates and metabolites from a layer to other layers [12]. A granular sludge having proper particle size, density and microfilm characteristics enhance reactor efficiency [14]. The knowledge of microbial community composition is fundamental to optimize granule performances: it has been recently proved that reactor operating parameters are correlated with granule microstructure [56]. In particular, the loading ratio was identified as the key parameter controlling granule transformations [56]. Beside upward velocity, intermittent gas sparging could be an effective way to promote liquid shear, altering granule surface structure [56]. The detailed microscopical analysis of granule structure demonstrated that larger granules (3–4 mm diameter) have multi-layered internal microstructures, with more consistent methanogenic activity than smaller granules (1–2 mm diameter) [57]. Granular sludge was shown to have superior performances than thickened digestate and anaerobic sewage in UASB reactors [35].

The microbial community in UASB reactors is capable to adapt even to sudden changes in operating parameters that can arise in continuous operations. Bacteria response to alkaline pH perturbations in a full-scale UASB reactor treating dairy wastewater showed that after an accidental pH increase (which affected microbial community structure) the microbial population was able to regain diversity and methanogenic activity [58]. Granulation fundamentals are not completely understood at the present, despite the number of existing granulation-based treatment plants [59]. Anaerobic granulation substantially consists of a microbial aggregation where communities are organized into dense and highly-structured aggregates, not necessitating any carrier media [59]. Microbial community composition was shown to modify in continuous operations, depending on the treated substrate [60]. As an example, in the work in [60], it was observed that *Methanosaeta* and *Methanobacterium*, despite being absent in the original inoculum, became dominant archaea in phenol UASB treatment, while the dominant bacteria were *Syntrophorhabdus* (known to degrade phenol to benzoate and then benzoate to acetate and H_2) and *Clostridium* [60].

A limit of the anaerobic trophic chain is the slow conversion metabolism of short-chain fatty acids and alcohols by syntrophic bacteria [61]. Recently, this bottleneck was solved by promoting the DIET mechanism, where electron transfer can directly proceed via bio-electrical connections or in combination with abiotic conductive materials such as biochar, activated carbon or magnetite [61]. DIET stimulates microbial communities, rapidly establishing biological electrical connections [62]. DIET was shown to reduce initial lag-phase, enhancing organic matter degradation and improving, at the same time, biogas production rate [63]. Ethanol could be used to stimulate DIET, favoring the formation of aggregates having higher conductivity and stability in response to environmental disturbances, with enrichment in *Petrimonas* and *Methanothrix* species [62]. Besides traditional conductive materials, in a recent literature study blast furnace dust was tested in UASB treatment of synthetic wastewater, highlighting an increased methane production due to an improved DIET mechanism [64].

4. Two-Stage UASB Anaerobic Digestion

An alternative to single stage digestion is two or multiple stage process, where in each phase the conditions are optimized for a different biomass type: typically, in the first step hydrolytic and acidogenic reactions are completed, followed by methanogenic reactions in the second stage. The microbial community operating in the AD process, in fact, can be broadly classified into acidogenic and methanogenic, whose optimal operational and environmental conditions do not coincide, mostly in terms of the pH where the maximum activity is observed [65]. Acidogenic bacteria, in fact, require a lower pH (4–6) and a shorter HRT than methanogenic bacteria [66].

The main advantages and drawbacks of two-stage AD are summarized in Table 2. Two-stage systems provide an optimal process stability with enhanced pollutant removal, an increased energy efficacy and an enhanced control of critical parameters, by reducing VFA and ammonia inhibition, controlling at the same time the production of harmful byproducts and providing sufficient buffering capacity [65]. On the other hand, co-digestion leads to increased capital costs and process control is not as simple as in mono-digestion operations, due to the not standardized operating conditions.

Recent studies claimed a 10–30% increase in methane yield via two-stage AD, even if this augment is typically not sufficient to sustain the costs of building a second digester in full-scale operations [67]. Through a techno-economic analysis, it was proved that two-stage AD is slightly more expensive than single-stage AD [67]. However, the economic profitability of two-stage AD should be assessed in each specific application, given the extreme variability in local market and substrate characteristics. Further work is required, in addition, to standardize two-stage AD to optimize OLR, HRT, total solids (TS)/ volatile solids (VS) balance [67].

Two-stage AD, with separation of acidogenesis from methanogenesis, has proved to be particularly effective in the treatment of readily biodegradable substrates such as food waste, where the second process stage can be accomplished using high-rate UASB reactor [65]. Generally, it should be reminded that acid fermentation reduces alkalinity and reactor instability is triggered when acid generation is faster than degradation through the methanogenic pathway [68]. Thus, the substrates which are known to undergo fast acidification, are the most suitable ones to apply two-stage treatment. Consistently, a two-stage system was applied to treat potato wastewater in the work in [68]. In the treatment of starch wastewater, pre-acidification was similarly shown to be beneficial for granular sludge stabilization, avoiding granules floatation and disintegration (triggered by extracellular polymeric substances, EPS), when compared to single-stage UASB system [69]. A combined two-stage CSTR-UASB digestion system achieved superior performances and an alkali-addition free cheese whey wastewater treatment, if compared to single UASB treatment [70].

Table 2. Main advantages and drawbacks of two-stage UASB treatment.

Advantages	Drawbacks
Improved process stability	Increased capital costs
Increased pollutant abatement	Not standardized operating conditions (Hydraulic Retention Time, Organic Loading Rate, total solids (TS)/ volatile solids (VS)
Augmented methane yield	Need for a second digester
Optimal operating conditions for diverse microorganism consortia	Biogas surplus typically not sufficient to cover expenses
Better process control	Need for an extra process control
Reduced fatty acid and ammonia inhibition	Economic sustainability not always favorable
Augmented buffering capacity	
Improved control of byproducts	
Enhanced sludge stabilization	
Reduced biomass floatation and disintegration	
Effective in readily biodegradable substrate treatment	

Two-stage UASB treatment is beneficial also when dealing with substrates containing a fraction of readily degradable matter together with a fraction of slowly degradable compounds. In fact, a two-stage process was proposed in [71] to treat purified terephthalic acid wastewater, with acidogenic and methanogenic reactors operating at different HRTs, obtaining an enhanced pollutant abatement. More complex schemes can be adopted in particular situations, operating with different COD loading rates in the different process stages. A three-stage UASB reactor with methanogen sludge recirculation was studied in [72] for H_2 and CH_4 production from cassava wastewater, with a significantly higher energy yield, if compared to single and two-stage UASB reactor [72]. Coherently, H_2 and CH_4 were produced in two-stage UASB reactor treating cassava wastewater with added cassava residue in [73]: at a thermophilic temperature (55 °C), by applying an OLR of 10.29 kg COD/m^3d, a favorable gas composition was highlighted both in the first (42.3% H_2, 55% CO_2, 2.7% CH_4) and second (70.5% CH_4, 28% CO_2, 1.5% H_2) reactor [73]. A two stage system, including a UASB reactor for H_2 production and an IC reactor for CH_4 production was proposed in [66] to treat medicine herbal wastewater, obtaining a H_2 yield of 3.0 L/L d in the UASB reactor (HRT = 6 h) and a CH_4 production of 2.54 L/L d in the IC reactor (HRT = 15 h) [66]. In optimal conditions, COD removal was 90%, with an energy conversion efficiency of 72.4% [66]. Modified UASB systems for hydrogen and VFA production are described more in depth in Section 7.

5. UASB Co-Digestion

Single substrate AD can lead to inhibitory phenomena, due for example to a lack in alkalinity or to an excessive ammonia concentration in the feeding stream [74]. The contemporary anaerobic treatment of multiple substrates, widely known as co-digestion process, is accepted to be beneficial for an enhanced process stability and an increased biogas generation, helping to balance macro- and micro-nutrient concentration [75].

Recent research focused on addition of micronutrients to enhance co-digestion performances, as well as on the contemporary treatment of substrates available in a local territory, contributing to circular economy and sustainability principles. The most recently reported UASB co-digestion results were summarized in Table 3. It can be seen that food waste and farm wastewater are commonly investigated streams [76–79], characterized by complementary properties, mostly in terms of carbon to nitrogen (C/N) ratio and nutrients, where co-digestion can lead to a significant increase in energy yield. Applied HRT in co-digestion studies is typically longer than that used for the treatment of single substrates (as reported in Table 1), while again mesophilic operations are claimed for most of the tests.

Micro-nutrient supplementation (in form of Fe, Co, Ni, Se, Mo) was proved to be crucial to enhance methanogenic activity, stimulating methane production [76]. The addition of metals and natural elements showed to have a positive effect both on COD removal and on biogas production in UASB co-digestion of high-loaded substrates [76,77]. Carbon-rich co-substrates can be strongly beneficial, instead, when working with sulfate rich-wastewater, given the competition between sulfate-reducing bacteria (SRB) and methanogenic biomass [80], which is reduced by C addition. On the other hand, nutrient and alkalinity can be provided by animal residues when working with highly degradable and acidifying substrates [81]. An integrated solution for the treatment of different farm wastes, such as the liquid fraction of manure and cheese whey, characterized by complementary properties, was recently proposed and could be applied in areas with intense presence of cheese factories and intensive livestock farming, reducing overall environmental pollution [82].

From the data reported in Table 3, a great effort in finding suitable substrates for co-digestion in UASB reactor emerges, even if most of the studies are conducted at laboratory phase and need a further deepening to prove up-scale feasibility. Moreover, the analysis of available waste and wastewater fluxes in a selected area is recommended, estimating the total potential energy production and allowing continuous operations throughout the year. In fact, not all the industrial waste and wastewater streams are produced in a continuous way: as an example, ethanol industry produces sugarcane molasses and vinasses, whose co-digestion proved to be highly beneficial, due to the factory batch operations mode [83].

Table 3. Recently reported UASB co-digestion results.

Substrate	Co-Substrate	Adjuvant	Temperature (°C)	Hydraulic Retention Time (d)	Organic Loading Rate (kg COD/m^{3}d)	Chemical Oxygen Demand Removal (%)	Yield	Reference
Food waste	Domestic wastewater	Cu^{2+}	35	10	3.8	>90	0.3 [1]	[76]
Food waste	Liquid slaughterhouse waste	Clinoptiolite	40	28	-	-	-	[77]
Cheese whey	Manure liquid fraction	None	35	2.2	19.4	95	6.4 [2]	[82]
Distilled gin	Swine wastewater	None	36	3.3	28.5	97	8.4 [2]	[81]
Sewage sludge-cow manure	Kitchen waste, yard waste, floral waste, dairy wastewater	None	36	1	-	78–86	4.5 [2]	[78]
Acid mine drainage	Cheese whey	None	30	1	-	68–84	-	[80]
Source-diverted blackwater	Food waste	None	35	2.6	10.0	82–84	2.42 [3]	[79]

[1] L CH$_4$/g COD$_{removed}$, [2] m^3 CH$_4$/m^{3}d, [3] L/L·d.

6. UASB Application as Anammox Process

Anammox process consists in the simultaneous abatement of ammonium and nitrite to nitrogen gas and is performed by anaerobic ammonium oxidation bacteria [84]. Anammox is being widely applied worldwide to remove high nitrogen loads, given that the total cost of partial nitrification-Anammox process is significantly lower than that of conventional nitrification-denitrification processes [84]. In fact, conventional biological nitrogen removal (BNR) processes are economically limited by the large amount of excess sludge that is inevitably produced [85]. It was recently proved that Anammox process is optimized by growing Anammox bacteria in granular form, enhancing biomass retention and shock resistance, as well as system resilience [84].

Landfill leachate is a complex wastewater with high levels of COD and ammonia, with a tendency to COD reduction and increase in NH_4^+-N over time: the treatment of mature landfill leachate is particularly cumbersome due to the low COD/N ratio [86]. In a recent literature work, a three-stage Simultaneous Ammonium oxidation Denitrifying (SAD) process was proposed to treat mature landfill leachate [85]. UASB was intended as Anammox reactor, efficiently removing ammonia and nitrite (formed in the previous process stage). An excellent TN removal (98.3%) was obtained in the integrated process, with reduction in oxygen consumption, sludge production and organic matter concentration, if compared to traditional aerobic treatment [85]. Similarly, in the work in [86] UASB reactor was used as Anammox process to treat landfill leachate after a pre-denitrification phase, fully degrading biodegradable COD in wastewater and reducing the need for oxygen supply in the aerobic process.

It is known that the activity of Anammox bacteria strongly depends on the operating temperature. An Anammox UASB system operating at ambient temperature (between 9 °C and 28 °C) was proposed in [87] for artificial wastewater treatment without temperature control, underlining high N removal ability (90%) during Summer period, with a maximum N removal rate up to 62.5 kg N/m^3d and an enrichment of Anammox bacteria in the UASB granular sludge [87]. An external nitrite source was used instead in an Anammox UASB reactor treating diluted chicken waste digestate, characterized by high presence of nitrogenous and organic compounds in the work in [88], obtaining a good pollutant abatement (respectively, 57% on total ammonia nitrogen and 80% on COD).

7. Modified UASB Systems for Bio-Hydrogen, Volatile Fatty Acids and Methane Production

The original UASB configuration was modified in a number of scientific studies to satisfy a particular purpose, such as producing hydrogen or VFA (rather than methane), or increasing reactor performances through the introduction of packing materials (Table 4). UASB reactor, in fact, can be used also for bio-hydrogen production by inhibiting methanogenic bacteria through sludge thermal pre-treatment, acid-basic procedures or headspace gas recirculation [89–91]. Different substrates have been tested for bio-hydrogen production, including palm oil mill effluent, winery wastewater and synthetic media [89–91]. Besides H_2, waste-derived VFA, especially acetate, are valuable bio-refinery products that can be used as precursors to fuels and chemicals in different industrial sectors [92]. In particular, foul condensate from a Kraft pulp mill was shown to be adapt for VFA production in a UASB reactor, due to its high methanol, ethanol and acetone content [92].

Table 4. Recently reported UASB modified systems.

Substrate	Reactor	Temperature (°C)	Product	Hydraulic Retention Time (d)	Organic Loading Rate (kg COD/m^{3}d)	Yield	Reference
Palm oil mill effluent	UASB	55	H$_2$	0.25	-	11.75 [1]	[89]
Winery wastewater	UASB	37	H$_2$	0.23	-	62 [2]	[91]
Synthetic media	UASB	-	H$_2$	0.25	-	4.34 [1]	[90]
Foul condensate from Kraft mill	UASB	22–55	VFA	3.13	8.6	52–70 [3]	[92]
Nitrobenzene	UASB with nanoscale zero-valent iron addition	35	CH$_4$	1	0.2 [4]	-	[93]
Petroleum wastewater	UASB with diatomite and maifanite addition	36	CH$_4$	10–20	11	1.61–2.2 [5]	[94]
Sewage water	Packed UASB reactor	Ambient	CH$_4$	0.21–0.25	1.8	-	[95]
Swine wastewater	UASB+ aerobic packed bed reactor	37	CH$_4$	0.79	3.26–10.14	0.26–0.81 [6]	[96]
Textile wastewater	Micro-aerated UASB	25	CH$_4$	-	1.27–1.5	-	[97]
Cattle slaughterhouse wastewater	UASB with synthetic grass packing	35	CH$_4$	1	10	2.0 [5]	[98]
Diluted food waste paste	UASB with biochar addition	30	CH$_4$	1	6.9–7.8	0.86 [7]	[99]

[1] L H$_2$/L$_{effluent}$ d, [2] mL H$_2$/L h, [3] % of utilized carbon, [4] kg NB/m^{3}d, [5] m^3/m^{3}d, [6] kg COD-CH$_4$/kg VSS d, [7] L biogas/g COD$_{removed}$ d.

The presence of support materials in a traditional UASB reactor for methane production triggers the formation of densely packed aggregates and a more consistent presence of large granular sludge ($\geq$0.6 mm), with an increase in EPS production, promotion of VFA degradation and methane yield stimulation, together with stabilized performances [93,94]. Different materials were proposed for packing UASB reactor, including metals, minerals, recycled plastic material, synthetic grass and biochar [93–95,98,99]. Biochar, in particular, is a biomass-derived carbonaceous material, obtained from pyrolysis process, with proved capability of enhancing AD process performances, by DIET stimulation, C/N ratio optimization, micro-pollutant adsorption [74]. Biochar is able to enhance UASB performances when treating highly soluble substrates, that rapidly produce VFA: in a recent study, a biochar-amended UASB reactor treating diluted food waste paste highlighted a significantly higher COD removal than control reactor (77% versus 47%), with improved biogas yield at an OLR of 6.9–7.8 kg COD/m^{3}d [99].

Hybrid aerobic-anaerobic reactors, coupling a UASB section and a packed bed reactor, were investigated in the work in [96] for swine wastewater treatment, with a progressive OLR increase, allowing nitrogen removal in the final aerobic phase. This solution would allow complying with current regulations for discharge to water bodies [96]. Micro-aerated UASB reactor can be a feasible solution to reduce effluent toxicity, in particular when treating complex industrial wastewater: micro-aeration allows removing the aromatic amines formed under anaerobic conditions [97].

8. UASB Treatment of Municipal Wastewater

The operating temperature is known to have a primary impact on anaerobic process kinetics (Section 2.2): low-temperature psychrophilic operations are particularly critical when treating diluted or refractory wastewater streams in a UASB reactor. In a consistent way, recently a low methanogenic and sulfidogenic activity was reported at temperatures of 17.5–25 °C in UASB treatment of sulfate-rich methanol wastewater (COD/SO$_4^{2-}$ ratio of 0.5), while an excellent COD removal (>90%) was obtained at 40–50 °C, with methanogens prevalence over sulfidogens [100].

In order to reduce UASB reactor heat requirement, in particular in cold climates, low-temperature digestion could be implemented by proper setting of the operating parameters, improving the overall energy balance. Considering a plant-wide perspective, in fact, in psychrophilic operations a higher amount of potential energy is available to be exploited for electricity or heat production. In recent literature studies, low-temperature (15 °C) UASB treatment of municipal sewage was proved to be technically feasible, in particular in co-digestion with rapidly degradable co-substrates (such as glucose), with moderate COD removal efficiency (23%) and triple specific methanogenic activity, if compared to sewage mono-digestion [101]. The main challenge in low-temperature UASB treatment of municipal wastewater consists in the slow hydrolysis kinetics of complex and suspended material, as well as in the slow methanogen growth [101].

In ambient temperature operations, a major impact is due to temperature fluctuations, that can be particularly limiting in the Winter period for biomass activity, and to reactor configuration, that has to be properly engineered to enhance applicable OLR and biogas yield [102,103]. A meaningful study [102] investigated low-temperature (10–20 °C) UASB treatment of domestic wastewater at short HRT (6 h). A stable COD removal (mean 60%) and a low effluent COD concentration (mean 90 mg/L) were claimed at temperatures ranging from 12.5 and 20 °C, with CH_4 production evaluated as 39.7% of the influent COD, while decreased performances were reported at 10 °C [102]. In another study the feasibility of low-temperature anaerobic operations still appeared somehow challenging: in dairy wastewater treatment through UASB and EGSB reactors (OLR of 7.5–9 kg COD/m^{3}d), it was shown that UASB performances were better than those of EGSB reactor [103].

Regarding low-temperature high-rate anaerobic treatment, it should be reminded, in addition, that lower operating temperatures increase methane solubility in water, leading to significant energy losses, due to the presence of dissolved methane in the effluent [104]. The use of sequential Down-flow Hanging Sponge (DHS) was recently suggested to recover a high fraction (57–88%) of dissolved CH_4 from UASB reactor effluent, improving at the same time treated water quality, due to 90% abatement of suspended solids and COD [104]. Membranes could be used, as well, to recover methane from the treated anaerobic effluent, even if a higher fouling propensity was claimed in UASB effluent (rather than AnMBR) in the work in [105], showing that the majority of foulants were protein-like substances.

A number of full-scale UASB plants are in operation worldwide without reactor heating, to reduce energy expenses and simplify daily operations. Regarding process efficiency, a long-term analysis of seven full-scale treatment plants operating at ambient temperature in India highlighted that UASB reactor, followed by conventional aerobic processes, failed to comply with required effluent standards, while UASB reactor, followed by a DHS system, allowed to obtain high COD, BOD, total suspended solids (TSS), ammonia and phosphate abatements [106] and thus could be considered a more environmentally friendly solution. Similarly, full-scale UASB treatment at ambient temperature (mean 17 °C) was reported in the work in [107] in Bolivia, with moderate COD and TSS abatements, even if biogas valorization was not done [106]. This proves that an advancement is needed in developing countries to efficiently exploit the high-energy content of biogas. In some cases, in fact, biogas is flared, without any energy recovery, or even dispersed in the atmosphere, with negative environmental effects due to huge greenhouse gases (GHG) emissions that contribute to global temperature increase.

The study and application of decentralized wastewater treatment systems is necessary to develop sustainable water management practices in small and remote communities, not served by public sewer networks. Among the available technologies, it is widely accepted that UASB process has significant advantages over aerobic treatment, including reduced capital investment, lower energy demand, reduced sludge production and biogas exploitation [108]. Energy recovery from decentralized wastewater streams provides a significant contribution to GHG emission reduction, promoting at the same time resource recovery [109]. Source-diverted blackwater, in particular, collected from vacuum toilets, is an ideal stream for AD application, given its high organic and solid content [109].

Recently, it was proved that the UASB reactor could be applied as a decentralized system, co-digesting separated blackwater and kitchen waste, with an overall reduction in environmental impact (considering water, fertilizers and emissions) if compared to different alternative scenarios, including septic tanks followed by composting and septic tanks followed by landfill [110]. Similarly, UASB reactor was tested to treat blackwater (COD concentration of 1.00 g/L) in the work in [111], by applying a HRT of 2.2 d at 35 °C temperature: COD removal efficiency was above 80%, even if some methanogenesis inhibition was observed due to SRB competition [111]. High efficiency blackwater treatment was reported also in [109], where an OLR of 4.1 kg COD/m^{3}d was applied (HRT of 2.6 d), obtaining a COD removal efficiency of 84% and a good methane production of 0.68 m^3 CH$_4$/m^{3}d. The positive reported outcome was related to the prevalence of H$_2$/CO$_2$ methanogenic pathway [109]. To compare the effect of diverse influent streams, UASB was used to treat both blackwater and sanitary wastewater in the work in [112], highlighting a higher COD abatement when treating blackwater (77% versus 60%). The good COD removal was maintained despite the fluctuating OLR, coupled with a moderate *E. coli* (1.96 log) and Total Coliforms (2.13 log) removal [112]. In a coherent way, a recent notable research proved that different blackwater types (coming from conventional and vacuum toilets) enriched distinct microbial consortia in UASB treatment, probably due to the diverse sulfate and ammonia loadings [113].

Another literature work showed that UASB reactor can be seen as a second treatment stage (after septic tank) for onsite wastewater treatment in developing countries, with high TSS (83%) and COD (88%) removal efficiencies [114]. Primary septic tank presence allows to reduce the required UASB start-up time [114]. A more complex scheme for raw municipal wastewater treatment, including a UASB reactor followed by a settler (having HRT of 3 h, where sludge was concentrated) and a digester, was proposed in the work in [115] to improve global energy efficiency of the system. The integrated process had total HRT of 6 h, achieving a mean COD removal of 49.2% [115]. Compared to single UASB reactor, the combined UASB-settler-digester system had similar COD abatement and methane production applying a reduced sludge recirculation rate, leading to 50% energy saving [115].

More advanced technological approaches, focused on resource recovery (beside energy recovery from biogas) involve the addition of calcium in a modified UASB treatment (with internal gas lift, GL) of vacuum collected blackwater, to increase total phosphorous (TP) retention [116]. An excellent COD (92%) and P (90%) removal was claimed, together with recovery, at the same time, of CaP granules (>0.4 mm diameter), containing an average of 7.8% P [116]. A potential 57% P recovery in source-separated sanitation was calculated in the combined UASB-GL process [116]. It should be reminded that P recovery is highly recommended to be implemented in wastewater treatment in the future, considering that P is an essential element for living organisms and that mineral-derived P is predicted to be depleted within the next 100 years [117].

It could be concluded that a number of research studies proved the feasibility of municipal wastewater treatment in UASB reactor, that is particularly efficient if innovative collecting techniques (such as vacuum sanitation) are applied, leading to an increased OLR and a consequently higher biogas generation, together with a reduced freshwater usage. Material recovery, particularly important in the case of phosphorous, can be coupled with energy recovery, enhancing the sustainability of wastewater treatment. Finally, in order to achieve 100% of clean and renewable energy, UASB process in decentralized areas could be coupled with solar thermal energy, in particular in hot climate areas. Through a modeling approach, it was recently demonstrated that a 100% energy saving could be obtained with an integrated UASB-solar system in Morocco for 10 months per year, while 70% energy saving could be reached in the residual two cold months per year [118]. In addition, it was proved that the operating temperature in the digester never dropped out of the mesophilic range, even in the absence of solar production (due to a poor irradiation) [118].

9. UASB Pre- and Post-Treatment

As previously mentioned, a number of different pre-treatments can be applied before AD to augment methane yield, depending on substrate characteristics and physicochemical composition: the applicable pre-treatments have been widely investigated in recent literature studies, both on traditional AD systems and advanced UASB reactors. The possible techniques to increase substrate biodegradability include physical (size reduction), chemical (acid, alkaline, organic solvent), thermophysical (hot water, steam explosion, ultrasound, microwave), thermochemical (wet oxidation, supercritical CO_2) and biological (microbial, enzymatic) methods [119].

Sewage contains 30–70% of particulate COD, degrading at a slower rate than soluble organic fraction: a pre-hydrolysis phase ensures better UASB performances in sewage treatment at an increased OLR, in particular if using two-stage systems [120]. Lignocellulosic streams can be efficiently pre-treated with acid and enzymatic methods, even if industrial scale application of these methods is still limited [4]. Temperature pre-treatment and temperature-phased advanced AD were lately studied to facilitate veterinary antibiotic degradation [121], that is characterized by a significant refractory fraction. In another work, chemical pre-treatment of onion waste was proposed in a packed bed reactor coupled with a UASB reactor to reduce VFA accumulation and improve biogas yield [122]. Sulfuric acid pre-treatment, in particular, allowed to decrease the required start-up time, with methane yield up to 0.41 m^3 CH_4/kg $VS_{removed}$ [122]. Mechanical solid-liquid separation of organic fraction of municipal solid waste (OFMSW), instead, allows to obtain a highly-degradable liquid fraction, called press water or leachate, able to produce consistent methane yields in high-rate systems, such as UASB reactor [16]. Ultrasound (US) and *Escherichia Coli/Aspergillus Niger* biodegradation pre-treatments were demonstrated to be efficient in reducing long-chain fatty acids (LCFA) concentration in UASB treatment of crude glycerol, reducing methanogenic bacteria inhibition, with methane yield increase up to 29% (with US pre-treatment) and 77% (with *A. Niger* pre-treatment) [123]. Furthermore, recently an emulsification pre-treatment was shown to be efficient for high-rate anaerobic treatment of an oleic acid based effluent [124].

Enzymatic pre-treatment is an environmental-friendly technique, if compared to chemical methods, and was proposed in the work in [125] to increase biogas yield from palm oil mill effluent (POME). A complex treatment scheme for oilfield wastewater treatment was instead proposed in the work in [126], including dissolved air floatation, yeast bioreactor, UASB reactor and biological aerated filter, showing excellent COD, suspended solids and oil removal. UASB reactor, in particular, enhanced effluent biodegradability for the final aerobic phase [126]. The most recent pre-treatment methods are basically aimed at increasing substrate biodegradability, in the case of lignocellulosic and refractory streams, or at reducing acid accumulation in the system, that leads to an imbalance in the anaerobic trophic chain, with acidogenesis prevalence over methanogenesis. Besides evaluating the effectiveness of the investigated pre-treatment method, the most important aspect to be assessed when upscaling a pre-treatment is to evaluate its economic and energy sustainability: the obtained augment in methane yield should be sufficient to cover the extra costs for mechanical equipment installation and energy (or chemical) consumption of the pre-treatment phase. Moreover, the additional environmental impact has to be studied, due to the usage of chemical compounds and to substrate degradation.

On the other hand, a post-treatment is typically required on effluents from UASB reactor to comply with the required legislative standards, especially regarding nutrients, pathogens and solids [12]. Recently, high-rate algal ponds (HRAP) were proposed as an innovative post-treatment for a UASB effluent operating at environmental temperature [15]. UASB co-treated raw sewage and harvested microbial biomass from HRAP reactor: overall COD and NH_4-N removals of 65% and 61% were claimed, respectively, in the co-digestion system, with 25% increase in CH_4 yield (from 156 to 211 NL CH_4/kg VS) if compared to UASB treatment of sewage alone [15]. UASB reactor was followed by an aerobic process in the work reported in [127] in recalcitrant azo-dye treatment, showing high COD (92.4%) and color (98.4%) removal at HRT of 6 h, corresponding to an OLR of 12.97 g COD/L·d, with maximum CH_4 yield of 13.3 mmol CH_4/g COD d [127].

Beside the traditional activated sludge process, an efficient post-treatment of anaerobic UASB effluents could be achieved by applying sequencing batch reactors (SBR) with granular aerobic sludge, even if particular attention has to be put in reactor start-up [128], similarly to what happens in anaerobic granular systems. A pilot system consisting of a UASB reactor and a successive microalgae post-treatment (*Chlorella sorokiniana* cultivated in three flat photobioreactors) was tested in the work reported in [129] to treat a mixture of municipal and piggery wastewater, obtaining a high organic removal efficiency, despite fluctuations in the influent characteristics. Microalgae abated dissolved inorganic carbon (46–56% removal), orthophosphate (40–60% abatement) and ammonia (100% efficiency) [129]. A simple surface aerator post-treatment was finally proposed in the work in [130] as an energy-saving technique after UASB treatment of municipal wastewater coming from an Indian WWTP. It could be concluded that UASB can be coupled with extremely different post-treatments, from conventional biological processes and simple aerators to advanced granular systems, depending on the required effluent quality; microalgae, in particular, have a high integration potential in WWTPs, due to the possibility to use the excess biomass for biogas production, leading to a circular cascade.

10. UASB Reactor and Wastewater Toxicity

Sewage sludge production is increasing worldwide due to the growing population and an improved wastewater treatment [131]. The applied techniques for sludge treatment substantially include agricultural application and thermochemical methods [131]. The environmental impact of water treatment residues should be thoroughly assessed through toxicity evaluation to allow a safe sludge application to the soil. Eco-toxicity of treated wastewater and sludge is being given attention: sewage sludge is reported to contain at hazardous concentrations heavy metals, pharmaceuticals and personal care products [131]. As previously stated, UASB treatment abates nutrients, biological agents and trace metals in a limited way [132], with accumulation in the residual solid fraction. Heavy metal pollution poses a serious threat for human and environmental health, if not effectively removed from wastewater [5]. Coherently, a recent study highlighted that UASB treatment, followed by an aerobic bioreactor, is not adequately efficient for acute toxicity abatement in municipal wastewater treatment, with final toxicity values ranging from non-toxic to moderately toxic [133]. The applied wastewater treatment processes should consider not only metal removal efficiency, but also the potential to promote the most preferable phase distributions [132]. Metals in UASB sludge demonstrated a high binding potential with coexisting anions, including carbonates, hydroxyls and bicarbonates, with possible negative environmental relapses [132].

The evaluation of trace metal composition is particularly important for sludge agricultural reutilization, considering that nowadays agricultural application is still a commonly applied technique for sludge coming from municipal wastewater treatment [5]. To this purpose, six different sewage sludges from UASB reactor were characterized in the work in [134], considering ten metals (Ni, Mn, Se, Co, Fe, Zn, K, Cu, Pb, Cr). Se, Zn, Ni and Fe were found at high percentage in the sulfide-organic matter fraction, while Co and K were mostly present in the exchangeable and carbonate fractions and Pb appeared in the residual fraction [134]. In a consistent way, a high capability of UASB reactor to remove selenate in presence of cadmium and zinc was reported in the work in [135] under psychrophilic and mesophilic conditions, with excellent removal efficiency, coupled with very high Cd(II) and Zn(II) abatements (particularly in the mesophilic range).

Another critical aspect to be considered in wastewater treatment, as previously mentioned, is the presence of pharmaceutical residuals, which pose severe risks to the receiving environment, if not removed from wastewater. Antibiotics, in particular, have been detected in water, soil, manure and sludge, due to the low removal efficiency commonly observed in conventional pharmaceutical WWTPs [136]. The application of advanced oxidation processes (AOPs) is being currently studied, considering that the highly reactive species produced in AOPs can destroy antibiotics and deactivate microbes, together with refractory compounds [136,137]. In this framework, a complex treatment scheme, including a UASB reactor, an anoxic-oxic tank and a final AOP (UV, ozonation, Fenton, Fenton/UV) was proposed in the work in [136] to simultaneously remove 18 antibiotics and 10 antibiotic resistant genes. Interestingly, UASB reactor gave the most significant contribution (85.8%) in pharmaceutical abatement, with mechanisms including both degradation and sorption to sludge [136]. A similar complex treatment scheme, with a UASB treatment followed by an aerobic sequencing batch reactor (SBR) and a final Fenton-like oxidation process, was proposed in [138] to abate antibiotics (highly concentrated in swine wastewater). UASB was efficient in abating organic contaminants (COD removal of 75%) while SBR and Fenton-like processes removed antibiotics (efficiencies, respectively, >95% and 74%), allowing for reutilization of treated wastewater as farm fertilizer [138]. It could be concluded that, while UASB treatment shows a good abatement even of hazardous and refractory emerging contaminants, an accumulation in the sludge fraction of these compounds is observed: a thorough characterization of the material is consequently required before land application, to avoid negative environmental effects.

11. Critical Aspects and Future Perspectives

Biogas from AD is a relevant renewable energy source that plays a significant role in environmental pollution mitigation and local electricity production [139]. While most of the available literature studies focus on UASB application to different substrates at laboratory or pilot scale, by analyzing process efficiency and biogas production, a broader focus is needed to evaluate techno-economic scale-up feasibility, together with the possibility to integrate biogas with other renewable energies, such as photovoltaic or wind energy. A more sustainable approach in the water-energy nexus is desirable, in particular in developing regions such as Latin America [140], with efficient biogas exploitation.

Methodologies based on life cycle assessment (LCA) and criteria indicators in sustainability studies allow to evaluate and compare different scenarios to optimize the energy and environmental aspects in biogas exploitation [140] and could be integrated with a detailed process modeling. The application of existing powerful modeling devices, in fact, helps to increase plant efficiency, fixing the most critical energy wastages. Nowadays, there exist hundreds of full-scale UASB reactors in various areas of the tropical world, in particular in India and Latin America (especially Brazil) [141]. However, operating problems, including scum accumulation in the settler compartment and in the gas-liquid-solid separator, as well as odor nuisance, often pose a significant threat to maximize reactor performances [141]. Harmful H_2S emissions due to the reduction of sulfate contained in municipal wastewater should be measured and abated [142]. An in-depth monitoring campaign was carried in full-scale UASB plants in Brazil in the work in [141] to assess these aspects, proposing a basic UASB operation optimization.

A simplified wastewater treatment scheme, including a UASB reactor followed by constructed wetlands, could be a feasible solution for developing countries, due to its simplicity and ease of operations [143]. To this purpose, a LCA approach was recently proposed to evaluate a Brazilian full-scale WWTP, consisting in a UASB reactor and a successive wetland treatment, including also reactor construction costs: a negative impact of air emissions from UASB reactor on global warming was again highlighted [143]. Consequently, a careful monitoring and a proper treatment unit of odor emissions from anaerobic reactors is recommended to avoid negative consequences also from a public perception perspective [16]. Different indicators (environmental, social and economic parameters) were proposed to assess the technical and economic suitability of a UASB reactor integrated with a down-flow hanging sponge (DHS) for sewage treatment, comparing this solution to other full-scale wastewater treatment schemes commonly found in India [144]. It was proved that UASB-DHS process had the highest value of the global sustainability indicator. In addition, a trade-off between an environmental-friendly wastewater treatment and the socio-economic aspects was highlighted [144].

Energy analysis is particularly important to assess the full-scale anaerobic treatment feasibility and the capability of the produced biogas to fulfil a significant share of plant energy requirement. The energy potential of sludge and biogas in a Brazilian WWTP (70,000 population equivalent) was studied in the work in [145], claiming that electricity produced from biogas could supply up to 57.6% of total WWTP energy demand [145]. Multivariate analysis in wastewater treatment can be a meaningful tool to reduce operating costs [146]. As an example, the application of Principal Component Analysis (PCA) in swine wastewater treatment (consisting in a UASB reactor followed by aerobic filters and wetland) allowed to increase COD removal efficiency from 45% to 67% [146].

The most important advantages and critical aspects for further UASB application, in light of the results of recent literature studies, were briefly summarized in Table 5. A notable interest was found in UASB reactor application to a pool of different high-loaded substrates, focusing on the treatment of recalcitrant wastewater (by selection of modified UASB systems or complex process schemes) and on low-temperature decentralized UASB treatment, to allow a better exploitation of this technology in remote and rural areas. Advanced high-rate systems, such as IC and ICX reactors, were specifically targeted to enhance treatable organic loads and to optimize the separation between sludge, effluent and biogas. Beside methane, also hydrogen and VFA can be produced in UASB systems, leading to biorefineries development. At the same time, an efficient and compact nitrogen removal can be performed through Anammox process. A significant number of studies related treatment efficiency and biogas production to microbial community composition, giving profound insights in the process understanding. The integration of biogas with other renewable energy sources (such as photovoltaic and wind energy), despite not being often considered, could lead to local 100% renewable energy communities. Biogas and solar energy integration is highly recommended in the future to allow a significant reduction in GHG emissions, mitigating climate change and following the proposed sustainable development goals (SDG). However, attention has to be put in the residual toxicity of sludge and treated wastewater, not to compromise the receiving environment, focusing both on traditional (heavy metals, microbial indicators) and emerging (pharmaceuticals, microplastic) contaminants.

Table 5. Advantages and critical aspects of UASB application, according to recent literature studies.

Advantages	Critical Aspects
Possibility to get clean energy in decentralized areas	Not efficient disinfection and nutrient removal
Proved efficiency on high-loaded biodegradable streams	Residual effluent toxicity in the sludge to be carefully evaluated
	Low-temperature operations not efficient on diluted streams
Modified UASB systems allow to efficiently treat refractory streams	Limited integration with other renewable energy sources
Co-digestion of complementary substrates in the same territory increases plant sustainability	Need for a post-treatment to abate pollutants under required law limits
Possibility to produce hydrogen and volatile fatty acids	Limited biogas valorization in developing countries
Possibility to abate N through granular Anammox process	UASB start-up phase is particularly critical
Significant reduction of excess sludge in comparison with traditional flocculent processes	Pre-treatments to increase biogas yield are not very applied at full-scale
Improved microbial understanding and use of modeling tools helps to optimize performances	Need for an efficient odor abatement

12. Conclusions

High-rate up-flow anaerobic sludge blanket (UASB) reactor is being increasingly applied worldwide to produce renewable energy in biogas form from a variety of wastewater streams, including both industrial and municipal substrates. In this review, the most recent advances regarding UASB application were critically presented and discussed, focusing on the most important engineering aspects for further full-scale development. It was demonstrated that UASB reactor is able to effectively treat not only highly biodegradable substrates, but also recalcitrant or diluted fractions, by selection of modified UASB systems, multi-stage anaerobic process, proper substrate pre-treatment. Anaerobic microbial community composition has been intensively studied, being sensitive to operating conditions and treated substrate. In addition, an improvement in understanding the complex hydrodynamic UASB behavior has been recently accomplished for powerful modeling purposes. Co-digestion between different substrates, available in the same area, allows increasing the obtainable methane yield, stabilizing reactor operations, due to an improved macro- and micro-nutrient balance. Furthermore, UASB has shown to be effective in nitrogen removal, when employed as Anammox treatment.

A particular deepening was made on low-temperature decentralized UASB treatment of municipal wastewater, which is able to locally integrate electricity with end users, creating efficient and sustainable renewable energy clusters. As for toxicity aspects, UASB was shown to be effective in reducing heavy metal concentration and in some cases antibiotics, even if its disinfection efficiency is known to be limited. Most of the recent studies report laboratory or pilot-scale trials and consequently a further effort is required to prove full-scale energy and economic sustainability of the proposed solutions. Finally, it was concluded that the integration of biogas with other renewable energies, such as wind and photovoltaic energy, needs to be encouraged, with an integration between process and energy models, in order to create 100% renewable communities and reduce greenhouse gases emissions.

Author Contributions: Conceptualization, M.M. and D.G.; methodology, M.M.; formal analysis, M.M.; investigation, M.M.; data curation, M.M.; writing—original draft preparation, M.M.; writing—review and editing, M.M. and M.B.; visualization, M.B.; supervision, D.G. All authors have read and agreed to the published version of the manuscript.

Funding: This research received no external funding.

Conflicts of Interest: The authors declare no conflict of interest.

References

1. Rao, P.V.; Baral, S.S.; Dey, R.; Mutnuri, S. Biogas generation potential by anaerobic digestion for sustainable energy development in India. *Renew. Sustain. Energy Rev.* **2010**, *14*, 2086–2094. [CrossRef]

2. Rasapoor, M.; Young, B.; Brar, R.; Sarmah, A.; Zhuang, W.-Q.; Baroutian, S. Recognizing the challenges of anaerobic digestion: Critical steps toward improving biogas generation. *Fuel* **2020**, *261*, 116497. [CrossRef]

3. Kumar, A.; Samadder, S.R. Performance evaluation of anaerobic digestion technology for energy recovery from organic fraction of municipal solid waste: A review. *Energy* **2020**, *197*, 117253. [CrossRef]

4. Gunes, B.; Stokes, J.; Davis, P.; Connolly, C.; Lawler, J. Pre-treatments to enhance biogas yield and quality from anaerobic digestion of whiskey distillery and brewery wastes: A review. *Renew. Sustain. Energy Rev.* **2019**, *113*, 109281. [CrossRef]

5. Misson, G.; Mainardis, M.; Incerti, G.; Goi, D.; Peressotti, A. Preliminary evaluation of potential methane production from anaerobic digestion of beach-cast seagrass wrack: The case study of high-adriatic coast. *J. Clean. Prod.* **2020**, *254*, 120131. [CrossRef]

6. Zhong, Y.; Chen, R.; Rojas-Sossa, J.-P.; Isaguirre, C.; Mashburn, A.; Marsh, T.; Liu, Y.; Liao, W. Anaerobic co-digestion of energy crop and agricultural wastes to prepare uniform-format cellulosic feedstock for biorefining. *Renew. Energy* **2020**, *147*, 1358–1370. [CrossRef]

7. Zhang, L.; Loh, K.-C.; Zhang, J. Food waste enhanced anaerobic digestion of biologically pretreated yard waste: Analysis of cellulose crystallinity and microbial communities. *Waste Manag.* **2018**, *79*, 109–119. [CrossRef] [PubMed]

8. Mainardis, M.; Flaibani, S.; Trigatti, M.; Goi, D. Techno-economic feasibility of anaerobic digestion of cheese whey in small Italian dairies and effect of ultrasound pre-treatment on methane yield. *J. Environ. Manag.* **2019**, *246*, 557–563. [CrossRef] [PubMed]

9. Bórawski, P.; Bełdycka-Bórawska, A.; Szymańska, E.J.; Jankowski, K.J.; Dubis, B.; Dunn, J.W. Development of renewable energy sources market and biofuels in the European Union. *J. Clean. Prod.* **2019**, *228*, 467–484. [CrossRef]

10. Chong, S.; Sen, T.K.; Kayaalp, A.; Ang, H.M. The performance enhancements of Upflow Anaerobic Sludge Blanket (UASB) reactors for domestic sludge treatment—A state-of-the-art review. *Water Res.* **2012**, *46*, 3434–3470. [CrossRef]

11. Latif, M.A.; Ghufran, R.; Wahid, Z.A.; Ahmad, A. Integrated application of upflow anaerobic sludge blanket reactor for the treatment of wastewaters. *Water Res.* **2011**, *45*, 4683–4699. [CrossRef]

12. Lim, S.J.; Kim, T.-H. Applicability and trends of anaerobic granular sludge treatment processes. *Biomass Bioenergy* **2014**, *60*, 189–202. [CrossRef]

13. Tauseef, S.M.; Abbasi, T.; Abbasi, S.A. Energy recovery from wastewaters with high-rate anaerobic digesters. *Renew. Sustain. Energy Rev.* **2013**, *19*, 704–741. [CrossRef]

14. Abbasi, T.; Abbasi, S.A. Formation and impact of granules in fostering clean energy production and wastewater treatment in Upflow Anaerobic Sludge Blanket (UASB) reactors. *Renew. Sustain. Energy Rev.* **2012**, *16*, 1696–1708. [CrossRef]

15. Vassalle, L.; Díez-Montero, R.; Machado, A.T.R.; Moreira, C.; Ferrer, I.; Mota, C.R.; Passos, F. Upflow anaerobic sludge blanket in microalgae-based sewage treatment: Co-digestion for improving biogas production. *Bioresour. Technol.* **2020**, *300*, 122677. [CrossRef] [PubMed]

16. Mainardis, M.; Goi, D. Pilot-UASB reactor tests for anaerobic valorisation of high-loaded liquid substrates in Friulian mountain area. *J. Environ. Chem. Eng.* **2019**, *7*, 103348. [CrossRef]

17. Enitan, A.M.; Kumari, S.; Odiyo, J.O.; Bux, F.; Swalaha, F.M. Principal component analysis and characterization of methane community in a full-scale bioenergy producing UASB reactor treating brewery wastewater. *Phys. Chem. Earth Parts A B C* **2018**, *108*, 1–8. [CrossRef]

18. Cruz-Salomón, A.; Meza-Gordillo, R.; Rosales-Quintero, A.; Ventura-Canseco, C.; Lagunas-Rivera, S.; Carrasco-Cervantes, J. Biogas production from a native beverage vinasse using a modified UASB bioreactor. *Fuel* **2017**, *198*, 170–174. [CrossRef]

19. Mainardis, M.; Cabbai, V.; Zannier, G.; Visintini, D.; Goi, D. Characterization and BMP tests of liquid substrates for high-rate anaerobic digestion. *Chem. Biochem. Eng. Q.* **2018**, *31*, 508–518. [CrossRef]

20. Prateep Na Talang, R.; Sirivithayapakorn, S. Environmental impacts and economic benefits of different wastewater management schemes for molasses-based ethanol production: A case study of Thailand. *J. Clean. Prod.* **2020**, *247*, 119141. [CrossRef]

21. Kong, Z.; Li, L.; Xue, Y.; Yang, M.; Li, Y.-Y. Challenges and prospects for the anaerobic treatment of chemical-industrial organic wastewater: A review. *J. Clean. Prod.* **2019**, *231*, 913–927. [CrossRef]

22. Liang, J.; Wang, Q.; Yoza, B.A.; Li, Q.X.; Chen, C.; Ming, J.; Yu, J.; Li, J.; Ke, M. Rapid granulation using calcium sulfate and polymers for refractory wastewater treatment in up-flow anaerobic sludge blanket reactor. *Bioresour. Technol.* **2020**, *305*, 123084. [CrossRef] [PubMed]

23. Li, L.; Kong, Z.; Xue, Y.; Wang, T.; Kato, H.; Li, Y.-Y. A comparative long-term operation using Up-flow Anaerobic Sludge Blanket (UASB) and Anaerobic Membrane Bioreactor (AnMBR) for the upgrading of anaerobic treatment of N, N-Dimethylformamide-containing wastewater. *Sci. Total Environ.* **2020**, *699*, 134370. [CrossRef] [PubMed]

24. Geißler, A.; Schwan, B.; Dornack, C. Developing a high-performance methane stage for biomass with high nitrogen loads. *Renew. Energy* **2019**, *143*, 1744–1754. [CrossRef]

25. Chen, H.; Wei, Y.; Xie, C.; Wang, H.; Chang, S.; Xiong, Y.; Du, C.; Xiao, B.; Yu, G. Anaerobic treatment of glutamate-rich wastewater in a continuous UASB reactor: Effect of hydraulic retention time and methanogenic degradation pathway. *Chemosphere* **2020**, *245*, 125672. [CrossRef]

26. Chen, H.; Wei, Y.; Liang, P.; Wang, C.; Hu, Y.; Xie, M.; Wang, Y.; Xiao, B.; Du, C.; Tian, H.; et al. Performance and microbial community variations of a Upflow Anaerobic Sludge Blanket (UASB) reactor for treating monosodium glutamate wastewater: Effects of organic loading rate. *J. Environ. Manag.* **2020**, *253*, 109691. [CrossRef]

27. Ribeiro, F.R.; Passos, F.; Gurgel, L.V.A.; Baêta, B.E.L.; de Aquino, S.F. Anaerobic digestion of hemicellulose hydrolysate produced after hydrothermal pretreatment of sugarcane bagasse in UASB reactor. *Sci. Total Environ.* **2017**, *584–585*, 1108–1113. [CrossRef]

28. Bakraoui, M.; Karouach, F.; Ouhammou, B.; Aggour, M.; Essamri, A.; El Bari, H. Biogas production from recycled paper mill wastewater by UASB digester: Optimal and mesophilic conditions. *Biotechnol. Rep.* **2020**, *25*, e00402. [CrossRef]

29. Liang, J.; Wang, Q.; Yoza, B.A.; Li, Q.X.; Ke, M.; Chen, C. Degradation of guar in an up-flow anaerobic sludge blanket reactor: Impacts of salinity on performance robustness, granulation and microbial community. *Chemosphere* **2019**, *232*, 327–336. [CrossRef]

30. Ni, C.-H.; Chang, C.-Y.; Lin, Y.-C.; Lin, J.C.-T. Simultaneous biodegradation of Tetrahydrofuran, 3-Buten-1-Ol and 1,4-Butanediol in real wastewater by a pilot high-rate UASB reactor. *Int. Biodeterior. Biodegrad.* **2019**, *143*, 104698. [CrossRef]

31. Gür, E.; Demirer, G.N. Anaerobic digestability and biogas production capacity of pistachio processing wastewater in UASB reactors. *J. Environ. Eng.* **2019**, *145*, 04019042. [CrossRef]

32. Han, Y.; Guo, J.; Zhang, Y.; Lian, J.; Guo, Y.; Song, Y.; Wang, S.; Yang, Q. Anaerobic granule sludge formation and perchlorate reduction in an Upflow Anaerobic Sludge Blanket (UASB) reactor. *Bioresour. Technol. Rep.* **2018**, *4*, 123–128. [CrossRef]

33. Vidal, J.; Carvajal, A.; Huiliñir, C.; Salazar, R. Slaughterhouse wastewater treatment by a combined anaerobic digestion/solar photoelectro-fenton process performed in semicontinuous operation. *Chem. Eng. J.* **2019**, *378*, 122097. [CrossRef]

34. Esparza-Soto, M.; Jacobo-López, A.; Lucero-Chávez, M.; Fall, C. Anaerobic treatment of chocolate-processing industry wastewater at different organic loading rates and temperatures. *Water Sci. Technol.* **2019**, *79*, 2251–2259. [CrossRef] [PubMed]

35. Rico, C.; Montes, J.A.; Rico, J.L. Evaluation of different types of anaerobic seed sludge for the high rate anaerobic digestion of pig slurry in UASB reactors. *Bioresour. Technol.* **2017**, *238*, 147–156. [CrossRef] [PubMed]

36. Li, J.; He, C.; Tian, T.; Liu, Z.; Gu, Z.; Zhang, G.; Wang, W. UASB-modified bardenpho process for enhancing bio-treatment efficiency of leachate from a municipal solid waste incineration plant. *Waste Manag.* **2020**, *102*, 97–105. [CrossRef] [PubMed]

37. Wang, W.; Wu, B.; Pan, S.; Yang, K.; Hu, Z.; Yuan, S. Performance robustness of the UASB reactors treating saline phenolic wastewater and analysis of microbial community structure. *J. Hazard. Mater.* **2017**, *331*, 21–27. [CrossRef]

38. Nakasaki, K.; Nguyen, K.K.; Ballesteros, F.C.; Maekawa, T.; Koyama, M. Characterizing the microbial community involved in anaerobic digestion of lipid-rich wastewater to produce methane gas. *Anaerobe* **2020**, *61*, 102082. [CrossRef]

39. Aziz, A.; Basheer, F.; Sengar, A.; Irfanullah; Khan, S.U.; Farooqi, I.H. Biological wastewater treatment (anaerobic-aerobic) technologies for safe discharge of treated slaughterhouse and meat processing wastewater. *Sci. Total Environ.* **2019**, *686*, 681–708. [CrossRef]

40. Mannacharaju, M.; Kannan Villalan, A.; Shenbagam, B.; Karmegam, P.M.; Natarajan, P.; Somasundaram, S.; Arumugam, G.; Ganesan, S. Towards sustainable system configuration for the treatment of fish processing wastewater using bioreactors. *Environ. Sci. Pollut. Res. Int.* **2020**, *27*, 353–365. [CrossRef]

41. Barrera, E.L.; Spanjers, H.; Romero, O.; Rosa, E.; Dewulf, J. A successful strategy for start-up of a laboratory-scale UASB reactor treating sulfate-rich sugar cane vinasse. *J. Chem. Technol. Biotechnol.* **2020**, *95*, 205–212. [CrossRef]

42. Lu, X.; Zhen, G.; Ni, J.; Hojo, T.; Kubota, K.; Li, Y.-Y. Effect of influent COD/SO42—Ratios on biodegradation behaviors of starch wastewater in an Upflow Anaerobic Sludge Blanket (UASB) reactor. *Bioresour. Technol.* **2016**, *214*, 175–183. [CrossRef] [PubMed]

43. Mora, M.; Lafuente, J.; Gabriel, D. Influence of crude glycerol load and pH shocks on the granulation and microbial diversity of a sulfidogenic upflow anaerobic sludge blanket reactor. *Process Saf. Environ. Prot.* **2020**, *133*, 159–168. [CrossRef]

44. Li, H.; Han, K.; Li, Z.; Zhang, J.; Li, H.; Huang, Y.; Shen, L.; Li, Q.; Wang, Y. Performance, granule conductivity and microbial community analysis of Upflow Anaerobic Sludge Blanket (UASB) reactors from mesophilic to thermophilic operation. *Biochem. Eng. J.* **2018**, *133*, 59–65. [CrossRef]

45. Tassew, F.A.; Bergland, W.H.; Dinamarca, C.; Bakke, R. Influences of temperature and substrate particle content on granular sludge bed anaerobic digestion. *Appl. Sci.* **2020**, *10*, 136. [CrossRef]

46. Zhang, L.; Ban, Q.; Li, J.; Wan, C. Functional bacterial and archaeal dynamics dictated by pH stress during sugar refinery wastewater in a UASB. *Bioresour. Technol.* **2019**, *288*, 121464. [CrossRef]

47. Vasconcelos, E.A.F.; Santaella, S.T.; Viana, M.B.; dos Santos, A.B.; Pinheiro, G.C.; Leitão, R.C. Composition and ecology of bacterial and archaeal communities in anaerobic reactor fed with residual glycerol. *Anaerobe* **2019**, *59*, 145–153. [CrossRef]

48. Debik, E.; Coskun, T. Use of the Static Granular Bed Reactor (SGBR) with anaerobic sludge to treat poultry slaughterhouse wastewater and kinetic modeling. *Bioresour. Technol.* **2009**, *100*, 2777–2782. [CrossRef]

49. Wang, T.; Huang, Z.; Ruan, W.; Zhao, M.; Shao, Y.; Miao, H. Insights into sludge granulation during anaerobic treatment of high-strength leachate via a full-scale IC reactor with external circulation system. *J. Environ. Sci.* **2018**, *64*, 227–234. [CrossRef]

50. Diamantis, V.; Aivasidis, A. Performance of an ECSB reactor for high-rate anaerobic treatment of cheese industry wastewater: Effect of pre-acidification on process efficiency and calcium precipitation. *Water Sci. Technol.* **2018**, *78*, 1893–1900. [CrossRef]

51. Hendrickx, T.L.G.; Pessotto, B.; Prins, R.; Habets, L.; Vogelaar, J. Biopaq®ICX: The next generation high rate anaerobic reactor proves itself at full scale. *Water Pract. Technol.* **2019**, *14*, 802–807. [CrossRef]

52. Parker, W.J. Application of the ADM1 model to advanced anaerobic digestion. *Bioresour. Technol.* **2005**, *96*, 1832–1842. [CrossRef] [PubMed]

53. Lorenzo-Llanes, J.; Pagés-Díaz, J.; Kalogirou, E.; Contino, F. Development and application in aspen plus of a process simulation model for the anaerobic digestion of vinasses in UASB reactors: Hydrodynamics and biochemical reactions. *J. Environ. Chem. Eng.* **2019**, 103540. [CrossRef]

54. Chen, Y.; He, J.; Mu, Y.; Huo, Y.-C.; Zhang, Z.; Kotsopoulos, T.A.; Zeng, R.J. Mathematical modeling of Upflow Anaerobic Sludge Blanket (UASB) reactors: Simultaneous accounting for hydrodynamics and bio-dynamics. *Chem. Eng. Sci.* **2015**, *137*, 677–684. [CrossRef]

55. Brito, M.G.S.L.; Nunes, F.C.B.; Magalhães, H.L.F.; Lima, W.M.P.B.; Moura, F.L.C.; Farias Neto, S.R.; Lima, A.G.B. Hydrodynamics of Uasb reactor treating domestic wastewater: A three-dimensional numerical study. *Water* **2020**, *12*, 279. [CrossRef]

56. Tsui, T.-H.; Ekama, G.A.; Chen, G.-H. Quantitative characterization and analysis of granule transformations: Role of intermittent gas sparging in a super high-rate anaerobic system. *Water Res.* **2018**, *139*, 177–186. [CrossRef]

57. Owusu-Agyeman, I.; Eyice, Ö.; Cetecioglu, Z.; Plaza, E. The study of structure of anaerobic granules and methane producing pathways of pilot-scale UASB reactors treating municipal wastewater under sub-mesophilic conditions. *Bioresour. Technol.* **2019**, *290*, 121733. [CrossRef]

58. Callejas, C.; Fernández, A.; Passeggi, M.; Wenzel, J.; Bovio, P.; Borzacconi, L.; Etchebehere, C. Microbiota adaptation after an alkaline pH perturbation in a full-scale UASB anaerobic reactor treating dairy wastewater. *Bioprocess Biosyst. Eng.* **2019**, *42*, 2035–2046. [CrossRef]

59. Show, K.-Y.; Yan, Y.; Yao, H.; Guo, H.; Li, T.; Show, D.-Y.; Chang, J.-S.; Lee, D.-J. Anaerobic granulation: A review of granulation hypotheses, bioreactor designs and emerging green applications. *Bioresour. Technol.* **2020**, *300*, 122751. [CrossRef]

60. Na, J.-G.; Lee, M.-K.; Yun, Y.-M.; Moon, C.; Kim, M.-S.; Kim, D.-H. Microbial community analysis of anaerobic granules in phenol-degrading UASB by next generation sequencing. *Biochem. Eng. J.* **2016**, *112*, 241–248. [CrossRef]

61. Tsui, T.-H.; Wu, H.; Song, B.; Liu, S.-S.; Bhardwaj, A.; Wong, J.W.C. Food waste leachate treatment using an Upflow Anaerobic Sludge Bed (UASB): Effect of conductive material dosage under low and high organic loads. *Bioresour. Technol.* **2020**, *304*, 122738. [CrossRef] [PubMed]

62. Zhao, Z.; Zhang, Y. Application of ethanol-type fermentation in establishment of direct interspecies electron transfer: A practical engineering case study. *Renew. Energy* **2019**, *136*, 846–855. [CrossRef]

63. Li, Y.; Chen, Y.; Wu, J. Enhancement of methane production in anaerobic digestion process: A review. *Appl. Energy* **2019**, *240*, 120–137. [CrossRef]

64. Yang, G.; Fang, H.; Wang, J.; Jia, H.; Zhang, H. Enhanced anaerobic digestion of Up-Flow Anaerobic Sludge Blanket (UASB) by Blast Furnace Dust (BFD): Feasibility and mechanism. *Int. J. Hydrogen Energy* **2019**, *44*, 17709–17719. [CrossRef]

65. Srisowmeya, G.; Chakravarthy, M.; Nandhini Devi, G. Critical considerations in two-stage anaerobic digestion of food waste—A review. *Renew. Sustain. Energy Rev.* **2020**, *119*, 109587. [CrossRef]

66. Sun, C.; Liu, F.; Song, Z.; Li, L.; Pan, Y.; Sheng, T.; Ren, G. Continuous hydrogen and methane production from the treatment of herbal medicines wastewater in the two-phase 'UASBH-ICM' system. *Water Sci. Technol.* **2019**, *80*, 1134–1144. [CrossRef]

67. Rajendran, K.; Mahapatra, D.; Venkatraman, A.V.; Muthuswamy, S.; Pugazhendhi, A. Advancing anaerobic digestion through two-stage processes: Current developments and future trends. *Renew. Sustain. Energy Rev.* **2020**, *123*, 109746. [CrossRef]

68. Kamyab, B.; Zilouei, H.; Rahmanian, B. Investigation of the effect of hydraulic retention time on anaerobic digestion of potato leachate in two-stage mixed-UASB system. *Biomass Bioenergy* **2019**, *130*, 105383. [CrossRef]

69. Wu, J.; Jiang, B.; Feng, B.; Li, L.; Moideen, S.N.F.; Chen, H.; Mribet, C.; Li, Y.-Y. Pre-acidification greatly improved granules physicochemical properties and operational stability of Upflow Anaerobic Sludge Blanket (UASB) reactor treating low-strength starch wastewater. *Bioresour. Technol.* **2020**, *302*, 122810. [CrossRef]

70. Diamantis, V.I.; Kapagiannidis, A.G.; Ntougias, S.; Tataki, V.; Melidis, P.; Aivasidis, A. Two-stage CSTR–UASB digestion enables superior and alkali addition-free cheese whey treatment. *Biochem. Eng. J.* **2014**, *84*, 45–52. [CrossRef]

71. Alpay, T.; Karabey, B.; Azbar, N.; Ozdemir, G. Purified terephthalic acid wastewater treatment using modified two-stage UASB bioreactor systems. *Curr. Microbiol.* **2020**. [CrossRef] [PubMed]

72. Jiraprasertwong, A.; Maitriwong, K.; Chavadej, S. Production of biogas from cassava wastewater using a three-stage Upflow Anaerobic Sludge Blanket (UASB) reactor. *Renew. Energy* **2019**, *130*, 191–205. [CrossRef]

73. Chavadej, S.; Wangmor, T.; Maitriwong, K.; Chaichirawiwat, P.; Rangsunvigit, P.; Intanoo, P. Separate production of hydrogen and methane from cassava wastewater with added cassava residue under a thermophilic temperature in relation to digestibility. *J. Biotechnol.* **2019**, *291*, 61–71. [CrossRef] [PubMed]

74. Mainardis, M.; Flaibani, S.; Mazzolini, F.; Peressotti, A.; Goi, D. Techno-economic analysis of anaerobic digestion implementation in small Italian breweries and evaluation of biochar and granular activated carbon addition effect on methane yield. *J. Environ. Chem. Eng.* **2019**, *7*, 103184. [CrossRef]

75. Siddique, N.I.; Wahid, Z.A. Achievements and perspectives of anaerobic co-digestion: A review. *J. Clean. Prod.* **2018**, *194*, 359–371. [CrossRef]

76. Chan, P.C.; Lu, Q.; de Toledo, R.A.; Gu, J.-D.; Shim, H. Improved anaerobic co-digestion of food waste and domestic wastewater by copper supplementation—Microbial community change and enhanced effluent quality. *Sci. Total Environ.* **2019**, *670*, 337–344. [CrossRef]

77. Loizia, P.; Neofytou, N.; Zorpas, A.A. The concept of circular economy strategy in food waste management for the optimization of energy production through anaerobic digestion. *Environ. Sci. Pollut. Res.* **2019**, *26*, 14766–14773. [CrossRef]

78. Kumari, K.; Suresh, S.; Arisutha, S.; Sudhakar, K. Anaerobic co-digestion of different wastes in a UASB reactor. *Waste Manag.* **2018**, *77*, 545–554. [CrossRef]

79. Gao, M.; Zhang, L.; Liu, Y. High-loading food waste and blackwater anaerobic co-digestion: Maximizing bioenergy recovery. *Chem. Eng. J.* **2020**, *394*, 124911. [CrossRef]

80. Sampaio, G.F.; Dos Santos, A.M.; da Costa, P.R.; Rodriguez, R.P.; Sancinetti, G.P. High rate of biological removal of sulfate, organic matter, and metals in UASB reactor to treat synthetic acid mine drainage and cheese whey wastewater as carbon source. *Water Environ. Res.* **2020**, *92*, 245–254. [CrossRef]

81. Montes, J.A.; Leivas, R.; Martínez-Prieto, D.; Rico, C. Biogas production from the liquid waste of distilled gin production: Optimization of UASB reactor performance with increasing organic loading rate for co-digestion with swine wastewater. *Bioresour. Technol.* **2019**, *274*, 43–47. [CrossRef] [PubMed]

82. Rico, C.; Muñoz, N.; Fernández, J.; Rico, J.L. High-load anaerobic co-digestion of cheese whey and liquid fraction of dairy manure in a one-stage UASB process: Limits in co-substrates ratio and organic loading rate. *Chem. Eng. J.* **2015**, *262*, 794–802. [CrossRef]

83. Junior, A.E.S.; Duda, R.M.; De Oliveira, R.A. Improving the energy balance of ethanol industry with methane production from vinasse and molasses in two-stage anaerobic reactors. *J. Clean. Prod.* **2019**, *238*, 117577. [CrossRef]

84. Ma, H.; Niu, Q.; Zhang, Y.; He, S.; Li, Y.-Y. Substrate inhibition and concentration control in an UASB-anammox process. *Bioresour. Technol.* **2017**, *238*, 263–272. [CrossRef]

85. Zhang, F.; Li, X.; Wang, Z.; Jiang, H.; Ren, S.; Peng, Y. Simultaneous ammonium oxidation denitrifying (SAD) in an innovative three-stage process for energy-efficient mature landfill leachate treatment with external sludge reduction. *Water Res.* **2020**, *169*, 115156. [CrossRef]

86. Li, X.; Lu, M.; Qiu, Q.; Huang, Y.; Li, B.; Yuan, Y.; Yuan, Y. the effect of different denitrification and partial nitrification-anammox coupling forms on nitrogen removal from mature landfill leachate at the pilot-scale. *Bioresour. Technol.* **2020**, *297*, 122430. [CrossRef]

87. Li, M.-C.; Song, Y.; Shen, W.; Wang, C.; Qi, W.-K.; Peng, Y.; Li, Y.-Y. The performance of an anaerobic ammonium oxidation upflow anaerobic sludge blanket reactor during natural periodic temperature variations. *Bioresour. Technol.* **2019**, *293*, 122039. [CrossRef]

88. Pekyavas, G.; Yangin-Gomec, C. Response of anammox bacteria to elevated nitrogen and organic matter in pre-digested chicken waste at a long-term operated UASB reactor initially seeded by methanogenic granules. *Bioresour. Technol. Rep.* **2019**, *7*, 100222. [CrossRef]

89. Mahmod, S.S.; Azahar, A.M.; Tan, J.P.; Jahim, J.M.; Abdul, P.M.; Mastar, M.S.; Anuar, N.; Mohammed Yunus, M.F.; Asis, A.J.; Wu, S.-Y.; et al. Operation performance of Up-flow Anaerobic Sludge Blanket (UASB) bioreactor for biohydrogen production by self-granulated sludge using pre-treated Palm Oil Mill Effluent (POME) as carbon source. *Renew. Energy* **2019**, *134*, 1262–1272. [CrossRef]

90. Moreno Dávila, I.M.M.; Tamayo Ordoñez, M.C.; Morales Martínez, T.K.; Soria Ortiz, A.I.; Gutiérrez Rodríguez, B.; Rodríguez de la Garza, J.A.; Ríos González, L.J. Effect of fermentation time/hydraulic retention time in a UASB reactor for hydrogen production using surface response methodology. *Int. J. Hydrogen Energy* **2020**. [CrossRef]

91. Buitrón, G.; Muñoz-Páez, K.M.; Quijano, G.; Carrillo-Reyes, J.; Albarrán-Contreras, B.A. Biohydrogen production from winery effluents: Control of the homoacetogenesis through the headspace gas recirculation. *J. Chem. Technol. Biotechnol.* **2020**, *95*, 544–552. [CrossRef]

92. Eregowda, T.; Kokko, M.E.; Rene, E.R.; Rintala, J.; Lens, P.N.L. Volatile fatty acid production from kraft mill foul condensate in upflow anaerobic sludge blanket reactors. *Environ. Technol.* **2020**, 1–14. [CrossRef] [PubMed]

93. Zhang, D.; Shen, J.; Shi, H.; Su, G.; Jiang, X.; Li, J.; Liu, X.; Mu, Y.; Wang, L. Substantially enhanced anaerobic reduction of nitrobenzene by biochar stabilized sulfide-modified nanoscale zero-valent iron: Process and mechanisms. *Environ. Int.* **2019**, *131*, 105020. [CrossRef] [PubMed]

94. Chen, C.; Liang, J.; Yoza, B.A.; Li, Q.X.; Zhan, Y.; Wang, Q. Evaluation of an Up-Flow Anaerobic Sludge Bed (UASB) reactor containing diatomite and maifanite for the improved treatment of petroleum wastewater. *Bioresour. Technol.* **2017**, *243*, 620–627. [CrossRef]

95. El-Khateeb, M.A.; Emam, W.M.; Darweesh, W.A.; El-Sayed, E.S.A. Integration of UASB and down flow hanging non-woven fabric (DHNW) reactors for the treatment of sewage water. *Desalin. Water Treat.* **2019**, *164*, 48–55. [CrossRef]

96. Gonzalez-Tineo, P.A.; Durán-Hinojosa, U.; Delgadillo-Mirquez, L.R.; Meza-Escalante, E.R.; Gortáres-Moroyoqui, P.; Ulloa-Mercado, R.G.; Serrano-Palacios, D. Performance improvement of an integrated anaerobic-aerobic hybrid reactor for the treatment of swine wastewater. *J. Water Process Eng.* **2020**, *34*, 101164. [CrossRef]

97. Carvalho, J.R.S.; Amaral, F.M.; Florencio, L.; Kato, M.T.; Delforno, T.P.; Gavazza, S. Microaerated UASB reactor treating textile wastewater: The core microbiome and removal of azo dye direct black 22. *Chemosphere* **2020**, *242*, 125157. [CrossRef]

98. Musa, M.A.; Idrus, S.; Harun, M.R.; Tuan Mohd Marzuki, T.F.; Abdul Wahab, A.M. A comparative study of biogas production from cattle slaughterhouse wastewater using conventional and modified Upflow Anaerobic Sludge Blanket (UASB) reactors. *Int. J. Environ. Res. Public Health* **2020**, *17*, 283. [CrossRef]

99. Wambugu, C.W.; Rene, E.R.; van de Vossenberg, J.; Dupont, C.; van Hullebusch, E.D. Role of biochar in anaerobic digestion based biorefinery for food waste. *Front. Energy Res.* **2019**, *7*. [CrossRef]

100. Wu, J.; Liu, Q.; Feng, B.; Kong, Z.; Jiang, B.; Li, Y.-Y. Temperature effects on the methanogenesis enhancement and sulfidogenesis suppression in the UASB treatment of sulfate-rich methanol wastewater. *Int. Biodeterior. Biodegrad.* **2019**, *142*, 182–190. [CrossRef]

101. Zhang, L.; Hendrickx, T.L.G.; Kampman, C.; Temmink, H.; Zeeman, G. Co-digestion to support low temperature anaerobic pretreatment of municipal sewage in a UASB–digester. *Bioresour. Technol.* **2013**, *148*, 560–566. [CrossRef]

102. Zhang, L.; De Vrieze, J.; Hendrickx, T.L.G.; Wei, W.; Temmink, H.; Rijnaarts, H.; Zeeman, G. Anaerobic treatment of raw domestic wastewater in a UASB-digester at 10 °C and microbial community dynamics. *Chem. Eng. J.* **2018**, *334*, 2088–2097. [CrossRef]

103. McAteer, P.G.; Christine Trego, A.; Thorn, C.; Mahony, T.; Abram, F.; O'Flaherty, V. Reactor configuration influences microbial community structure during high-rate, low-temperature anaerobic treatment of dairy wastewater. *Bioresour. Technol.* **2020**, 123221. [CrossRef] [PubMed]

104. Crone, B.C.; Garland, J.L.; Sorial, G.A.; Vane, L.M. Significance of dissolved methane in effluents of anaerobically treated low strength wastewater and potential for recovery as an energy product: A review. *Water Res.* **2016**, *104*, 520–531. [CrossRef] [PubMed]

105. Rongwong, W.; Goh, K.; Sethunga, G.S.M.D.P.; Bae, T.-H. Fouling formation in membrane contactors for methane recovery from anaerobic effluents. *J. Membr. Sci.* **2019**, *573*, 534–543. [CrossRef]

106. Hasan, M.N.; Khan, A.A.; Ahmad, S.; Lew, B. Anaerobic and aerobic sewage treatment plants in Northern India: Two years intensive evaluation and perspectives. *Environ. Technol. Innov.* **2019**, *15*, 100396. [CrossRef]

107. Saavedra, O.; Escalera, R.; Heredia, G.; Montoya, R.; Echeverría, I.; Villarroel, A.; Brito, L.L. Evaluation of a domestic wastewater treatment plant at an intermediate city in Cochabamba, Bolivia. *Water Pract. Technol.* **2019**, *14*, 908–920. [CrossRef]

108. Lijó, L.; Malamis, S.; González-García, S.; Moreira, M.T.; Fatone, F.; Katsou, E. Decentralised schemes for integrated management of wastewater and domestic organic waste: The case of a small community. *J. Environ. Manag.* **2017**, *203*, 732–740. [CrossRef]

109. Gao, M.; Zhang, L.; Guo, B.; Zhang, Y.; Liu, Y. Enhancing biomethane recovery from source-diverted blackwater through hydrogenotrophic methanogenesis dominant pathway. *Chem. Eng. J.* **2019**, *378*, 122258. [CrossRef]

110. Prado, L.O.; Souza, H.H.S.; Chiquito, G.M.; Paulo, P.L.; Boncz, M.A. A comparison of different scenarios for on-site reuse of blackwater and kitchen waste using the life cycle assessment methodology. *Environ. Impact Assess. Rev.* **2020**, *82*, 106362. [CrossRef]

111. Gao, M.; Guo, B.; Zhang, L.; Zhang, Y.; Yu, N.; Liu, Y. Biomethane recovery from source-diverted household blackwater: Impacts from feed sulfate. *Process Saf. Environ. Prot.* **2020**, *136*, 28–38. [CrossRef]

112. Slompo, N.D.M.; Quartaroli, L.; Zeeman, G.; da Silva, G.H.R.; Daniel, L.A. Black water treatment by an Upflow Anaerobic Sludge Blanket (UASB) reactor: A pilot study. *Water Sci. Technol.* **2019**, *80*, 1505–1511. [CrossRef] [PubMed]

113. Gao, M.; Guo, B.; Zhang, L.; Zhang, Y.; Liu, Y. Microbial community dynamics in anaerobic digesters treating conventional and vacuum toilet flushed blackwater. *Water Res.* **2019**, *160*, 249–258. [CrossRef] [PubMed]

114. Adhikari, J.R.; Lohani, S.P. Design, installation, operation and experimentation of septic tank—UASB wastewater treatment system. *Renew. Energy* **2019**, *143*, 1406–1415. [CrossRef]

115. Xu, S.; Zhang, L.; Huang, S.; Zeeman, G.; Rijnaarts, H.; Liu, Y. Improving the energy efficiency of a pilot-scale UASB-digester for low temperature domestic wastewater treatment. *Biochem. Eng. J.* **2018**, *135*, 71–78. [CrossRef]

116. Cunha, J.R.; Schott, C.; van der Weijden, R.D.; Leal, L.H.; Zeeman, G.; Buisman, C. Recovery of calcium phosphate granules from black water using a hybrid upflow anaerobic sludge bed and gas-lift reactor. *Environ. Res.* **2019**, *178*, 108671. [CrossRef]

117. Li, B.; Boiarkina, I.; Yu, W.; Huang, H.M.; Munir, T.; Wang, G.Q.; Young, B.R. Phosphorous recovery through struvite crystallization: Challenges for future design. *Sci. Total Environ.* **2019**, *648*, 1244–1256. [CrossRef]

118. Ouhammou, B.; Aggour, M.; Frimane, Â.; Bakraoui, M.; El Bari, H.; Essamri, A. A new system design and analysis of a solar bio-digester unit. *Energy Convers. Manag.* **2019**, *198*, 111779. [CrossRef]

119. Kainthola, J.; Kalamdhad, A.S.; Goud, V.V. A review on enhanced biogas production from anaerobic digestion of lignocellulosic biomass by different enhancement techniques. *Process Biochem.* **2019**, *84*, 81–90. [CrossRef]

120. Rajagopal, R.; Choudhury, M.R.; Anwar, N.; Goyette, B.; Rahaman, M.S. Influence of pre-hydrolysis on sewage treatment in an Up-Flow Anaerobic Sludge BLANKET (UASB) reactor: A review. *Water* **2019**, *11*, 372. [CrossRef]

121. Gurmessa, B.; Pedretti, E.F.; Cocco, S.; Cardelli, V.; Corti, G. Manure anaerobic digestion effects and the role of pre- and post-treatments on veterinary antibiotics and antibiotic resistance genes removal efficiency. *Sci. Total Environ.* **2020**, *721*, 137532. [CrossRef] [PubMed]

122. Domínguez-Maldonado, J.A.; Alzate-Gaviria, L.; Milquez-Sanabria, H.A.; Tapia-Tussell, R.; Leal-Bautista, R.M.; España-Gamboa, E.I. Chemical pretreatments to enrich the acidogenic phase in a system coupled packed bed reactor with a UASB reactor using peels and rotten onion waste. *Waste Biomass Valorization* **2019**. [CrossRef]

123. Paulista, L.O.; Boaventura, R.A.R.; Vilar, V.J.P.; Pinheiro, A.L.N.; Martins, R.J.E. Enhancing methane yield from crude glycerol anaerobic digestion by coupling with ultrasound or A. Niger/E. Coli biodegradation. *Environ. Sci. Pollut. Res.* **2020**, *27*, 1461–1474. [CrossRef] [PubMed]

124. Eftaxias, A.; Diamantis, V.; Michailidis, C.; Stamatelatou, K.; Aivasidis, A. The role of emulsification as pre-treatment on the anaerobic digestion of oleic acid: Process performance, modeling, and sludge metabolic properties. *Biomass Convers Biorefinery* **2020**. [CrossRef]

125. Uddin, M.N.; Rahman, M.A.; Taweekun, J.; Techato, K.; Mofijur, M.; Rasul, M. Enhancement of biogas generation in Up-Flow Sludge Blanket (UASB) bioreactor from Palm Oil Mill Effluent (POME). *Energy Procedia* **2019**, *160*, 670–676. [CrossRef]

126. Zhang, L. Advanced treatment of oilfield wastewater by a combination of DAF, yeast bioreactor, UASB, and BAF processes. *Sep. Sci. Technol.* **2020**, 1–10. [CrossRef]

127. Gadow, S.I.; Li, Y.-Y. Development of an integrated anaerobic/aerobic bioreactor for biodegradation of recalcitrant azo dye and bioenergy recovery: HRT effects and functional resilience. *Bioresour. Technol. Rep.* **2020**, *9*, 100388. [CrossRef]

128. Owaes, M.; Gaur, R.Z.; Hasan, M.N.; Gani, K.M.; Kumari, S.; Bux, F.; Khan, A.A.; Kazmi, A. Performance assessment of aerobic granulation for the post treatment of anaerobic effluents. *Environ. Technol. Innov.* **2020**, *17*, 100588. [CrossRef]

129. Leite, L.d.S.; Hoffmann, M.T.; Daniel, L.A. Microalgae cultivation for municipal and piggery wastewater treatment in Brazil. *J. Water Process Eng.* **2019**, *31*, 100821. [CrossRef]

130. Walia, R.; Kumar, P.; Mehrotra, I. Post-treatment of effluent from UASB reactor by surface aerator. *Int. J. Environ. Sci. Technol.* **2020**, *17*, 983–992. [CrossRef]

131. Tarpani, R.R.Z.; Alfonsín, C.; Hospido, A.; Azapagic, A. Life cycle environmental impacts of sewage sludge treatment methods for resource recovery considering ecotoxicity of heavy metals and pharmaceutical and personal care products. *J. Environ. Manag.* **2020**, *260*, 109643. [CrossRef] [PubMed]

132. Kumar, M.; Gogoi, A.; Mukherjee, S. Metal removal, partitioning and phase distributions in the wastewater and sludge: Performance evaluation of conventional, upflow anaerobic sludge blanket and downflow hanging sponge treatment systems. *J. Clean. Prod.* **2020**, *249*, 119426. [CrossRef]

133. de Souza Celente, G.; Colares, G.S.; da Silva Araújo, P.; Machado, Ê.L.; Lobo, E.A. Acute ecotoxicity and genotoxicity assessment of two wastewater treatment units. *Environ. Sci. Pollut. Res. Int.* **2020**. [CrossRef] [PubMed]

134. Braga, A.F.M.; Zaiat, M.; Silva, G.H.R.; Fermoso, F.G. Metal fractionation in sludge from sewage UASB treatment. *J. Environ. Manag.* **2017**, *193*, 98–107. [CrossRef] [PubMed]

135. Zeng, T.; Rene, E.R.; Hu, Q.; Lens, P.N.L. Continuous biological removal of selenate in the presence of cadmium and zinc in UASB reactors at psychrophilic and mesophilic conditions. *Biochem. Eng. J.* **2019**, *141*, 102–111. [CrossRef]

136. Hou, J.; Chen, Z.; Gao, J.; Xie, Y.; Li, L.; Qin, S.; Wang, Q.; Mao, D.; Luo, Y. Simultaneous removal of antibiotics and antibiotic resistance genes from pharmaceutical wastewater using the combinations of up-flow anaerobic sludge bed, anoxic-oxic tank, and advanced oxidation technologies. *Water Res.* **2019**, *159*, 511–520. [CrossRef] [PubMed]

137. Mainardis, M.; Buttazzoni, M.; De Bortoli, N.; Mion, M.; Goi, D. Evaluation of ozonation applicability to pulp and paper streams for a sustainable wastewater treatment. *J. Clean. Prod.* **2020**, *258*, 120781. [CrossRef]

138. Qian, M.; Yang, L.; Chen, X.; Li, K.; Xue, W.; Li, Y.; Zhao, H.; Cao, G.; Guan, X.; Shen, G.; et al. The treatment of veterinary antibiotics in swine wastewater by biodegradation and fenton-like oxidation. *Sci. Total Environ.* **2020**, *710*, 136299. [CrossRef]

139. Freitas, F.F.; De Souza, S.S.; Ferreira, L.R.A.; Otto, R.B.; Alessio, F.J.; De Souza, S.N.M.; Venturini, O.J.; Ando Junior, O.H. The Brazilian market of distributed biogas generation: Overview, technological development and case study. *Renew. Sustain. Energy Rev.* **2019**, *101*, 146–157. [CrossRef]

140. Meneses-Jácome, A.; Diaz-Chavez, R.; Velásquez-Arredondo, H.I.; Cárdenas-Chávez, D.L.; Parra, R.; Ruiz-Colorado, A.A. Sustainable energy from agro-industrial wastewaters in Latin-America. *Renew. Sustain. Energy Rev.* **2016**, *56*, 1249–1262. [CrossRef]

141. Bressani-Ribeiro, T.; Chamhum-Silva, L.A.; Chernicharo, C.A.L. Constraints, performance and perspectives of anaerobic sewage treatment: Lessons from full-scale sewage treatment plants in Brazil. *Water Sci. Technol.* **2019**, *80*, 418–425. [CrossRef] [PubMed]

142. Gaur, R.Z.; Khan, A.A.; Lew, B.; Diamantis, V.; Kazmi, A.A. Performance of full-scale UASB reactors treating low or medium strength municipal wastewater. *Environ. Process.* **2017**, *4*, 137–146. [CrossRef]

143. Lopes, T.A.S.; Queiroz, L.M.; Torres, E.A.; Kiperstok, A. Low complexity wastewater treatment process in developing countries: A LCA approach to evaluate environmental gains. *Sci. Total Environ.* **2020**, *720*, 137593. [CrossRef]

144. Maharjan, N.; Nomoto, N.; Tagawa, T.; Okubo, T.; Uemura, S.; Khalil, N.; Hatamoto, M.; Yamaguchi, T.; Harada, H. Assessment of UASB-DHS technology for sewage treatment: A comparative study from a sustainability perspective. *Environ. Technol.* **2019**, *40*, 2825–2832. [CrossRef] [PubMed]

145. Rosa, A.P.; Chernicharo, C.A.L.; Lobato, L.C.S.; Silva, R.V.; Padilha, R.F.; Borges, J.M. Assessing the potential of renewable energy sources (biogas and sludge) in a full-scale UASB-based treatment plant. *Renew. Energy* **2018**, *124*, 21–26. [CrossRef]
146. Oliveira, J.F.d.; Fia, R.; Fia, F.R.L.; Rodrigues, F.N.; Matos, M.P.d.; Siniscalchi, L.A.B. Principal component analysis as a criterion for monitoring variable organic load of swine wastewater in integrated biological reactors UASB, SABF and HSSF-CW. *J. Environ. Manag.* **2020**, *262*, 110386. [CrossRef]

 *bioengineering*

Review

Technologies for Biogas Upgrading to Biomethane: A Review

Amir Izzuddin Adnan [1], Mei Yin Ong [1], Saifuddin Nomanbhay [1,*], Kit Wayne Chew [2] and Pau Loke Show [2]

[1]   Institute of Sustainable Energy, Universiti Tenaga Nasional, Kajang 43000, Selangor, Malaysia; izzuddin.amir95@gmail.com (A.I.A.); me089475@hotmail.com (M.Y.O.)

[2]   Department of Chemical and Environmental Engineering, Faculty of Science and Engineering, University of Nottingham Malaysia, Jalan Broga, Semenyih 43500, Selangor, Malaysia; kitwayne.chew@gmail.com (K.W.C.); pauloke.show@nottingham.edu.my (P.L.S.)

*   Correspondence: saifuddin@uniten.edu.my; Tel.: +6-03-8921-7285

Received: 17 September 2019; Accepted: 30 September 2019; Published: 2 October 2019

Abstract: The environmental impacts and high long-term costs of poor waste disposal have pushed the industry to realize the potential of turning this problem into an economic and sustainable initiative. Anaerobic digestion and the production of biogas can provide an efficient means of meeting several objectives concerning energy, environmental, and waste management policy. Biogas contains methane (60%) and carbon dioxide (40%) as its principal constituent. Excluding methane, other gasses contained in biogas are considered as contaminants. Removal of these impurities, especially carbon dioxide, will increase the biogas quality for further use. Integrating biological processes into the bio-refinery that effectively consume carbon dioxide will become increasingly important. Such process integration could significantly improve the sustainability of the overall bio-refinery process. The biogas upgrading by utilization of carbon dioxide rather than removal of it is a suitable strategy in this direction. The present work is a critical review that summarizes state-of-the-art technologies for biogas upgrading with particular attention to the emerging biological methanation processes. It also discusses the future perspectives for overcoming the challenges associated with upgradation. While biogas offers a good substitution for fossil fuels, it still not a perfect solution for global greenhouse gas emissions and further research still needs to be conducted.

Keywords: anaerobic digestion; biogas upgrading; biomethane; bio-succinic acid; CO_2 utilization; feasibility assessment

---

1. Introduction

In the last decades, fossil fuels have been utilized at a high rate as the main energy source for the industrial process as well as daily usage. The result is the increasing crisis of global energy and environmental problems. It has been predicted that the global consumption of energy will increase nearly threefold in the next thirty years [1]. Massive carbon dioxide (CO_2) emission during fossil fuel combustion has raised the concern on energy sustainability and environmental protection issues. The rate of CO_2 that is presently being released at a global scale is more than 1000 kg/s, although it is the imbalance between emissions and sinks that is responsible for the increasing CO_2 concentration in the atmosphere [2]. The reductions of CO_2 emission into the atmosphere can only be achieved by either reducing the CO_2 emissions from the sources or increasing the usage of CO_2. A wide-ranging research plan is needed to develop a variety of carbon utilization technologies suitable for utilizing the abundance of carbon waste in the atmosphere, integrating enabling technologies and resources, and producing a wide range of carbon-based products. Therefore, extensive research needs to be conducted to address the knowledge gaps throughout the carbon utilization landscape in order to reduce greenhouse gas emissions (GHG) while generating economic value. The conversion of CO_2 into added-value chemicals and fuels is considered as one of the great challenges of the 21st century.

To achieve sustainable development, energy resources with low environmental impact should be utilized. Besides petroleum, biomass is the largest source of carbon-rich material available on Earth [3]. Biorefineries represent tremendous potential for the efficient utilization of renewable resources. A biorefinery can be described as a facility that integrates biomass conversion processes and technologies in a sustainable and efficient way to produce a variety of marketable products (food, feed, chemicals, and materials) and energy (biofuels, power, and/or heat) from biomass. Biogas is a well-established renewable energy source for combined heat and power (CHP) generation. Biogas production is a treatment technology that generates renewable energy and recycles organic waste into a digested biomass, which can be used as fertilizer and soil amendment. Biogas is considered a renewable energy source due to the fact that the organic waste has consumed carbon dioxide in the photosynthesis process, and as such can be described as carbon-neutral [4]. The amount of wastes and residues generated has led to the demand for technologies and processes that can help to reduce these residues, which can help achieve the ambitious objective of "zero-waste" targets (or, at least, waste minimization) while obtaining valuable commodities, including renewable-based methane-rich product gas streams. In these regards, waste management technologies based on the anaerobic digestion of different residual streams, such as municipal solid wastes in landfills, agriculture crops, and urban wastewaters that allow the production of biogas, have played a significant role in the last decades. To date, efforts have been made to improve the methane (CH_4) yield during anaerobic digestion. Feedstock selection, process design and operation, digestion enhancement, and co-digestion with multiple substrates have been extensively studied, and several reviews are available [5–9].

Commercial biogas production has increased since it can be used as fuel or energy production while contributes to a lower GHG concentration when it is collected in a closed process and not emitted to the atmosphere. Depending on the nature of the substrate and pH of the reactor, biogas produced consists of CH_4 in a range of 50–70% and CO_2 at a concentration of 30–50%, with the addition of minor components such as hydrogen sulfide (H_2S), nitrogen (N_2), oxygen (O_2), siloxanes, volatile organic compounds (VOCs), carbon monoxide (CO), and ammonia (NH_3). It is estimated that biogas usage in the world will be doubled in the coming years, increasing from 14.5 GW in 2012 to 29.5 GW in 2022 [10,11]. Apart from CH_4, the remaining components in biogas are undesirable and considered as impurities. Basically, there are two steps involved in biogas treatment, cleaning (removal of minor unwanted components of biogas), and upgrading (removal of CO_2 content) [10,11]. After the processes, the final product is called biomethane which composed of CH_4 (95–99%) and CO_2 (1–5%), with no trace of H_2S. Biogas cleaning is usually considered the first step for biogas applications and is an energy-demanding process. The second treatment is called "biogas upgrading" and aims to increase the low calorific value of the biogas, and thus, to convert it to a higher fuel standard [12]. Nowadays, there are different treatments targeted at removing the undesired compounds from the biogas, thus expanding its range of applications. High CH_4 purity biogas has the same properties as natural gas, especially in terms of heating value, therefore, this clean biogas is qualified to be injected into a natural gas grid [13]. An early notable review report on biogas upgrading was published in 2009, providing a complete view on the situation of biogas upgrading at that time, however, the topic on CO_2 removal was only briefly discussed [14]. More review reports on biogas purification and upgrading had appeared recently. The first of them was by Ryckebosch and others (2011) [15] discussing the state of affairs of different techniques for biogas transformation and their functions, efficiency, and barriers. Next, Bauer et al. (2013) [16] reviewed and compared the commercial technologies on biogas upgrading. In later years, Sun et al. (2015) [12] had come out with a more detailed review on biogas upgrading technology, focusing on biogas purity and impurities, CH_4 recovery and loss, upgrading efficiency, investment, and operating cost. These were among the many reviews that were conducted on the topic of biogas upgrading involving CO_2 removal. Therefore, in this review, an attempt is made to present new technologies for biogas upgrading via the utilization and conversion of CO_2 rather than the removal of CO_2. The already matured technologies will only be briefly summarized.

2. Biogas Upgrading via Carbon Dioxide Removal Technologies

As a means to upgrade biogas to a higher fuel standard, that is, to remove unwanted components such as CO_2 and H_2S thus increasing its specific caloric value, several different approaches have been proposed [17,18]. The mature technologies that are today currently applied for biogas upgrading are illustrated in Figure 1. The focus of this section is to summarize the important details regarding current CO_2 removal technologies rather than going into details on it.

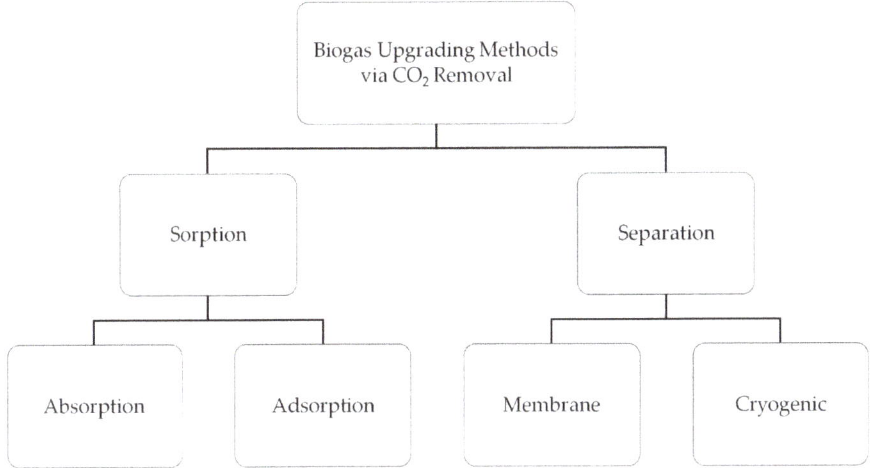

Figure 1. Technologies for biogas upgrading via CO_2 removal route.

The gas sorption is divided into two categories: physical and chemical scrubbing. Physical scrubbing and chemical scrubbing processes were summarized in Figures 2 and 3 respectively. Next, the adsorption method was usually done in a process called pressure swing adsorption and can be seen as a summarized point in Figure 4. Then, the term separation is applied in membrane technology and cryogenic separation and depicted as in Figures 5 and 6, respectively.

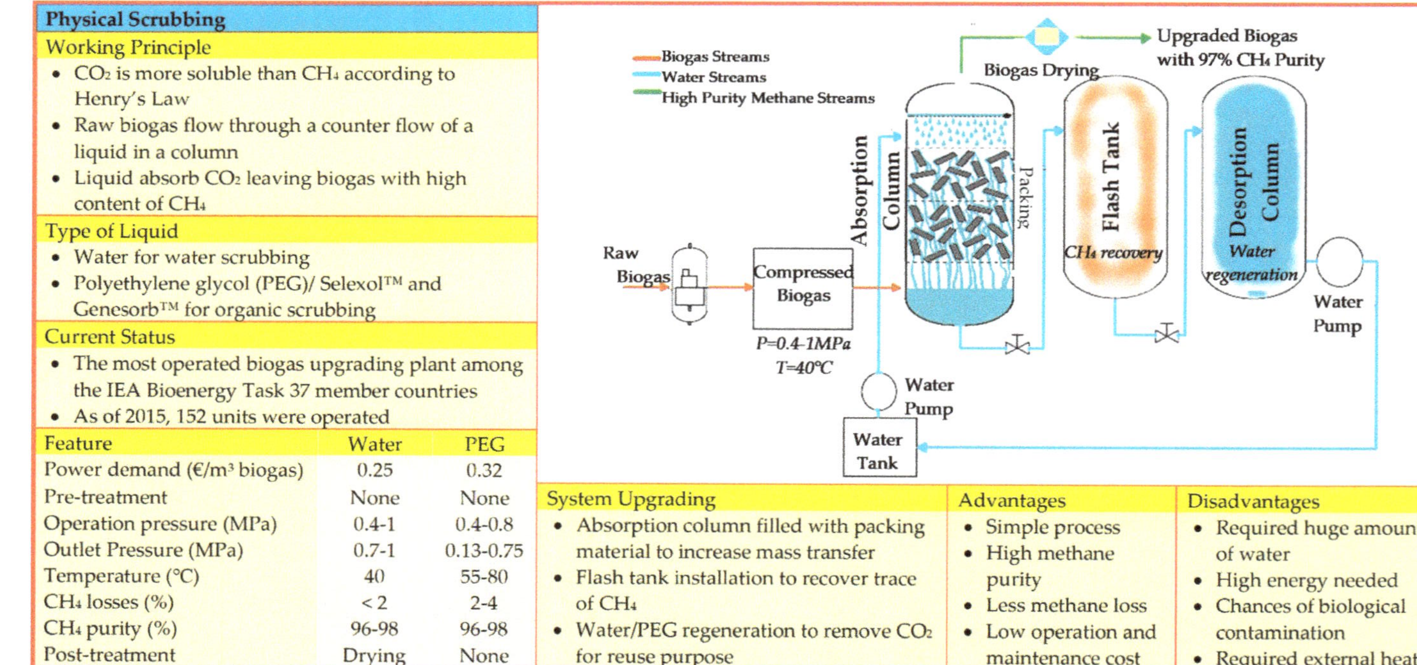

Feature	Water	PEG
Power demand (€/m³ biogas)	0.25	0.32
Pre-treatment	None	None
Operation pressure (MPa)	0.4-1	0.4-0.8
Outlet Pressure (MPa)	0.7-1	0.13-0.75
Temperature (°C)	40	55-80
CH₄ losses (%)	< 2	2-4
CH₄ purity (%)	96-98	96-98
Post-treatment	Drying	None

System Upgrading	Advantages	Disadvantages
• Absorption column filled with packing material to increase mass transfer • Flash tank installation to recover trace of CH₄ • Water/PEG regeneration to remove CO₂ for reuse purpose	• Simple process • High methane purity • Less methane loss • Low operation and maintenance cost	• Required huge amount of water • High energy needed • Chances of biological contamination • Required external heat

Figure 2. Summary of fundamental knowledge on physical scrubbing technology [15,19–28].

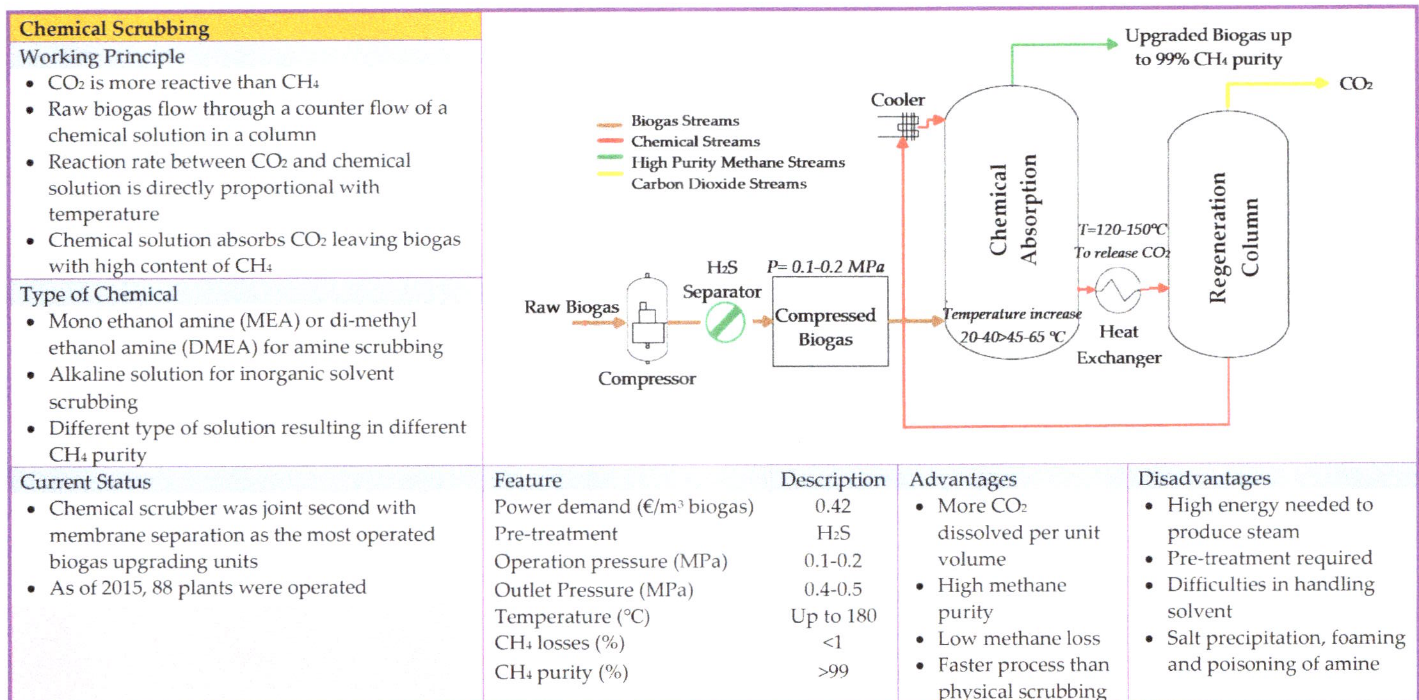

Figure 3. Summary of basic information on chemical scrubbing technology [15,19,20,25,27–32].

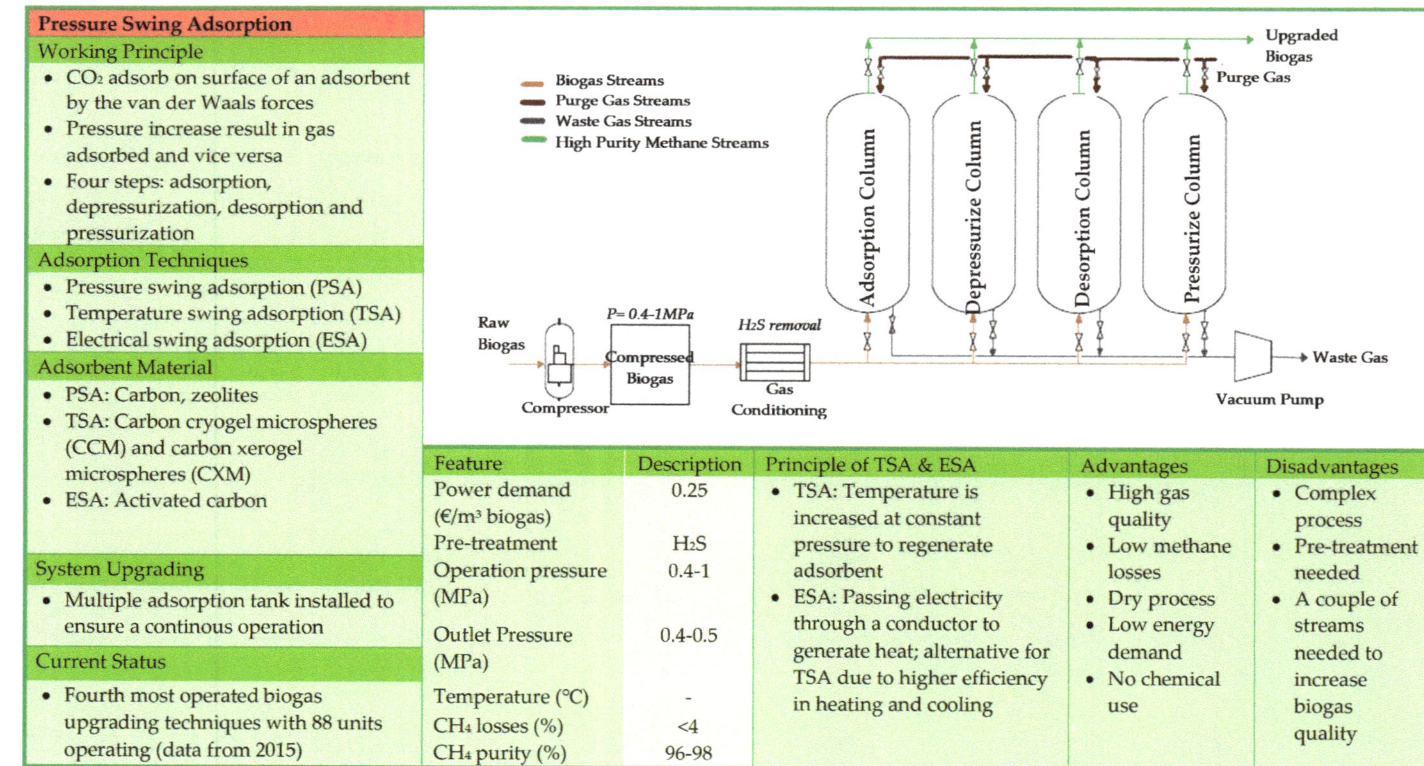

Feature	Description	Principle of TSA & ESA	Advantages	Disadvantages
Power demand (€/m³ biogas)	0.25	• TSA: Temperature is increased at constant pressure to regenerate adsorbent • ESA: Passing electricity through a conductor to generate heat; alternative for TSA due to higher efficiency in heating and cooling	• High gas quality • Low methane losses • Dry process • Low energy demand • No chemical use	• Complex process • Pre-treatment needed • A couple of streams needed to increase biogas quality
Pre-treatment	H2S			
Operation pressure (MPa)	0.4-1			
Outlet Pressure (MPa)	0.4-0.5			
Temperature (°C)	-			
CH4 losses (%)	<4			
CH4 purity (%)	96-98			

Figure 4. Depicts information on pressure swing adsorption technology [19,20,25,28,33–37].

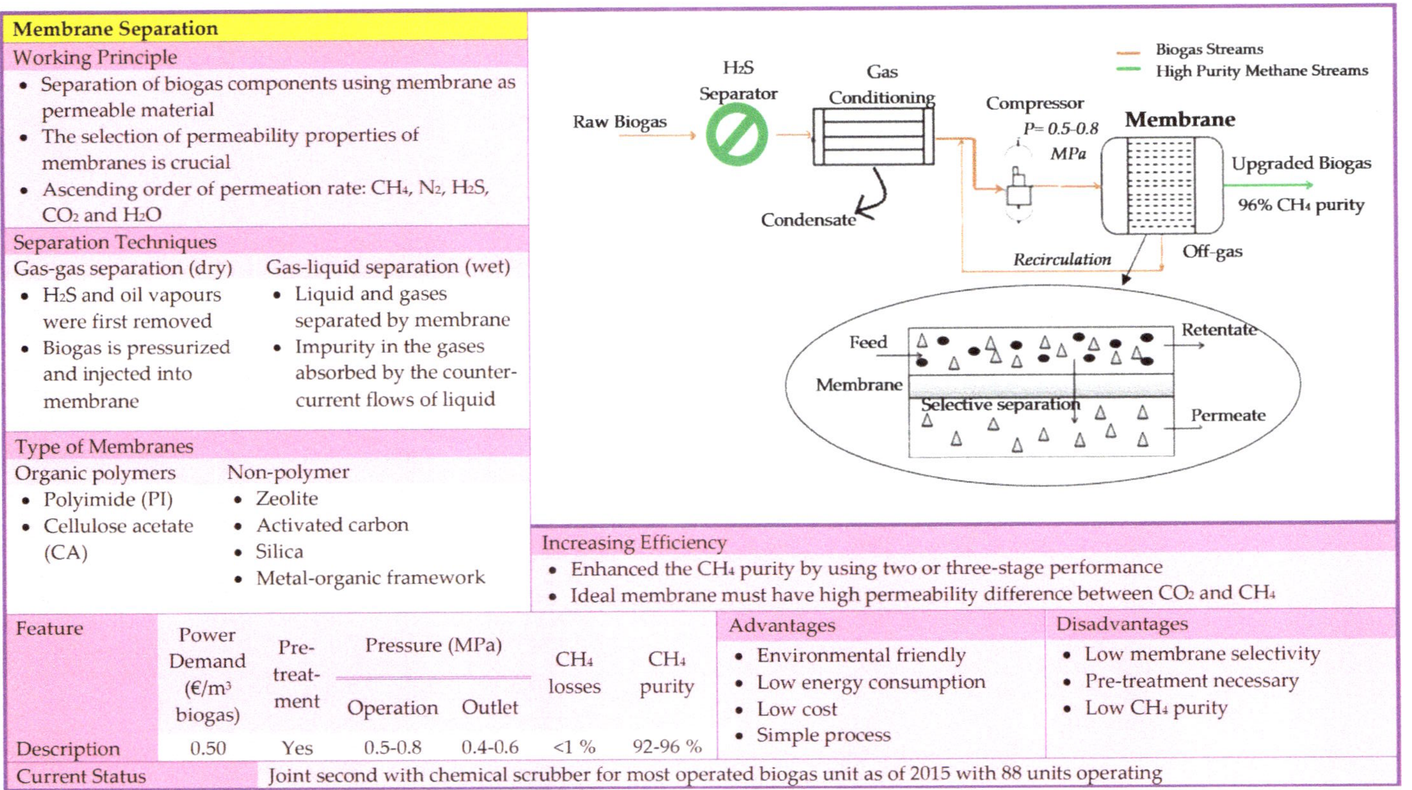

Feature	Power Demand (€/m³ biogas)	Pre-treatment	Pressure (MPa) Operation	Pressure (MPa) Outlet	CH₄ losses	CH₄ purity
Description	0.50	Yes	0.5-0.8	0.4-0.6	<1 %	92-96 %
Current Status	Joint second with chemical scrubber for most operated biogas unit as of 2015 with 88 units operating					

Figure 5. Summary of base knowledge of membrane separation technology [20,21,25,27–29,38,39].

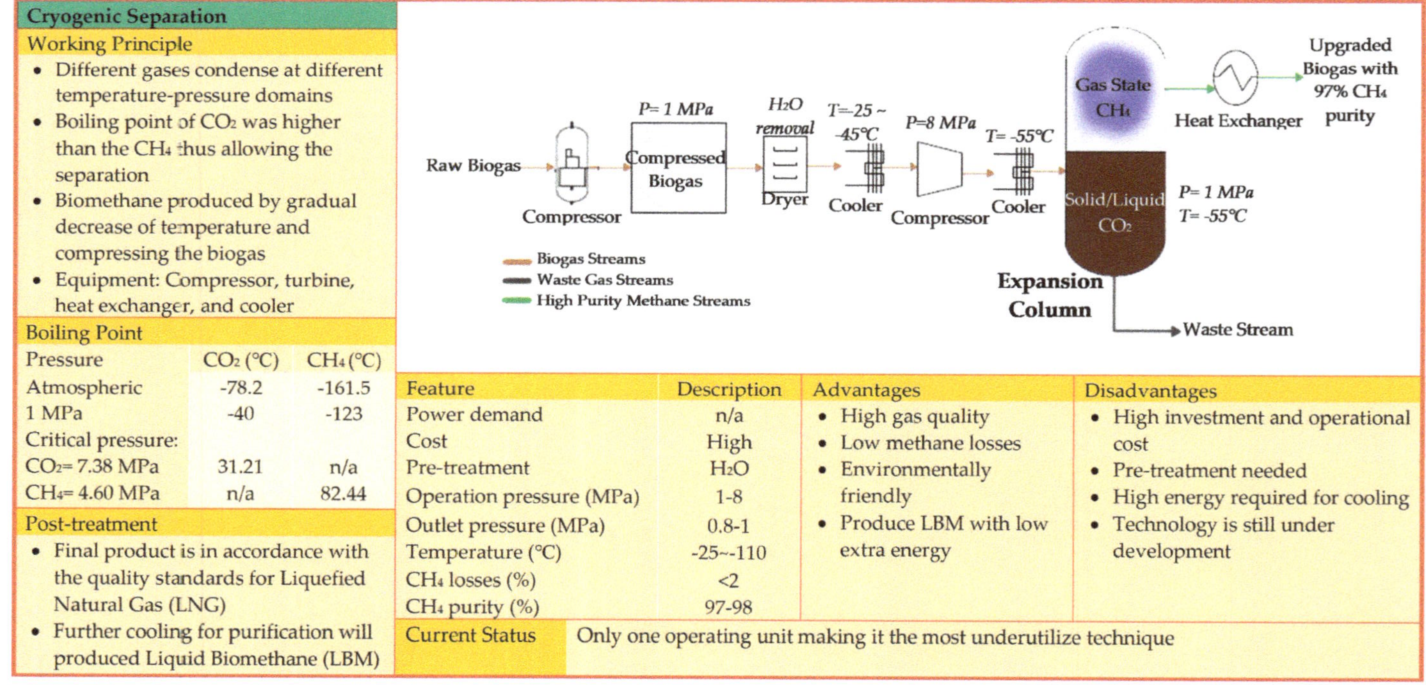

Boiling Point		
Pressure	CO₂ (°C)	CH₄ (°C)
Atmospheric	-78.2	-161.5
1 MPa	-40	-123
Critical pressure:		
CO₂= 7.38 MPa	31.21	n/a
CH₄= 4.60 MPa	n/a	82.44

Post-treatment
- Final product is in accordance with the quality standards for Liquefied Natural Gas (LNG)
- Further cooling for purification will produced Liquid Biomethane (LBM)

Feature	Description	Advantages	Disadvantages
Power demand	n/a	• High gas quality	• High investment and operational cost
Cost	High	• Low methane losses	
Pre-treatment	H₂O	• Environmentally friendly	• Pre-treatment needed
Operation pressure (MPa)	1-8		• High energy required for cooling
Outlet pressure (MPa)	0.8-1	• Produce LBM with low extra energy	• Technology is still under development
Temperature (°C)	-25~-110		
CH₄ losses (%)	<2		
CH₄ purity (%)	97-98		
Current Status	Only one operating unit making it the most underutilize technique		

Figure 6. Depicts fundamental knowledge on cryogenic separation [15,19,20,25,28,29,40–43].

The benefits of biogas to the environment are often discussed as a sustainable source of fuels [44]. However, some biogas components released from biogas upgrading are associated with GHG, especially CO_2. The direct impacts of excessive CO_2 emission are global warming, ocean acidification, and carbon fertilization. The released CO_2 needs to be disposed of. It includes the processes of CO_2 liquifying and injection into underground aquifers. The drawback of this process is the possibility of CO_2 leaking and returning to the surface. Furthermore, the cost of CO_2 disposal is very high and uncertain (among the factors that contribute to cost are the size of the plant and the distance). Thus, a possible solution for this problem is through CO_2 utilization technology. This technology holds big potential for a new way of upgrading biogas, since the benefits of utilizing CO_2 could potentially overcome the cost of CO_2 disposal and reduce the cost of biogas upgrading. The next section of this review will focus on the discussion of various techniques for the utilization of CO_2 as reported in the literature.

3. Biogas Upgrading via Carbon Dioxide Utilization Technologies

In the previous section, biogas was upgraded to enrich the methane content and treated directly as fuel without essential chemical changes. The technologies are always changing, and researchers have developed methods to further explore the value of raw biogas. In recent years, biogas has been used as feedstock in producing chemical material by utilizing the CO_2 content in the biogas [45]. In addition, this low-grade biogas will benefit society by the production of high-quality products instead of inefficient heat supply that results in higher pollution. This section will discuss the state-of-art of emerging technologies for biogas upgrading through CO_2 utilization.

3.1. Chemical Processes

It is well known that using CO_2 as a feedstock for the synthesis of commodity chemicals and fuels has the potential to be beneficial for the economy and environment [46]. CO_2 with the molecular weight of 44.01 and critical density of 468 kg/m^3 can be in a liquid state at a pressure below 415.8 kPa and in the form of solid under $-78\ °C$. It is a massively produced waste and the main contributor to global warming. Despite the potential, the challenges that arise from the utilization of CO_2 are the need for large inputs of energy and the strong bonds that are not particularly reactive due to its kinetic and thermodynamic stability. For instance, it is not affected by heat under normal conditions until the temperature reaches about 2000 °C [47]. Consequently, the process of converting CO_2 requires stoichiometric amounts of energy-intensive reagents that lead to the generation of other waste and increasing GHG footprints. Thus, the main challenge is to develop a new technology that can reduce the use of non-renewable energy and reduce GHG emissions.

Methanation reaction, also called a Sabatier reaction is a reaction between CO_2 and H_2 to produce CH_4 and water (H_2O). Although the reaction is between CO_2 and H_2, there is the potential of using biogas directly as feedstock for CO_2 methanation as CH_4 content in the biogas has only a little influence on the reaction at high pressure [48]. The research has found that the methanation of CO_2 above 0.8 MPa will be ideal to decrease the effect of CH_4 on the conversion process [49]. CH_4 is consumed by the consumer widely as a fuel in 2014 (3500 billion cubic meters) [50]. The main source of CH_4 is natural gas, and occasionally as a result of synthetization. The process of hydrogenation of CO_2 to CH_4 using Ni catalyst is explained by Sabatier reaction in Equation (1) [51].

$$CO_2 + 4H_2 \rightarrow CH_4 + 2H_2O \qquad \Delta H = -165\ kJ/mol \qquad (1)$$

The research in the improvement of catalysts is still developing. Challenges that need to be confronted include the catalysts that can operate at lower temperatures where the reaction more promising and preventing the deactivation of nickel-based catalysts due to sintering and oxidation. Sintering occurs due to the high temperature and water while oxidation is due to the presence of H_2 [52,53]. The improvement of catalysts and processes that have been recently discovered are simplified in Table 1.

Table 1. Improvement of catalysts in methane production.

Modification	Description/Results	Reference
Ruthenium *Electrochemical*	More advanced than nickel but costly	[54]
N-doped carbon Copper-on-carbon	Using the standard three-electrode or H cells Faradaic efficiencies 80% to 94%	[55–57]
Copper	Electrodeposited on a carbon gas diffusion electrode 38 mA/cm^2 densities of methane formation	[56]

On the other hand, by changing the nature of catalysts to less reactive catalysts result in the production of methanol. In 2015, approximately 70 billion kg of methanol (CH_3OH) was produced worldwide from the synthetization of syngas ($H_2 + CO_2$) obtained directly from fossil fuels [58–61]. The mechanism of methanol production, seen in Equation (2), involves a side reaction between CO_2 and H_2 to produce CO and H_2O based on water gas-shift reaction as shown by Equation (3).

$$CO_2 + 3H_2 \leftrightarrow CH_3OH + H_2O \qquad \Delta H_{298K} = -90.70 \text{ kJ/mol} \qquad (2)$$

$$H_2 + CO_2 \leftrightarrow CO + H_2O \qquad \Delta H_{298K} = 41.19 \text{ kJ/mol} \qquad (3)$$

The methanol formation here is an exothermic reaction and the molecular weight of molecules with carbon decrease. Thus, there will be an increase in pressure and a decrease in temperature for selectivity. But, as mentioned earlier, CO_2 is not very reactive and needs a high reaction temperature (>513 K) for CO_2 conversion to occur. In recent years, a lot of research has been done on the catalysts used for direct hydrogenation of CO_2 to methanol, and the results have shown that high pressure is needed to achieve high methanol selectivity [58,62,63]. The most suitable catalyst is not yet available in the current industry. Two challenges for catalyst development are the huge amount of water produced by both reactions that inhibit the product and the undesirable reverse water gas–shift reaction that consumes hydrogen, thus results in a decrease in the yield for methanol. Copper-zinc-aluminum oxide catalyst is often used in CO_2 hydrogenation. The process is run at 5.0–10.0 MPa and 473–523 K. But, the catalyst is not effective again for hydrogenating pure CO_2 [64]. Significant amounts of research into the direct hydrogenation of CO_2 to methanol is continuing. Some of the researches are simplified in Table 2.

Table 2. Modification of direct hydrogenation of CO_2 to methanol.

Modification	Description/Result	Reference
Transition metal carbides: 1. Molybdenum carbide (Mo_2C) and cementite (Fe_3C) 2. Tantalum carbide (TaC) and Silicon carbide (SiC)	High CO_2 conversion and good methanol selectivity Almost inactive	[65]
Two-stage bed system	Higher performance	[66]
Heterogeneous copper-based catalysts	Based on CO hydrogenation	[59]
Molybdenum-bismuth bimetallic chalcogenide electrocatalyst	Produce methanol with 70% of Faradaic efficiency with requirement of acetonitrile/ionic liquid electrolyte solution	[67]

Another product that can be obtained from the methanation of CO_2 is carbon monoxide. CO is usually obtained through partial oxidation of hydrocarbons or coal at high temperatures around 800 °C. CO is a valuable feedstock in the synthesis of different commodities such as methanol and other higher-order hydrocarbons. The method of obtaining CO from CO_2 from the methanation process is the reverse water–gas shift reaction (shown in Equation (3)) as the major by-product [68]. The reaction is endothermic and requires s high temperature (~500 °C). A wide range of heterogeneous catalysts often used are copper-, iron-, or ceria-based systems for the reverse water–gas shift reaction. The problems of these catalysts are poor thermal stability and undesired side product often formed. Due to this thermodynamic constraint, it is unlikely for the research on converting CO_2 to CO using reverse water–gas shift reaction to advance beyond this stage. Furthermore, there are other potential routes to generate CO from CO_2 at a significantly more advanced state. To directly reduce CO_2 to CO and O_2, the use of electrochemical splitting provides an alternative way. Unfortunately, the subject will not be discussed further in this paper, but information on the process can be obtained here [69,70].

3.2. Biological Processes

Biological processes complement chemical options due to its uniqueness of carbon utilization resource requirements and product opportunities. It focuses on the aptitude of microorganisms to convert CO_2 into useful products. Biological fixation of CO_2 is a sustainable solution to reduce CO_2 content in biogas due to its nature which is environmentally-friendly and eliminates the step of captured CO_2 disposal [71]. One of the biological methods to utilize CO_2 in biogas relies on the utilization of H_2 for the conversion of CO_2 to CH_4 based on the action of hydrogenotrophic methanogens. The reaction is shown in Equation (4).

$$4H_2 + CO_2 \rightarrow CH_4 + 2H_2O \qquad \Delta G° = -130.7 \text{ kJ/mol} \qquad (4)$$

The source of H_2 is the hydrolysis of water. To ensure the method is sustainable, electricity needed in the hydrolysis process came from renewable sources, such as solar and wind. One of the disadvantages of H_2 was its low volumetric energy density, resulting in storage difficulties [72]. This H_2 assisted biogas upgrading can occur in a so-called in-situ and ex-situ biological biogas upgrading. Ex-situ upgrading had been discussed in previous sections and includes absorption, adsorption, membrane separation, and cryogenic methods. It requires the CO_2 to be removed first, thus defeating the purpose of utilizing the CO_2 in biogas, which is the focus of this topic. Ex-situ upgrading will not be discussed further but the review can be found here [73]. Meanwhile, the process of in-situ upgrading does not require the CO_2 to be removed first, rather it will be converted into CH_4 leading to a significant increment in biogas purity [13].

In-situ biological biogas upgrading uses the injection of H_2 inside a biogas reactor during anaerobic digestion to react with CO_2, resulting in CH_4 production by the action of autochthonous methanogenic archaea [13]. This can be operated through two different pathways: hydrogenotrophic methanogenesis and Wood–Ljungdahl [74]. Hydrogenotrophic methanogenesis performs direct conversion of CO_2 to CH_4 with the addition of H_2 as a source of electrons, according to Equation (4). Meanwhile the Wood–Ljungdahl pathway indirectly converts CO_2 to CH_4 via two reactions according to Equations (5) and (6).

$$4H_2 + 2CO_2 \rightarrow CH_3COOH + 2H_2O \qquad \Delta G° = -104.5 \text{ kJ/mol} \qquad (5)$$

$$CH_3COOH \rightarrow CH_4 + CO_2 \qquad \Delta G° = -31.0 \text{ kJ/mol} \qquad (6)$$

The CO_2 is converted to acetate acid with the help of homoacetogenic bacteria. Then the acetate acid is converted into CH_4 with the present of acetoclastic methanogenic archaea. H_2 plays a crucial role in the whole process of anaerobic digestion. Exogenous addition of H_2 results in the increase of both hydrogenotrophic methanogens and homoacetogenic species, producing acetate from H_2 and CO_2 [75]. The downside of adding H_2 to the process is the inhibition of syntrophic acetogens which are involved in propionate and butyrate degradation and syntrophic acetate oxidizers (SAO) [76]. It is important to control the concentration of H_2 to ensure the equilibrium of biochemical reactions. The process is illustrated in Figure 7.

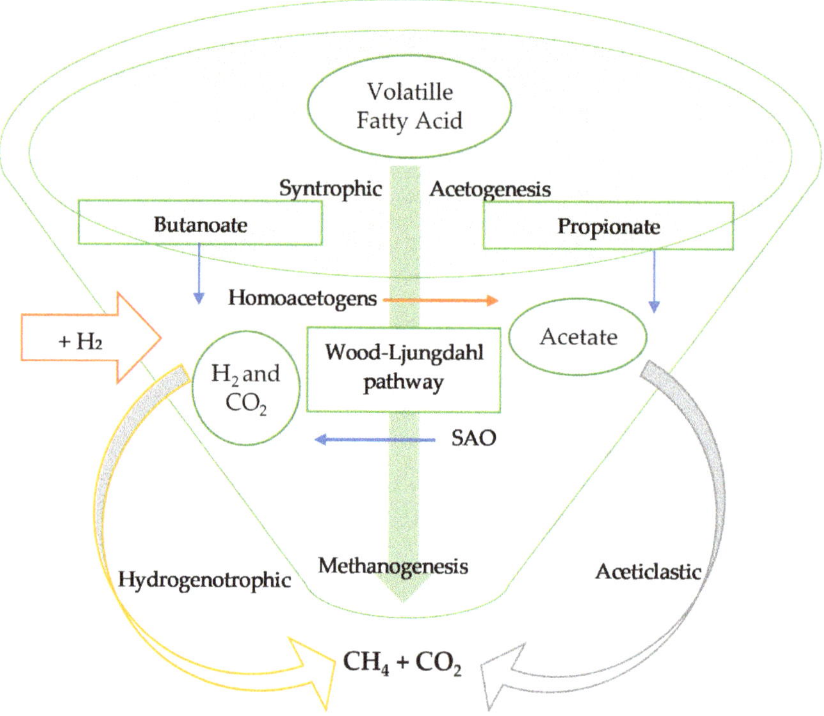

Figure 7. Metabolic pathways for hydrogen assisted methanogenesis [25].

One type of biogas reactor often used in this process is called "continuous stirred tank reactor" (CSTR). The process is heavily connected to the pH level in the reactor. The main challenge is to prevent a pH value above 8.5 because it will lead to methanogenesis inhibition [77,78]. Another challenge arises from the oxidation of the volatile fatty acid (VFA) and alcohols associated with the concentration of the injected hydrogen. To prevent the increasing of the pH level and VFA oxidation, co-digestion with acidic waste [79] and injection of high H_2 concentrations in reactor [80] were proposed to solve the problems, respectively. Additionally, a ton of research had been done on how to increase the efficiency of the process. A select few of these are listed in Table 3.

Table 3. In-situ enriched H_2 upgrading technologies.

Reactor Type	Upgrading Technology	Substrate	Temperature (°C)	HRT (days)	H_2 Flow (L/L-days)	pH	CH₄ (%)	CO₂ (%)	Reference
1.5 (R1) and 2L (R2) CSTR	a) Mesophilic digester with external H_2 addition b) Thermophilic digester with external H_2 addition	Cattle ma-nure	35–55	R1 = 25 R2 = 20	R1 = 0.192 R2 = 0.510	R1 = 7.78 R2 = 7.95	89 85	7 9	[78]
120 mL Batch bottle	Exogenous H_2 addition	Maize Leaf	52	24	0.04–0.10	7–8	88–89	10–12	[81]
Two 600 mL CSTR	Co-digested substrates with exogenous H_2 addition	Cattle ma-nure and whey	55	15	1.5–1.7	7.7–7.9	53–75	6.6–13	[79]
Two 3.5 L CSTR	H_2 addition	Cattle ma-nure	55	14	28.6 mL/L/h	8.3	68	12	[82]

3.3. Assessment on Feasibility of Biogas Upgrading

In methanation and biological reaction, costs that need to be considered are investment and operational costs, on top of costs associated with H_2 electrolysis and methanation. Assumptions made were that a large-scale plant for conversion was constructed and that the declining future cost for H_2 electrolysis was achieved due to the higher market penetration rate.

3.3.1. Cost Estimation

H_2 electrolysis involves the production of H_2 and O_2 from electricity (renewable) and water. There are two techniques that can carry out hydrolysis, the low-temperature process, and the high-temperature process. However, the lack of flexibility of high-temperature electrolysis had impaired the use of it [83]. Thus, a further assumption was made based on the low-temperature process. Based on these assumptions, investment costs obtained were in the range of 656–768 €/kW; the operating costs were about four percent of it; efficiency was 67%; and electricity consumption was 4.1 kWh/m^3 [84,85]. The cost of water supply is negligible because it was considered less relevant and can be obtained from the methanation reaction.

For the methanation reaction, besides investment and operating cost, there were costs for capturing CO_2 from biogas and H_2 storage. Assuming the implementation of the system was at well-established biogas upgrading units, the cost can be neglected. During methanation, heat was released and will be used to capture the CO_2 from the biogas, resulting in zero cost on heat generation. The water obtained can be used for H_2 hydrolysis. The storage of H_2 in steel tanks is a well-established technology and can be put at 27 €/kWh as investment costs [86]. The investment cost for the methanation plant can be assumed in the range of 652–785 €/kW; and the operating costs were about four percent of it [85]. However, for biological process, the technique is still under development and the cost cannot be estimated.

In addition, estimation of producing methanol from biogas was done by Zhang et al. (2017) [87]. In the literature, different analyses are taken to calculate the cost. For a plant scale of 5×10^6 kg/day methanol, the total cost will be in range of USD 827 million to USD 1036 million. For comparison, capital cost for fossil fuel-based methanol was around USD 480 million [88]. From an economical point of view, it can be concluded that industrial exploitation of biogas has a long path ahead of them to be on the same level with current fossil fuel-based processes. For sure, by upgrading biogas by converting CO_2 to methane and methanol is relevant but is now not a viable short-term benefit when compared to already established technologies.

3.3.2. Advantages and Disadvantages

The created mixture in the form of biomethane has a strong resemblance to natural gas. Thus, the distribution of biomethane can be done from existing gas pipelines. This displays a major advantage, as the infrastructure for transporting the biomethane already exists. In contrast to H_2, new distribution network is needed if it became the main energy carrier. Second, production of biomethane can help balance the electric grid. For example, renewables energy such as solar and wind are intermittent and not flexible enough. By producing biomethane, it helps to make use of excess electricity produced whenever the demand is low. On the other hand, biomethane can be used as fuel in a power plant when the demand is high and exceeding the limit of produced electricity. Finally, unlike electricity, biomethane is carbon neutral and can be stored efficiently for future use.

One of the drawbacks of the technique is low efficiency. When converting biogas into biomethane using H_2, the efficiency is only 60%. In addition, if the biomethane produced was to be used to produce electricity, the efficiency drops to 36%. After analyzing the cost, a question is raised: is this technique economically viable? At the moment, the technique is not viable. However, it is likely to be possible in the future when a system with a large share of intermittent renewables are available.

4. Novel Technologies in Carbon Dioxide Conversion

In recent years, the development of new technologies has resulted in the production of a useful commodity by the discovery of new converting processes of CO_2 from waste and atmosphere. These efforts led to the limiting of GHG emissions to the atmosphere of climate-altering pollutants. While CO_2 has been safely used for enhanced oil and carbon feedstock, there is an increased focus on identifying options for re-use of CO_2 for other purposes. There were three stages of development in CO_2 conversion technologies, which can be classified as past, present, and future [65]. In the past, CO_2 conversion technologies focused on producing urea, methanol, cyclic carbonate, and salicylic acid. Then its focus shifted to the making of CO_2 based polymers, fuels, and reactions such as methanation and dry reforming. Meanwhile, CO_2 conversion technologies in the future are predicted to be focusing on production of carboxylic and succinic acid (SA). Thus, this section will be focusing on the possibility of producing SA from CO_2 components in biogas.

SA ($C_4H_6O_4$), also known as butanedioic acid is a four-carbon diacid used as a platform for synthesis of various commodities as shown in Figure 8. It is mostly produced from LPG or petroleum oil through specific chemical process. Although, recent analysis revealed that production of bio-SA from bacterial fermentation, which is a renewable source, can be more cost-effective than the traditional processes [89]. In recent years, the advancement of bio-based production of SA was very significant, and as a consequence, a variety of microorganisms has been engineered for the synthesis of SA from sugars, glycerol, or acetate [90]. Furthermore, the CO_2 is fixed into the bacteria reducing the greenhouse gas emission that lead to pollution. In fact, carbon footprint of bio-SA production is 0.85 kg CO_2 eq/kg compared to 1.8 kg CO_2 eq/kg of carbon footprint by petroleum-based SA [91]. One way to operate a CO_2 fixation process is through reductive tricarboxylic acid (TCA) cycle. In this anaerobic SA production which fully operated under pure CO_2 condition, 1 mol CO_2 can produce 1 mol of SA [92]. However, to establish a truly circular bio-economy and utilizing the abundant industrial by-product of CO_2, valorization of CO_2 as a substitute to the sugar-based substrates is today of particular relevance [93]. Moreover, if the off-gas from biogas industries could be effectively utilized as a CO_2 source for SA fermentation, it will simultaneously decrease the cost of the whole process while meeting the commercial-scale requirements for natural gas grid [94].

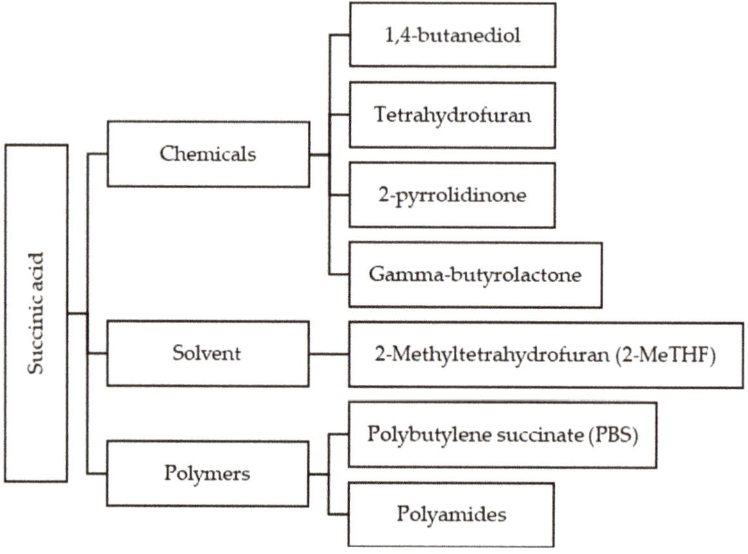

Figure 8. Potential products by using succinic acid (SA) as feedstock [95].

4.1. Simultaneous Biogas Upgrading and Bio-Succinic Acid Production

As mentioned earlier, biogas consists of 60% CH_4 and 40% CO_2. The presence of CO_2 limits the use of biogas. In 2014, Gunnarsson et al. (2014) [96] had come out with a novel approach for converting the CO_2 component in biogas into SA through a biological process. The study demonstrates a new biogas upgrading technology, which makes use of bacterial fermentation to simultaneously produce high-purity CH_4 and bio-SA. The microorganism used was a strain of *Actinobacillus succinogenes* 130Z (DSM 22257). Application properties are as follows: Substrate: Glucose 30–32 g/L; reactors: 3-L; T: 37 °C; pH: 6.75; ω: 200 rpm; t: 24 h; P: 101.325 and 140 kPa; gas–liquid ratio: 8.3:1 and 5:1. The results of the study are tabulated in Table 4. Stages of the processes are simplified in Equation (7).

$$\text{Substrate } (\textit{Anaerobic digestion}) \rightarrow \text{Biogas } (60\% \text{ CH}_4|40\% \text{ CO}_2) \text{ } (\textit{Fermentation}) \\ \rightarrow \text{Natural Gas } (95\% \text{ CH}_4) + \text{Succinate} \tag{7}$$

Based on Table 4, slight over-pressure during fermentation was ideal for the solubility of CO_2, thus increasing the CH_4 content in biogas. Increasing the pressure while reducing the ratio also affects other parameters, as CO_2 consumption rate increased by 16.4%, SA concentration increased by 6.2%, and SA yield increased by 13.8%. The final 95% CH_4 purity produced was similar to that of commercial biogas upgrading technologies (95–98%) [21]. This study sparks vast potentials for future investigation on the large-scale implementation for practical application in industries. Then in 2018, a group of inspired researchers from Germany, led by Patrick Ballmann, provided a plan to further study this new concept of simultaneous upgrading by replacing the glucose with lignocelluloses from straw [97]. A further modification was done on the straw to provide a suitable strain for SA production while reducing the by-products. To this state, only *A. succinogenes* has been used for SA production coupled with biogas upgrading [25]. That remained the case for a few years until Babaei et al. (2019) [98] conducted an experiment using *Basfia succiniciproducens* (DSM 22022) as a bacterial strain for the fermentation of SA.

Table 4. Performance of the system at different pressure and gas–liquid ratio.

| | Pressure (kPa) | | | |
| | 101.325 | | 140 | |
Gas-liquid ratio	8.3:1	5:1	8.3:1	5:1
CO_2 solubility (mM)	9.15	9.15	16.7	16.7
CO_2 fixation rate (L CO_2/L-d)	1.35	1.52	2.59	1.77
CH_4 purity (%)	76.4	85.2	91.1	95.4
SA yield (g/g)	0.60	0.56	0.62	0.63
SA productivity (g/L-h)	0.53	0.53	0.60	0.56
SA concentration (g/L)	12.85	12.74	14.39	13.53
By-products concentration (g/L)	9.5	11.63	8.65	9.96

The experiment conducted by Babaei et al. (2019) [98] was to determine the possibilities of expanding the simultaneous SA production with a biogas upgrading process by using organic fraction of household kitchen waste (OFHKW) as substrate, replacing the common use of glucose while comparing the performance of *A. succinogenes* and *B. succiniciproducens* in producing SA. OFHKW was broken down by enzymatic hydrolysis to produce monomeric fermentable sugars prior to the fermentation process. The experiment was divided by two major parts: The first was to determine the condition for *B. succiniciproducens* to produce SA, the second was to prove the ability of *B. succiniciproducens* to conduct a simultaneous biogas upgrading with SA production. Application properties, results, and discussion of the study are simplified in Table 5.

Table 5. Summary of the fermentation process using either *B. succiniciproducens* or *A. succinogenes* as bacterial strain.

Task		Application Properties	Results	Discussion
SA Production		Carbon source: $MgCO_3$ 5–100 g/L; Substrate: OFHKW 17, 25, 35 & 60 g/L; Serum bottles: 250-mL; T: 37 °C; pH: 6.7 ± 0.1; ω: 150 rpm	*B. succiniciproducens* SA concentration: Maximum titer of 17.9 ± 0.43 g/L; Overall reaction: Substrate + $2\,CO_2 \rightarrow$ 2 lactate + 2 acetate + 2 formate *A. succinogenes* SA concentration: Maximum titer of 21.1 ± 3.5 g/L	Higher substrate concentration results in higher SA production; *B. succiniciproducens* is preferred for SA fermentation due to better performances at lower concentration, whereas the by-products were lower
Simultaneous Upgrading	*B. succiniciproducens*	Carbon source: Biogas; Substrate: OFHKW 17 g/L; Reactors: 3-L; T: 37 °C; pH: 6.7; ω: 200rpm; t: 8 h; P: 130 & 140 kPa	SA concentration: 3.8 ± 0.8 g/L (0.25 $g_{SA}/g_{glucose}$); CO_2 content: 8.0% (v/v) reduction; CH_4 content: 4.7% (v/v) increase	In term of duration and sugar consumption rate, *B. succiniciproducens* (8 h) is still superior than *A. succinogenes* (24 h); The best way to conduct fermentation process was by gradual additional of substrate instead of starting with high substrate concentration
	A. succinogenes	Carbon source: Biogas; Substrate: Glucose 32 g/L; Reactors: 3-L; T: 37 °C; pH: 6.75; ω: 200 rpm; t: 24 h; P: 101.325 & 140 kPa	SA concentration: 14.39 g/L; CH_4 content: 31% (v/v) increase	

This novel approach of using household waste as a substrate to produce SA provides the information on how to accomplish a fermentation process using either *A. succinogenes* or *B. succiniciproducens*. The research will be a benchmark for fellow researchers to utilize other home-grown or local products in the production of SA. Additionally, this study proves the ability of *B. succiniciproducens* to be an alternative as a bacterium capable of converting CO_2 content in biogas into SA. Nevertheless, further studies still need to be done on other bacteria to identify the possibilities of upgrading biogas while producing SA.

4.2. Future Perspective of Succinic Acid Production

These studies proved that both biomethane and biochemical (SA) can be produced by utilizing unconventional biomasses. To further improve the utilization of CO_2 in biogas, research can be done on metabolic engineering of some other bacteria to produce higher SA titer with no by-products. On top of using *A. succinogenes* 130Z (DSM 22257) and *B. succiniciproducens* (DSM 22022), other bacterial strains had been identified that hold a potential to convert CO_2 in biogas into SA. Fermentation techniques are also a factor in increasing the SA titer. Some of the bacterial strain and fermentation techniques that can possibly be integrated into SA fermentation technique are listed in Table 6. Although these studies were aimed at the direct conversion of CO_2 into SA, it will set a base for further research on integrating it in simultaneous biogas upgrading.

Additionally, to implement this technology on a larger scale, further improvement of the simultaneous biogas upgrading, and succinic acid production technology is required. Because there is still no available matured technology in the market, cost breakdown cannot be conducted. Nevertheless, the demand for bio-SA has been increasing over the years. By selling the produced SA, it will reduce the cost of whole operation. Market forecast of bio-SA was conducted by different researchers and can be seen in a simplified form in Figure 9. This reflects the relevance of producing bio-SA in the future.

Table 6. Summary of performances of succinic acid fermentation studies by various microorganisms.

Microorganism	Reactor Type/ Fermentation Technique	Substrate	Titer (g/L)	Yield (g/g)	Reference
A. *succinogenes*	Repeat-batch	Glucose	33.9	0.86	[99]
A. *succinogenes* 130Z	Suspended cell	Glucose	10.4	0.27–0.73	[99]
A. *succinogenes* 130Z	Recycled cell	Glucose	18.6	0.50–0.59	[100]
A. *succinogenes* 130Z	Batch	Whey	21.5	0.57	[101]
A. *succinogenes* 130Z	Continuous	Corn	39.6	0.78	[102]
A. *succinogenes* FZ53	Batch	Glucose	105.8	0.8	[103]
M. *succiniciproducens*	Batch	Glucose	14	0.7	[104]
M. *succiniciproducens*	Batch	Whey	13.5	0.72	[105]
M. *succiniciproducens* MBEL55E	Suspended cell	Lactose	10.3	0.63–0.69	[105]
M. *succiniciproducens* MBEL55E	Suspended cell	Glucose Xylose	14.1	0.34–0.61	[100]
M. *succiniciproducens* MBEL55E	Recycled cell	Glucose	12.8	0.48–0.64	[100]
M. *succiniciproducens* LPK7	Recycled cell	Glucose	12.9	0.10–0.71	[106]
A. *succiniciproducens*	Continuous	Whey	24	0.72	[103]
A. *succiniciproducens* ATCC No. 29305	Suspended cell	Lactose	24.0	0.62–0.72	[107]
A. *succiniciproducens* ATCC No. 29305	Suspended cell	Lactose	14.0	0.81–0.94	[108]
A. *succiniciproducens* ATCC No. 29305	Suspended cell	Glucose	29.6	0.73–0.82	[109]
A. *succiniciproducens* ATCC No. 29305	Suspended cell	Glycerol	16.1	1.23–1.50	[110]
A. *succiniciproducens* ATCC No. 53488	Recycled cell	Glucose	16.5	0.74–0.83	[111]

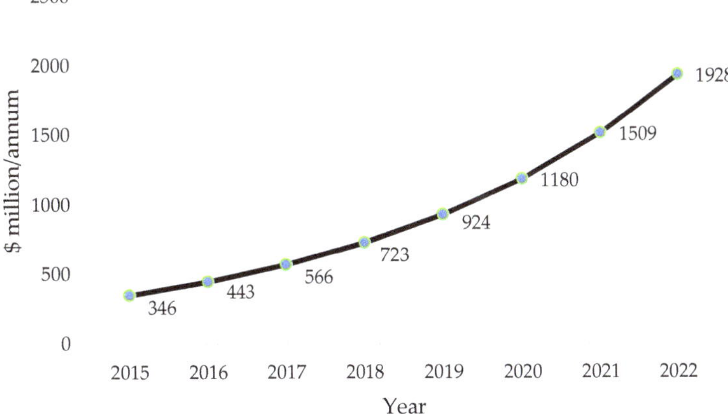

Figure 9. Market forecast of bio-SA volume from the year 2015 to 2022 [112].

The evaluation of the performance of microbial conversion of CO_2 into SA is an important step in providing practical solutions, knowledge, and addressing the gaps in the current understanding [113]. While SA is still widely produced from petrochemical and glucose because of ubiquitous substrate availability, simple process design and high productivities, effort toward producing SA from CO_2

as sustainable source is still growing and will be applicable if technical barriers that needed to be identified such as limiting gas transfer rates can be overcome [114].

5. Conclusions

Global industrial emission of carbon dioxide had risen to an all-time high in 2018 and it is unlikely to reduce soon [115,116]. Growing demand for oil and natural gas globally overshadowed the effort in the development of renewable energy. Furthermore, fossil-fuel infrastructure is still expanding, particularly in developing countries. If current trends continue, the fear of the worst effects of global warming—extreme weather, rising sea levels, plant and animal extinctions, ocean acidification, major shifts in climate, and unprecedented social upheaval—will be inevitable. One of the solutions for these problems is the utilization of bio-natural gas as the substitutes for fossil fuels. In fact, biogas reduces the emission of carbon dioxide while capturing methane, ensuring a cleaner environment. While these are major leaps toward cleaner fuels, still there is room for improvement. Major research had been made from time to time on the techniques to upgrade the biogas to a higher degree. Throughout the years, various technologies and techniques had been developed on how to fully utilize biogas and its by-product so there is no waste release into the environment. One major hurdle for biogas implementation is the cost which hurts its potential employment. While biogas is not the perfect solution for global greenhouse gas emissions, its place in the world of waste management has been very much solidified and will continue to evolve in the coming years.

Author Contributions: Conceptualization—S.N. and A.I.A.; writing, original draft preparation—S.N. and A.I.A.; writing, review, and editing—S.N., K.W.C., and M.Y.O.; writing, proofreading—A.I.A., K.W.C., and P.L.S.; funding acquisition—S.N.

Funding: The authors (SN and AIA) would like to thank the Ministry of Education (MOE) Malaysia for the financial support through the Fundamental Research Grant Scheme (MOHE Project Ref. No.: FRGS/1/2018/STG01/UNITEN/01/1). A note of appreciation to iRMC UNITEN for supporting the publication of this paper through publication fund BOLD 2025 (RJO10436494). AIA also wishes to express his gratitude to UNITEN for providing the UNITEN Postgraduate Scholarship 2019.

Acknowledgments: The authors would like to express their special thanks to Universiti Tenaga Nasional for providing facilities and equipment to ensure the accomplishment of this project.

Conflicts of Interest: The authors declare that they have no competing interests.

References

1. Höök, M.; Tang, X. Depletion of fossil fuels and anthropogenic climate change—A review. *Energy Policy* **2013**, *52*, 797–809. [CrossRef]
2. Goede, A.; van de Sanden, R. CO_2-Neutral fuels. *Europhys. News* **2016**, *47*, 22–26. [CrossRef]
3. Ragauskas, A.J.; Williams, C.K.; Davison, B.H.; Britovsek, G.; Cairney, J.; Eckert, C.A.; Frederick, W.J.J.; Hallett, J.P.; Leak, D.J.; Liotta, C.L.; et al. The path forward for biofuels and biomaterials. *Science* **2006**, *311*, 484–489. [CrossRef] [PubMed]
4. Masse, D.I.; Talbot, G.; Gilbert, Y. A scientific review of the agronomic, environmental and social benefits of anaerobic digestion. In *Anaerobic Digestions*; Caruana, D.J., Olsen, E.A., Eds.; Nova Science Publishers: New York, NY, USA, 2012; pp. 109–131. ISBN 978-1-62081-610-3.
5. Mata-Alvarez, J.; Macé, S.; Llabrés, P. Anaerobic digestion of organic solid wastes: An overview of research achievements and perspectives. *Bioresour. Technol.* **2000**, *74*, 3–16. [CrossRef]
6. Muhammad Nasir, I.; Mohd Ghazi, T.I.; Omar, R. Production of biogas from solid organic wastes through anaerobic digestion: A review. *Appl. Microbiol. Biotechnol.* **2012**, *95*, 321–329. [CrossRef]
7. Carlsson, M.; Lagerkvist, A.; Morgan-Sagastume, F. The effects of substrate pre-treatment on anaerobic digestion systems: A review. *Waste Manag.* **2012**, *32*, 1634–1650. [CrossRef]
8. Long, J.H.; Aziz, T.N.; Reyes, F.L.; de los Ducoste, J.J. Anaerobic co-digestion of fat, oil, and grease (FOG): A review of gas production and process limitations. *Process Saf. Environ. Prot.* **2012**, *90*, 231–245. [CrossRef]
9. Montalvo, S.; Guerrero, L.; Borja, R.; Sánchez, E.; Milán, Z.; Cortés, I.; de la la Rubia, M.A. Application of natural zeolites in anaerobic digestion processes: A review. *Appl. Clay Sci.* **2012**, *58*, 125–133. [CrossRef]

10. Raboni, M.; Urbini, G. Production and use of biogas in Europe: A survey of current status and perspectives. *Ambient. Agua* **2014**, *9*. [CrossRef]
11. Kárászová, M.; Sedláková, Z.; Izák, P. Gas permeation processes in biogas upgrading: A short review. *Chem. Pap.* **2015**, *69*, 1277–1283. [CrossRef]
12. Sun, Q.; Li, H.; Yan, J.; Liu, L.; Yu, Z.; Yu, X. Selection of appropriate biogas upgrading technology—A review of biogas cleaning, upgrading and utilisation. *Renew. Sustain. Energy Rev.* **2015**, *51*, 521–532. [CrossRef]
13. Kougias, P.G.; Treu, L.; Benavente, D.P.; Boe, K.; Campanaro, S.; Angelidaki, I. Ex-situ biogas upgrading and enhancement in different reactor systems. *Bioresour. Technol.* **2017**, *225*, 429–437. [CrossRef] [PubMed]
14. Abatzoglou, N.; Boivin, S. A review of biogas purification processes. *Biofuels Bioprod. Biorefin.* **2009**, *3*, 42–71. [CrossRef]
15. Ryckebosch, E.; Drouillon, M.; Vervaeren, H. Techniques for transformation of biogas to biomethane. *Biomass Bioenergy* **2011**, *35*, 1633–1645. [CrossRef]
16. Bauer, F.; Persson, T.; Hulteberg, C.; Tamm, D. Biogas upgrading—Technology overview, comparison and perspectives for the future. *Biofuels Bioprod Biorefin.* **2013**, *7*, 499–511. [CrossRef]
17. Aziz, N.I.H.A.; Hanafiah, M.M.; Gheewala, S.H. A review on life cycle assessment of biogas production: Challenges and future perspectives in Malaysia. *Biomass Bioenergy* **2019**, *122*, 361–374. [CrossRef]
18. García-Gutiérrez, P.; Jacquemin, J.; McCrellis, C.; Dimitriou, I.; Taylor, S.F.R.; Hardacre, C.; Allen, R.W.K. Techno-economic feasibility of selective CO_2 capture processes from biogas streams using ionic liquids as physical absorbents. *Energy Fuels* **2016**, *30*, 5052–5064. [CrossRef]
19. Andriani, D.; Wresta, A.; Atmaja, T.D.; Saepudin, A. A review on optimization production and upgrading biogas through CO_2 removal using various techniques. *Appl. Biochem. Biotechnol.* **2014**, *172*, 1909–1928. [CrossRef]
20. Awe, O.W.; Zhao, Y.; Nzihou, A.; Minh, D.P.; Lyczko, N. A Review of biogas utilisation, purification and upgrading technologies. *Waste Biomass Valoriz.* **2017**, *8*, 267–283. [CrossRef]
21. Leonzio, G. Upgrading of biogas to bio-methane with chemical absorption process: Simulation and environmental impact. *J. Clean. Prod.* **2016**, *131*, 364–375. [CrossRef]
22. Xia, A.; Cheng, J.; Murphy, J.D. Innovation in biological production and upgrading of methane and hydrogen for use as gaseous transport biofuel. *Biotechnol. Adv.* **2016**, *34*, 451–472. [CrossRef] [PubMed]
23. Patterson, T.; Esteves, S.; Dinsdale, R.; Guwy, A. An evaluation of the policy and techno-economic factors affecting the potential for biogas upgrading for transport fuel use in the UK. *Energy Policy* **2011**, *39*, 1806–1816. [CrossRef]
24. Angelidaki, I.; Treu, L.; Tsapekos, P.; Luo, G.; Campanaro, S.; Wenzel, H.; Kougias, P.G. Biogas upgrading and utilization: Current status and perspectives. *Biotechnol. Adv.* **2018**, *36*, 452–466. [CrossRef] [PubMed]
25. Tock, L.; Gassner, M.; Maréchal, F. Thermochemical production of liquid fuels from biomass: Thermo-economic modeling, process design and process integration analysis. *Biomass Bioenergy* **2010**, *34*, 1838–1854. [CrossRef]
26. Zhao, Q.; Leonhardt, E.; MacConnell, C.; Frear, C.; Chen, S. Purification technologies for biogas generated by anaerobic digestion. In *Climate Friendly Farming—Final Report*; Center for Sustaining Agriculture and Natural Resources: Washington, DC, USA, 2010; p. 24.
27. Hoyer, K.; Hulteberg, C.; Svensson, M.; Jenberg, J.; NØrregÅrd, Ø. *Biogas Upgrading: A Technical Review*; Energiforsk: Stockholm, Sweden, 2016; ISBN 9789176732755.
28. Zhou, K.; Chaemchuen, S.; Verpoort, F. Alternative materials in technologies for biogas upgrading via CO_2 capture. *Renew. Sustain. Energy Rev.* **2017**, *79*, 1414–1441. [CrossRef]
29. Ullah Khan, I.; Hafiz Dzarfan Othman, M.; Hashim, H.; Matsuura, T.; Ismail, A.F.; Rezaei-DashtArzhandi, M.; Wan Azelee, I. Biogas as a renewable energy fuel—A review of biogas upgrading, utilisation and storage. *Energy Convers. Manag.* **2017**, *150*, 277–294. [CrossRef]
30. Lasocki, J.; Kołodziejczyk, K.; Matuszewska, A. Laboratory-scale investigation of biogas treatment by removal of hydrogen sulfide and carbon dioxide. *Polish J. Environ. Stud.* **2015**, *24*, 1427–1434. [CrossRef]
31. Georgiou, D.; Petrolekas, P.D.; Hatzixanthis, S.; Aivasidis, A. Absorption of carbon dioxide by raw and treated dye-bath effluents. *J. Hazard. Mater.* **2007**, *144*, 369–376. [CrossRef]
32. Deng, L.; Hägg, M.B. Techno-economic evaluation of biogas upgrading process using CO_2 facilitated transport membrane. *Int. J. Greenh. Gas Control* **2010**, *4*, 638–646. [CrossRef]

33. Ho, M.T.; Allinson, G.W.; Wiley, D.E. Reducing the cost of CO_2 capture from flue gases using pressure swing adsorption. *Ind. Eng. Chem. Res.* **2008**, *47*, 4883–4890. [CrossRef]

34. Augelletti, R.; Conti, M.; Annesini, M.C. Pressure swing adsorption for biogas upgrading. A new process configuration for the separation of biomethane and carbon dioxide. *J. Clean. Prod.* **2017**, *140*, 1390–1398. [CrossRef]

35. An, H.; Feng, B.; Su, S. CO_2 capture by electrothermal swing adsorption with activated carbon fibre materials. *Int. J. Greenh. Gas Control* **2011**, *5*, 16–25. [CrossRef]

36. Plaza, M.G.; García, S.; Rubiera, F.; Pis, J.J.; Pevida, C. Post-combustion CO_2 capture with a commercial activated carbon: Comparison of different regeneration strategies. *Chem. Eng. J.* **2010**, *163*, 41–47. [CrossRef]

37. Petersson, A.; Wellinger, A. *Biogas Upgrading Technologies—Developments and Innovations*; IEA Bioenergy: Paris, France, 2009; p. 20.

38. Bauer, F.; Hulteberg, C.; Persson, T.; Tamm, D.; Granskning, B. *Biogas Upgrading—Review of Commercial Technologies*; Svenskt Gastekniskt Center AB: Malmo, Sweeden, 2013.

39. Harasimowicz, M.; Orluk, P.; Zakrzewska-Trznadel, G.; Chmielewski, A.G. Application of polyimide membranes for biogas purification and enrichment. *J. Hazard. Mater.* **2007**, *144*, 698–702. [CrossRef]

40. Munoz, R.; Meier, L.; Diaz, I.; Jeison, D. A review on the state-of-the-art of physical/chemical and biological technologies for biogas upgrading. *Rev. Environ. Sci. Biotechnol.* **2015**, *14*, 727–759. [CrossRef]

41. Green, D.W.; Perry, R.H. *Perry's Chemical Engineers' Hand Book*, 8th ed.; McGraw-Hill: New York, NY, USA, 2008.

42. Grande, C.A.; Blom, R. Cryogenic adsorption of methane and carbon dioxide on zeolites 4A and 13X. *Energy Fuels* **2014**, *28*, 6688–6693. [CrossRef]

43. Marsh, M.; Officer, C.E.; Krich, K.; Krich, K.; Augenstein, D.; Benemann, J.; Rutledge, B.; Salour, D. *Biomethane from Dairy Waste: A Sourcebook for the Production and Use of Renewable Natural Gas in California*; Suscon: San Francisco, CA, USA, 2005.

44. Cecchi, F.; Cavinato, C. Anaerobic digestion of bio-waste: A mini-review focusing on territorial and environmental aspects. *Waste Manag. Res.* **2015**, *33*, 429–438. [CrossRef]

45. Gao, Y.; Jiang, J.; Meng, Y.; Yan, F.; Aihemaiti, A. A review of recent developments in hydrogen production via biogas dry reforming. *Energy Convers. Manag.* **2018**, *171*, 133–155. [CrossRef]

46. Aresta, M. *Carbon Dioxide as Chemical Feedstock*, 1st ed.; Wiley-VCH: Weinheim, Germany, 2010.

47. Williams, M. *The merck index: An encyclopedia of chemicals, drugs, and biologicals*, 14th ed.; Merck Inc.: Rahway, NJ, USA, 2006; 2564p.

48. Stangeland, K.; Kalai, D.; Li, H.; Yu, Z. CO_2 Methanation: The effect of catalysts and reaction conditions. *Energy Procedia* **2017**, *105*, 2022–2027. [CrossRef]

49. Jürgensen, L.; Augustine, E.; Born, J.; Holm-nielsen, J.B. Bioresource technology dynamic biogas upgrading based on the Sabatier process: Thermodynamic and dynamic process simulation. *Bioresource Technol.* **2015**, *178*, 323–329.

50. U.S. Energy Information Administration (EIA). *International Energy Outlook 2017*; EIA: Washington, DC, USA, 2017.

51. Su, X.; Xu, J.; Liang, B.; Duan, H.; Hou, B.; Huang, Y. Catalytic carbon dioxide hydrogenation to methane: A review of recent studies. *J. Energy Chem.* **2016**, *25*, 553–565. [CrossRef]

52. Hashemnejad, S.M.; Parvari, M. Deactivation and regeneration of nickel-based catalysts for steam-methane reforming. *Chin. J. Catal.* **2011**, *32*, 273–279. [CrossRef]

53. Mutz, B.; Gänzler, A.M.; Nachtegaal, M.; Müller, O.; Frahm, R.; Kleist, W.; Grunwaldt, J.-D. Surface oxidation of supported ni particles and its impact on the catalytic performance during dynamically operated methanation of CO_2. *Catalysts* **2017**, *7*, 279. [CrossRef]

54. National Academies of Sciences Engineering and Medicine Gaseous Carbon Waste Streams Utilization. In *Gaseous Carbon Waste Streams Utilization: Status and Research Needs*; The National Academies Press: Washington, DC, USA, 2019; pp. 63–95. ISBN 978-0-309-48336-0.

55. Manthiram, K.; Beberwyck, B.J.; Alivisatos, A.P. Enhanced electrochemical methanation of carbon dioxide with a dispersible nanoscale copper catalyst. *J. Am. Chem. Soc.* **2014**, *136*, 13319–13325. [CrossRef]

56. Qiu, Y.-L.; Zhong, H.-X.; Zhang, T.-T.; Xu, W.-B.; Li, X.-F.; Zhang, H.-M. Copper electrode fabricated via pulse electrodeposition: Toward high methane selectivity and activity for CO_2 electroreduction. *ACS Catal.* **2017**, *7*, 6302–6310. [CrossRef]

57. Sun, X.; Kang, X.; Zhu, Q.; Ma, J.; Yang, G.; Liu, Z.; Han, B. Very highly efficient reduction of CO_2 to CH_4 using metal-free N-doped carbon electrodes. *Chem. Sci.* **2016**, *7*, 2883–2887. [CrossRef]

58. Álvarez, A.; Bansode, A.; Urakawa, A.; Bavykina, A.V.; Wezendonk, T.A.; Makkee, M.; Gascon, J.; Kapteijn, F. Challenges in the greener production of formates/formic acid, methanol, and DME by heterogeneously catalyzed CO_2 hydrogenation processes. *Chem. Rev.* **2017**, *117*, 9804–9838. [CrossRef]

59. Ganesh, I. Conversion of carbon dioxide into methanol—A potential liquid fuel: Fundamental challenges and opportunities (a review). *Renew. Sustain. Energy Rev.* **2014**, *31*, 221–257. [CrossRef]

60. Pérez-Fortes, M.; Schöneberger, J.C.; Boulamanti, A.; Tzimas, E. Methanol synthesis using captured CO_2 as raw material: Techno-economic and environmental assessment. *Appl. Energy* **2016**, *161*, 718–732. [CrossRef]

61. Saeidi, S.; Aishah, N.; Amin, S.; Reza, M. Hydrogenation of CO_2 to value-added products—A review and potential future developments. *Biochem. Pharmacol.* **2014**, *5*, 66–81. [CrossRef]

62. Klankermayer, J.; Wesselbaum, S.; Beydoun, K.; Leitner, W. Selective catalytic synthesis using the combination of carbon dioxide and hydrogen: Catalytic chess at the interface of energy and chemistry. *Angew. Chemie Int. Ed.* **2016**, *55*, 7296–7343. [CrossRef] [PubMed]

63. Wang, W.-H.; Himeda, Y.; Muckerman, J.T.; Manbeck, G.F.; Fujita, E. CO_2 Hydrogenation to formate and methanol as an alternative to photo- and electrochemical CO_2 reduction. *Chem. Rev.* **2015**, *115*, 12936–12973. [CrossRef] [PubMed]

64. Liu, Y.; Zhang, Y.; Wang, T.; Tsubaki, N. Efficient conversion of carbon dioxide to methanol using copper catalyst by a new low-temperature hydrogenation process. *Chem. Lett.* **2007**, *36*, 1182–1183. [CrossRef]

65. Alper, E.; Yuksel Orhan, O. CO_2 Utilization: Developments in conversion processes. *Petroleum* **2017**, *3*, 109–126. [CrossRef]

66. Rahimpour, M.R. A two-stage catalyst bed concept for conversion of carbon dioxide into methanol. *Fuel Process. Technol.* **2008**, *89*, 556–566. [CrossRef]

67. Sun, X.; Zhu, Q.; Kang, X.; Liu, H.; Qian, Q.; Zhang, Z.; Han, B. Molybdenum–Bismuth bimetallic chalcogenide nanosheets for highly efficient electrocatalytic reduction of carbon dioxide to methanol. *Angew. Chemie Int. Ed.* **2016**, *55*, 6771–6775. [CrossRef] [PubMed]

68. Wang, W.; Wang, S.; Ma, X.; Gong, J. Recent advances in catalytic hydrogenation of carbon dioxide. *Chem. Soc. Rev.* **2011**, *40*, 3703–3727. [CrossRef]

69. Qiao, J.; Liu, Y.; Hong, F.; Zhang, J. A review of catalysts for the electroreduction of carbon dioxide to produce low-carbon fuels. *Chem. Soc. Rev.* **2014**, *43*, 631–675. [CrossRef]

70. White, J.L.; Baruch, M.F.; Pander, J.E.; Hu, Y.; Fortmeyer, I.C.; Park, J.E.; Zhang, T.; Liao, K.; Gu, J.; Yan, Y.; et al. Light-Driven heterogeneous reduction of carbon dioxide: Photocatalysts and photoelectrodes. *Chem. Rev.* **2015**, *115*, 12888–12935. [CrossRef]

71. Wu, M.; Zhang, W.; Ji, Y.; Yi, X.; Ma, J.; Wu, H.; Jiang, M. Coupled CO_2 fixation from ethylene oxide off-gas with bio-based succinic acid production by engineered recombinant *Escherichia coli*. *Biochem. Eng. J.* **2017**, *117*, 1–6. [CrossRef]

72. Jürgensen, L.; Ehimen, E.A.; Born, J.; Holm-Nielsen, J.B. Utilization of surplus electricity from wind power for dynamic biogas upgrading: Northern Germany case study. *Biomass Bioenergy* **2014**, *66*, 126–132. [CrossRef]

73. Singhal, S.; Agarwal, S.; Arora, S.; Sharma, P.; Singhal, N. Upgrading techniques for transformation of biogas to bio-CNG: A review. *Int. J. Energy Res.* **2017**, *41*, 1657–1669. [CrossRef]

74. Stams, A.J.M.; Plugge, C.M. Electron transfer in syntrophic communities of anaerobic bacteria and archaea. *Nat. Rev. Microbiol.* **2009**, *7*, 568. [CrossRef] [PubMed]

75. Schuchmann, K.; Müller, V. Autotrophy at the thermodynamic limit of life: A model for energy conservation in acetogenic bacteria. *Nat. Rev. Microbiol.* **2014**, *12*, 809. [CrossRef] [PubMed]

76. Demirel, B.; Scherer, P. The roles of acetotrophic and hydrogenotrophic methanogens during anaerobic conversion of biomass to methane: A review. *Rev. Environ. Sci. Biotechnol.* **2008**, *7*, 173–190. [CrossRef]

77. Luo, G.; Angelidaki, I. Integrated biogas upgrading and hydrogen utilization in an anaerobic reactor containing enriched hydrogenotrophic methanogenic culture. *Biotechnol. Bioeng.* **2012**, *109*, 2729–2736. [CrossRef]

78. Bassani, I.; Kougias, P.G.; Treu, L.; Angelidaki, I. Biogas upgrading via hydrogenotrophic methanogenesis in two-stage continuous stirred tank reactors at mesophilic and thermophilic conditions. *Environ. Sci. Technol.* **2015**, *49*, 12585–12593. [CrossRef] [PubMed]

79. Luo, G.; Angelidaki, I. Co-digestion of manure and whey for in situ biogas upgrading by the addition of H2: Process performance and microbial insights. *Appl. Microbiol. Biotechnol.* **2013**, *97*, 1373–1381. [CrossRef]

80. Batstone, D.J.; Keller, J.; Angelidaki, I.; Kalyuzhnyi, S.V.; Pavlostathis, S.G.; Rozzi, A.; Sanders, W.T.; Siegrist, H.; Vavilin, V.A. The IWA anaerobic digestion model No 1 (ADM1). *Water Sci. Technol.* **2002**, *45*, 65–73. [CrossRef]

81. Mulat, D.G.; Mosbæk, F.; Ward, A.J.; Polag, D.; Greule, M.; Keppler, F.; Nielsen, J.L.; Feilberg, A. Exogenous addition of H2 for an in situ biogas upgrading through biological reduction of carbon dioxide into methane. *Waste Manag.* **2017**, *68*, 146–156. [CrossRef]

82. Luo, G.; Johansson, S.; Boe, K.; Xie, L.; Zhou, Q.; Angelidaki, I. Simultaneous hydrogen utilization and in situ biogas upgrading in an anaerobic reactor. *Biotechnol. Bioeng.* **2012**, *109*, 1088–1094. [CrossRef] [PubMed]

83. Fasihi, M.; Bogdanov, D.; Breyer, C. Techno-Economic assessment of power-to-liquids (PtL) fuels production and global trading based on hybrid *pv*-wind power plants. *Energy Procedia* **2016**, *99*, 243–268. [CrossRef]

84. Caldera, U. Role of seawater desalination in the management of an integrated water and 100% renewable energy based power sector in Saudi Arabia. *Water* **2018**, *10*, 3. [CrossRef]

85. Fasihi, M.; Bogdanov, D.; Breyer, C. Long-Term hydrocarbon trade options for the Maghreb Region and Europe—Renewable energy based synthetic fuels for a net zero emissions world. *Sustainability* **2017**, *9*, 306. [CrossRef]

86. Lövenich, A.; Fasihi, M.; Graf, A.; Kasten, P.; Langenheld, A.; Meyer, K.; Peter, F.; Podewils, C. *The Future Cost of Electricity-Based Synthetic Fuels*; Frontier Economics Ltd.: Cologne, Germany, 2018.

87. Zhang, C.; Jun, K.-W.; Gao, R.; Kwak, G.; Park, H.-G. Carbon dioxide utilization in a gas-to-methanol process combined with CO_2/Steam-mixed reforming: Techno-economic analysis. *Fuel* **2017**, *190*, 303–311. [CrossRef]

88. Stokes, H.C. *The Economics of Methanol Production*; The Stokes Consulting Group: Naples, FL, USA, 2002.

89. Mazière, A.; Prinsen, P.; García, A.; Luque, R.; Len, C. A review of progress in (bio)catalytic routes from/to renewable succinic acid. *Biofuels Bioprod. Biorefin.* **2017**, *11*, 908–931. [CrossRef]

90. Valderrama-Gomez, M.A.; Kreitmayer, D.; Wolf, S.; Marin-Sanguino, A.; Kremling, A. Application of theoretical methods to increase succinate production in engineered strains. *Bioprocess Biosyst. Eng.* **2017**, *40*, 479–497. [CrossRef]

91. Adom, F.; Dunn, J.B.; Han, J.; Sather, N. Life-Cycle fossil energy consumption and greenhouse gas emissions of bioderived chemicals and their conventional counterparts. *Environ. Sci. Technol.* **2014**, *48*, 14624–14631. [CrossRef]

92. Cheng, K.-K.; Zhao, X.-B.; Zeng, J.; Zhang, J.-A. Biotechnological production of succinic acid: Current state and perspectives. *Biofuels Bioprod. Biorefin.* **2012**, *6*, 302–318. [CrossRef]

93. Mohan, S.V.; Modestra, J.A.; Amulya, K.; Butti, S.K.; Velvizhi, G. A circular bioeconomy with biobased products from CO_2 sequestration. *Trends Biotechnol.* **2016**, *34*, 506–519. [CrossRef]

94. Cao, W.; Wang, Y.; Luo, J.; Yin, J.; Xing, J.; Wan, Y. Effectively converting carbon dioxide into succinic acid under mild pressure with Actinobacillus succinogenes by an integrated fermentation and membrane separation process. *Bioresour. Technol.* **2018**, *266*, 26–33. [CrossRef] [PubMed]

95. Cukalovic, A.; Stevens, C. V Feasibility of production methods for succinic acid derivatives: A marriage of renewable resources and chemical technology. *Biofuels Bioprod. Biorefin.* **2008**, *2*, 505–529. [CrossRef]

96. Gunnarsson, I.B.; Alvarado-Morales, M.; Angelidaki, I. Utilization of CO_2 fixating bacterium *Actinobacillus succinogenes* 130Z for simultaneous biogas upgrading and biosuccinic acid production. *Environ. Sci. Technol.* **2014**, *48*, 12464–12468. [CrossRef] [PubMed]

97. Ballmann, P.; Dröge, S.; Wilkens, M. Integrated succinic acid production using lignocellulose and carbon dioxide from biogas plants. *Chemie Ing. Tech.* **2018**, *90*, 1253. [CrossRef]

98. Babaei, M.; Tsapekos, P.; Alvarado-Morales, M.; Hosseini, M.; Ebrahimi, S.; Niaei, A.; Angelidaki, I. Valorization of organic waste with simultaneous biogas upgrading for the production of succinic acid. *Biochem. Eng. J.* **2019**, *147*, 136–145. [CrossRef]

99. Urbance, S.E.; Pometto, A.L., III; Dispirito, A.A.; Denli, Y. Evaluation of succinic acid continuous and repeat-batch biofilm fermentation by *Actinobacillus succinogenes* using plastic composite support bioreactors. *Appl. Microbiol. Biotechnol.* **2004**, *65*, 664–670. [CrossRef]

100. Kim, D.Y.; Yim, S.C.; Lee, P.C.; Lee, W.G.; Lee, S.Y.; Chang, H.N. Batch and continuous fermentation of succinic acid from wood hydrolysate by *Mannheimia succiniciproducens* MBEL55E. *Enzyme Microb. Technol.* **2004**, *35*, 648–653. [CrossRef]

101. Wan, C.; Li, Y.; Shahbazi, A.; Xiu, S. Succinic acid production from cheese whey using *Actinobacillus succinogenes* 130 Z. *Appl. Biochem. Biotechnol.* **2008**, *145*, 111–119. [CrossRef]
102. Bradfield, M.F.A.; Mohagheghi, A.; Salvachúa, D.; Smith, H.; Black, B.A.; Dowe, N.; Beckham, G.T.; Nicol, W. Continuous succinic acid production by *Actinobacillus succinogenes* on xylose-enriched hydrolysate. *Biotechnol. Biofuels* **2015**, *8*, 181. [CrossRef]
103. Song, H.; Lee, S.Y. Production of succinic acid by bacterial fermentation. *Enzyme Microb. Technol.* **2006**, *39*, 352–361. [CrossRef]
104. Lee, P.C.; Lee, S.Y.; Hong, S.H.; Chang, H.N. Isolation and characterization of a new succinic acid-producing bacterium, *Mannheimia succiniciproducens* MBEL55E, from bovine rumen. *Appl. Microbiol. Biotechnol.* **2002**, *58*, 663–668. [PubMed]
105. Lee, P.C.; Lee, S.Y.; Hong, S.H.; Chang, H.N. Batch and continuous cultures of *Mannheimia succiniciproducens* MBEL55E for the production of succinic acid from whey and corn steep liquor. *Bioprocess Biosyst. Eng.* **2003**, *26*, 63–67. [CrossRef] [PubMed]
106. Jae Oh, I.; Lee, H.; Park, C.; Lee, S.Y.; Lee, J. Succinic acid production by continuous fermentation process using *Mannheimia succiniciproducens* LPK7. *J. Microbiol. Biotechnol.* **2008**, *18*, 908–912.
107. Samuelov, N.S.; Datta, R.; Jain, M.K.; Zeikus, J.G. Whey fermentation by *Anaerobiospirillum succiniciproducens* for production of a succinate-based animal feed additive. *Appl. Environ. Microbiol.* **1999**, *65*, 2260–2263. [PubMed]
108. Lee, P.C.; Lee, W.G.; Kwon, S.; Lee, S.Y.; Chang, H.N. Batch and continuous cultivation of *Anaerobiospirillum succiniciproducens* for the production of succinic acid from whey. *Appl. Microbiol. Biotechnol.* **2000**, *54*, 23–27. [CrossRef] [PubMed]
109. Lee, P.C.; Lee, S.Y.; Chang, H.N. Kinetic study of organic acid formations and growth of *Anaerobiospirillum succiniciproducens* during continuous cultures. *J. Microbiol. Biotechnol.* **2009**, *19*, 1379–1384. [CrossRef]
110. Lee, P.C.; Lee, S.Y.; Chang, H.N. Kinetic study on succinic acid and acetic acid formation during continuous cultures of *Anaerobiospirillum succiniciproducens* grown on glycerol. *Bioprocess Biosyst. Eng.* **2010**, *33*, 465–471. [CrossRef] [PubMed]
111. Lee, P.-C.; Lee, S.-Y.; Chang, H.-N. Cell recycled culture of succinic acid-producing *Anaerobiospirillum succiniciproducens* using an internal membrane filtration system. *J. Microbiol. Biotechnol.* **2008**, *18*, 1252–1256.
112. Tan, J.P.; Jahim, J.; Harun, S.; Wu, T.Y. Overview of the Potential of Bio-Succinic Acid Production from Oil Palm Fronds. *J. Phys. Sci.* **2017**, *28*, 53–72.
113. Köhler, K.A.K.; Rühl, J.; Blank, L.M.; Schmid, A. Integration of biocatalyst and process engineering for sustainable and efficient n-butanol production. *Eng. Life Sci.* **2015**, *15*, 4–19. [CrossRef]
114. Liebal, U.W.; Blank, L.M.; Ebert, B.E. CO_2 to succinic acid—Estimating the potential of biocatalytic routes. *Metab. Eng. Commun.* **2018**, *7*, 1–10. [CrossRef] [PubMed]
115. Figueres, C.; Le Quéré, C.; Mahindra, A.; Bäte, O.; Whiteman, G.; Peters, G.; Guan, D. Emissions are still rising: Ramp up the cuts. *Nature* **2018**, *564*, 27–30. [PubMed]
116. Le Quéré, C.; Andrew, R.M.; Friedlingstein, P.; Sitch, S.; Hauck, J.; Pongratz, J.; Pickers, P.; Korsbakken, J.I.; Peters, G.P.; Canadell, J.G. Global carbon budget 2018. *Earth Syst. Sci. Data* **2018**, *10*, 2141–2194. [CrossRef]

Article

Economic Perspectives of Biogas Production via Anaerobic Digestion

Arpit H. Bhatt [1] and Ling Tao [2],*

[1] Strategic Energy Analysis Center, National Renewable Energy Laboratory, Golden, CO 80401, USA; Arpit.Bhatt@nrel.gov
[2] National Bioenergy Center, National Renewable Energy Laboratory, Golden, CO 80401, USA
* Correspondence: Ling.Tao@nrel.gov

Received: 31 May 2020; Accepted: 8 July 2020; Published: 14 July 2020

Abstract: As the demand for utilizing environment-friendly and sustainable energy sources is increasing, the adoption of waste-to-energy technologies has started gaining attention. Producing biogas via anaerobic digestion (AD) is promising and well-established; however, this process in many circumstances is unable to be cost competitive with natural gas. In this research, we provide a technical assessment of current process challenges and compare the cost of biogas production via the AD process from the literature, Aspen Plus process modeling, and CapdetWorks software. We also provide insights on critical factors affecting the AD process and recommendations on optimizing the process. We utilize four types of wet wastes, including wastewater sludge, food waste, swine manure, and fat, oil, and grease, to provide a quantitative assessment of theoretical energy yields of biogas production and its economic potential at different plant scales. Our results show that the cost of biogas production from process and economic models are in line with the literature with a potential to go even lower for small-scale plants with technological advancements. This research illuminates potential cost savings for biogas production using different wastes and guide investors to make informed decisions, while achieving waste management goals.

Keywords: biogas; anaerobic digestion; waste-to-energy; wet waste; bioenergy; techno-economic analysis

1. Introduction

The demand for energy continues to rise with global population growth and rise in urbanization [1,2]. This has led to increased energy requirements and fossil energy utilization, increasing pollution across the globe [3,4]. With the limited supply of conventional fossil raw materials and their adverse environmental implications on air quality, research efforts on utilizing environmentally friendly renewable alternative sources have gained significant importance [5]. In addition, continuous research coupled with technological developments in this field have helped the investment and deployment of clean energy technologies to surge across the world. One plausible and established waste-to-energy (WTE) option that has been widely adopted is the production of biogas from organic-rich waste streams via anaerobic digestion (AD) process or technology. As waste is an increasing issue worldwide, proper utilization of its energy potential via economically viable and technically feasible technologies can help promote sustainability and meet global renewable energy demand, while limiting risks pertaining to landfilling.

The total wet waste feedstocks in the United States (U.S.) present an annual energy resource of 42.9 billion liter gasoline equivalent (11.3 billion gallon gas equivalent [GGE]), which includes wastewater residuals, food waste, animal waste, and fats, oils and greases (FOG) [6]. Conversion of these wet waste feedstocks into useful products, such as biogas (methane [CH_4] and carbon dioxide [CO_2]), represents a significant opportunity for additional expansion of transforming under-utilized resources into renewable energy. Different types of organic waste include wastewater treatment (WWT) plant sludge (primary and secondary sludge), agri-food industry waste (part of municipal solid waste including fruit and vegetable by-products, canteen waste, kitchen waste), green waste (waste from shearing of grass, sheets), animal waste (swine, dairy manures), and FOG (animal fats, used cooking oils, restaurant vats for degreasing). Figure 1 shows the wet waste availability in the U.S. along with its energy potential that could be utilized for WTE technologies.

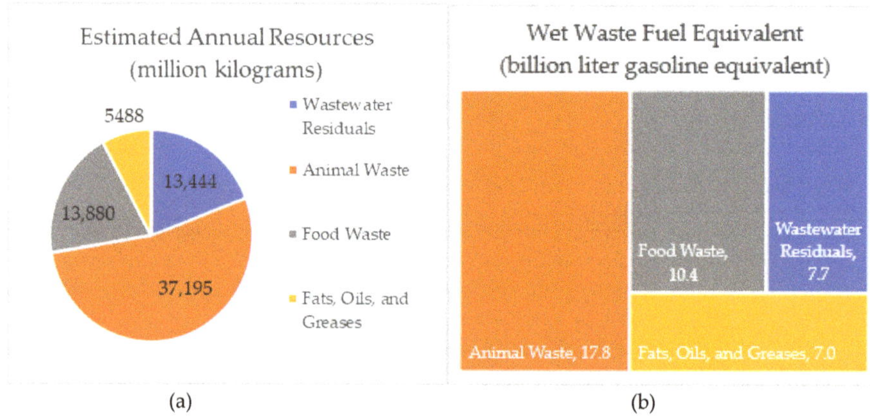

Figure 1. (**a**) Availability of wet wastes in the U.S.; (**b**) Wet waste energy potential in the U.S. [6]. 1 kg = 0.0011 U.S. tons; 1 L = 0.264 U.S. liquid gallons.

As shown in the U.S. Department of Energy's (DOE) Bioenergy Technologies Office (BETO) report [6], these wastes are geographically located in areas with high population density. Although the aggregation of wastes provides benefits with lower transportation costs, they are more vulnerable to stringent and cost-intensive disposal regulations owing to landfill constraints. Thus, novel waste management solutions should be considered for this huge waste energy potential that can help with disposal problems. Current and widely applied waste management practices include direct disposal or landfilling and direct combustion or incineration [7,8], which leads to emissions of hazardous pollutants harmful to human health as well as greenhouse gas (GHG) emissions [8–10]. Other sustainable practices that are accepted include composting, recycle/reuse of organic matter, and animal feed production, but yield low energy [9]. Moreover, many waste processing technologies are tailored to a specific family of feedstock that has significant compositional, temporal, and geographic variability. Thus, for sustainable energy production, there is a global need to adopt technologies that can utilize a wide variety of environmentally friendly feedstocks with high energy density.

A widely popular and cost-effective technology currently employed at several WWT facilities is AD, which effectively treats a variety of biodegradable streams rich in organic carbon, such as sludge, food waste, yard waste, wood, process residues, animal manure [11], algae biomass from sludge obtained from phototrophic recovery [12,13], etc., and convert them into CH_4 containing biogas. Biogas contains 40–75% CH_4 and 15–60% CO_2 (by volume), with small amounts of hydrogen (H_2), nitrogen (N_2), hydrogen sulfide (H_2S), oxygen (O_2), and water (H_2O). Biogas has a wide variety of applications including as a substitute for natural gas and heating oil, an upgrade for utilization as a transport fuel, and use in the production of heat and electricity using combined heat and power (CHP) technology [14]. Moreover, the odors associated with manures along with emissions of several pollutants (e.g., ammonia, CH_4, CO_2, nitrous oxide, and methyl-mercaptanes) can be reduced through use of AD technology, enhancing the agricultural sustainability [15]. Although production of biogas using the AD technology provides an environmentally sustainable approach, there are issues that could affect the process and biogas yields depending on the type of wet waste, further explained in the Discussion section.

This study provides a quantitative analysis on theoretical biogas energy yields and offers viewpoints on both technical and economic perspectives of wet wastes to biogas production. This paper provides a framework to establish the cost of biogas production from AD, utilizing a variety of wet wastes at a wide range of plant scales as compared to costs obtained from the literature. We utilize two process and economic models: (1) Aspen Plus coupled with techno-economic analysis (TEA), and (2) CapdetWorks, to estimate and compare the production cost of biogas in U.S. dollars per gigajoule (USD/GJ). The insights on the current technology status, process challenges, energy yields, and economic potentials are summarized in the study in order to not only provide advancements with waste-to-biogas conversion technologies but also to provide recommendations for optimizing the AD process. This analysis helps illuminate the potential cost savings for biogas production and guide investors and AD developers to make informed decisions.

2. Current Practices—Anaerobic Digestion for Biogas Production

A series of biological processing steps with the core conversion using the AD technology, converts the complex organic matter in waste products to simple monomers by using a consortia of microorganisms. The AD technology facilitates organic decomposition and reduces inorganic matter in the absence of O_2 [16]. These organic materials are converted to final products, which are mainly biogas, digestate (a liquid form in most cases), and a combination of solid and liquid effluents derived from digestate treatments [17]. Biogas can be used to produce electricity and heat, while the effluent liquid rich in crop nutrients is used as agricultural fertilizer depending on the amount of nutrient loading, especially nitrogen [17]. The solid fraction from solid/liquid separation of the digestate can be used as dairy bedding or converted to potting soil mixes. Generally, the feed to the digester is pretreated using different physical operations to reduce maintenance issues. For example, the solids from wastewater primarily include primary and secondary sludge which are grinded, shredded, or screened for efficient operations. Additionally, the accumulation of grits inside the tanks reduces the working volume, which affects biogas production and increases maintenance and cleaning [18]. Therefore, degritting is applied to prevent accumulation, which helps to improve process efficiency as well. Similarly, plastics, stones, and metals are removed from food waste during pretreatment depending on the collection methods.

2.1. Steps Involved in AD Process

AD includes four biological steps, as shown in Figure 2 [17]; (1) *Hydrolysis* to breakdown insoluble organic matter to simple monomers via hydrolytic microorganisms. Proteins are converted to amino acids, lipids into fatty acids, starch to glucose, and carbohydrates to monomeric sugars. (2) *Acidogenesis* to convert simple sugars and acids to volatile fatty acids (VFAs) and alcohols via acidogens. This is the fastest step in the AD process which involves complex consortia of bacteria, including bacteriocides, clostridia, bifidobacterial, streptococci, and Enterobacteriaceae [19]. (3) *Acetogenesis* to further convert VFAs to acetate, CO_2, and H_2, and (4) *Methanogenesis* to convert acetate and H_2 to CH_4 and CO_2 via methanogenic bacteria. The consortia of microbial populations need to be maintained to stabilize the AD process and increase biogas production efficiency.

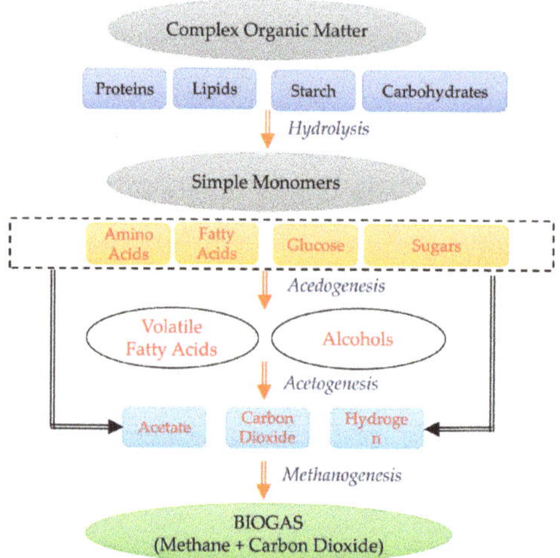

Figure 2. Steps in the Anaerobic Digestion Process.

2.2. Factors Affecting the AD Process

There are several operating factors affecting the production of biogas in the AD process. These mainly include temperature, hydraulic retention time (HRT), organic loading rate (OLR), and pH. Other factors affecting the gas production also include tank volume, feedstock type, feeding pattern, and carbon to nitrogen (C/N) ratio.

Temperature is a critical parameter for the AD process for survival of microbial consortia and to consistently produce biogas, as for each 6 °C decrease (20 °F), the biogas production falls by 50% [17]. Two temperature ranges are most suitable for biogas production—thermophilic and mesophilic. Thermophilic bacteria operate at high temperature conditions (48.9–60.0 °C or 120–140 °F), thus, reducing the retention time to decompose more substrate and produce more biogas. However, these systems are costly to operate as energy is required to maintain higher operating temperature, and they are prone to acidification and are easily influenced by toxins [20,21]. Alternatively, mesophilic bacteria functioning at lower temperatures (32.2–43.3 °C, or 90–110 °F) produce less biogas as compared to thermophilic but are easy to operate, low in investment costs, and more stable towards environmental changes. However, they have poor biodegradability and are susceptible to nutrient imbalance [22,23].

In addition to temperature, HRT also affects biogas production. HRT is the average time (usually, a few days to 40 days, depending on the type of organic waste and digester) feedstocks reside in the digester to decompose based on chemical oxygen demand (COD)/biological oxygen demand. Longer retention times provide enough time for organic matter to degrade depending on the microbial consortia present in the digester at different rates and times. Shorter retention times would inhibit methanogenesis while longer retention times lead to insufficient utilization of components [20,24]. Similarly, the amount of volatile solids (VS) fed to the digester every day (OLR) is also an important parameter affecting biogas yield. Biogas production increases with higher OLR; however, it disrupts the bacterial population, leading to higher hydrolytic bacteria and acidogens. This would lead to lower methanogen population required for biogas production. The literature contains maximum OLRs for various organic feedstocks to avoid irreversible acidification and high biogas yields (9.2 kg VS/m^3/day for sludge, 10.5 kg VS/m^3/day for food waste) [25,26], with an optimal range between 1.5 and 6 kg VS/m^3/day for all waste [15].

pH is another important factor affecting the bacterial activity, and thus, biogas production. Methanogens are highly sensitive to acidic environment (pH < 7), while acidogens are inhibited leading to a rapid increase in methanogens at higher pH levels. The optimal pH for acidogenesis is between pH 5.5 and 6.5 [27], while methanogenesis is most efficient between pH 6.5 and 8.2 [28]. Thus, it is important to maintain pH between 6.5 and 7.5 to sustain an optimal concentration of acidogens and methanogens in the digester for higher biogas yields. Other factors affecting the AD process include type of feedstock for predicting the composition and rate of reaction, tank volume for determining the retention time, and C/N ratio, replicating the amount of nutrient levels in the digester required for AD steps affecting the biogas yield [20].

3. Methods

For this analysis, we consider four different types of feed for biogas production—wastewater sludge, food waste, swine manure, and FOG. Table 1 shows the major data assumptions and correlations obtained from the literature for different wastes along with biogas parameters, including mass yield and gas composition analyzed in this study. These assumptions are based on the data obtained from the literature to understand wet organic waste feed characteristics, biogas production, and composition [6,29–55].

Table 1. Composition of Wet Wastes and Physical Parameters for Biogas Production via Anaerobic Digestion.

Parameters		Wastewater Sludge	Food Waste	Swine Manure	FOG
Composition (Dry Weight%)					
Ash		7.5%	5.0%	15.2%	0%
Lipids		18.0%	21.0%	3.8%	78.0%
Proteins		24.0%	19.0%	20.0%	7.0%
Fermentable Carbohydrates		16.0%	55.0%	36.5%	15.0%
Lignin		0%	0%	21.0%	0%
Extractives (all non-fermentable components)		34.5%	0%	3.5%	0%
Component Parameters					
Energy Density	MMBtu/t TS	19.5	22.9	17.1	39.0
	MJ/kg TS	20.6	24.2	18.0	41.1
Moisture Content (%)		96%	75%	93%	6–95%
TS (%)		Primary—2–6% / Secondary—2–10%	25%	7%	5–94%
COD (mg/L)	Range	47,200–140,000	39,800–350,000	20,600–35,000	92,000–149,000
	Mean	135,711	154,000	28,430	120,500
COD Reduction		55.5%	65.0%	55.0%	82.0%
Biogas Yield	m³/t TS	500–600	646	566	1168–1422
	L/kg TS	500–600	646	565	1169–1422
	MMBtu/t TS	11–13	14	12	20–25
	MJ/kg TS	12–14	15	13	21–27
Typical Scale	wet US tons/day, unless noted	1–300 MGD	1–250	1–250	1–200
	wet metric tons/day, unless noted	1–300 MGD	0.9–227	0.9–227	0.9–181
	kg/day, unless noted	3785–1,135,500 m³/day	907–227,000	907–227,000	907–181,000

FOG = fat, oil, and grease, MMBtu = million British thermal units, t = metric ton, MJ = megajoule, TS = total solids, COD = chemical oxygen demand, CO_2 = carbon dioxide, CH_4 = methane, m³ = cubic meter, MGD = million gallons per day, 1 metric ton = 1000 kg, I US ton = 0.907185 metric tons, 1 MGD = 3785 m³.

3.1. Feedstock Composition

The feedstock composition can play an important role in the overall yields and process economics of the system. For wastewater sludge, the moisture content varies greatly (>80–99 wt%, mass basis unless otherwise specified) depending on the dewatering techniques employed prior to reaching the AD reactors. Here, we assume that the stream entering the AD reactor has been dewatered to 4% solids content, which is consistent with previous modeled efforts [56]. The ash content of wastewater sludge can also vary greatly depending on the sources, assumed at 7.5% in this study [29]. The nonvolatile solids, other than ash, account for 17.5% of the dry mass such that only 75% of the total dry solids are volatile [30,31]. Proteins have the highest concentration in the wastewater sludge at 24% followed by 18% lipids, 16% fermentable carbohydrates, and 17% of other material set as extractives.

Food waste can vary widely in composition depending on the source and usage. Based on the literature, the moisture content in food waste ranges from 54–87%, ash content ranges from 1–15%, proteins range from 3–44%, lipids range from 4–43%, and carbohydrates range from 10–76% [32–38,57]. We assume that the food waste entering the AD unit contains 25% solids, with 55% carbohydrates as the largest percent of the biomass on dry basis, followed by 21% lipids, 19% proteins, and 5% ash.

For swine manure, the moisture content is assumed 93% with an ash content of 15% of total solids (TS) [39]. The TS have 37% carbohydrates, 20% proteins, 21% lignin, 4% lipids, and 4% extractives [40]. Swine manure can have a high concentration of nutrients after AD, specifically phosphorous. These biosolids with high nutrient contents can be a revenue generating stream. Further research would be needed to understand the nutrient production amounts and balances to valorize this stream.

Finally, FOGs are predominantly lipids with a small protein and carbohydrate content compared to the other three feedstocks. FOG is a combination of fats, oils, and greases, mainly produced from cooking and the food producing industry. For this study, we consider fat as a primary source of the FOG stream, which is assumed to have 90% moisture, and the dry weight solids have 78% lipids, 7% proteins, and 15% carbohydrates [41].

3.2. Biogas Composition and Energy Yield

The composition of CH_4 and CO_2 in the biogas produced would vary depending on the waste characteristics and operating conditions of the AD process. A summary of the percentage of initial COD values [44–53,58–62], COD reduction, and typical biogas production facility scales are shown in Table 1. The fractional conversions of fermentable components are based on the COD reductions from the literature, i.e., 55.5% for sludge, 65% for food waste, 55% for swine manure, and 82% for FOG. It should be noted that depending on the composition and material that comprises the food waste stream, the COD reduction can be as high as 90%. The feedstock composition and the conversion of fermentable components (carbohydrates, proteins, and lipids) into biogas based on waste-specific COD reduction rates forms the basis for estimating the CH_4 and CO_2 content in biogas.

The energy yield is another important parameter to estimate the biogas production values for a given amount of waste (on a dry or wet basis). There is a wide range of biogas production values reported in the literature for different types of wet waste. Based on the literature values of biogas production (mass basis), energy content or heating value of the substrate, COD values, and VS and TS in the feed, we estimate the biogas energy yield for sludge, food waste, swine manure, and FOG [6,29,30,41–54,58–72]. We also estimate the theoretical energy yield of biogas by dividing the energy content of biogas production (in terms of MJ/kg using the volumetric biogas yield in m^3/kg and calorific value of biogas in MJ/m^3) by the feedstock energy inputs (in terms of heat content, MJ/kg) to the process, without accounting for heat loss due to process operation. Refer to Section 4.1 for detailed results on each wet waste feedstock.

3.3. Biogas Cost Analysis

3.3.1. Cost Data from Open Literature

To compare the cost of biogas production from various models, we first collect the data from the literature for various AD processes utilizing different types of organic wastes for different plant scales. In addition, we also summarize data for organics present in the feed (COD), VS. and TS content, and biogas production values. We use these values to determine the biogas energy yields for each wet waste feedstock and estimate the biogas production (in GJ) for different feed rates. When data on biogas production rates are not available, we use the theoretical energy yields for each feed type to estimate the costs.

In addition to direct capital costs associated with the AD reactor obtained from literature, we add indirect capital costs, which include additional piping (4.5% of inside battery limits (ISBL]), project contingency (10% of total direct costs [TDC], combination of total installed capital costs and piping costs), and other costs (10% of TDC). We also consider working capital, which is 5% of fixed capital investment (FCI) (combination of TDC and other capital costs). The capital costs are amortized annually considering a 30-year straight line depreciation. We also consider operating expenses, including electricity import, heat, water for the process, and additional nutrients (including ammonia) provided to the AD microbes. Other operating costs include maintenance and property insurance, which is 3% of the plant capital costs and 0.7% of the FCI, respectively. Capital and operating costs are then converted to USD/GJ based on biogas production rates from the literature.

All costs are converted to 2016 USD using the Plant Cost Index from Chemical Engineering Magazine [73], the Industrial Inorganic Chemical Index from SRI Consulting [74], and the labor indices provided by the U.S. Department of Labor Bureau of Labor Statistics [75]. For comparison, we convert the annualized total costs (capital and operating) into USD/GJ based on respective energy yields.

3.3.2. Cost Data from Aspen Plus Simulation

We model an AD system using Aspen Plus for four feedstocks at various plant scales (see Table 1) analyzed in this paper. Aspen Plus is a chemical process simulator which includes unit operations to build a process model for simulating complex calculations for integrated batch and continuous processes. First, the process simulation calculates material and energy balances using Aspen Plus [76] based on the block flow diagram shown in Figure 3. Then, an in-house spreadsheet calculates capital and operating expenses for each case. The final effluent out of the AD unit goes through solid-liquid separation to produce AD sludge waste and nutrient rich liquor. Utilities for heat and power requirements are also included in the model. It should be noted that we do not include biogas clean-up in this analysis as most of the literature data do not report this cost.

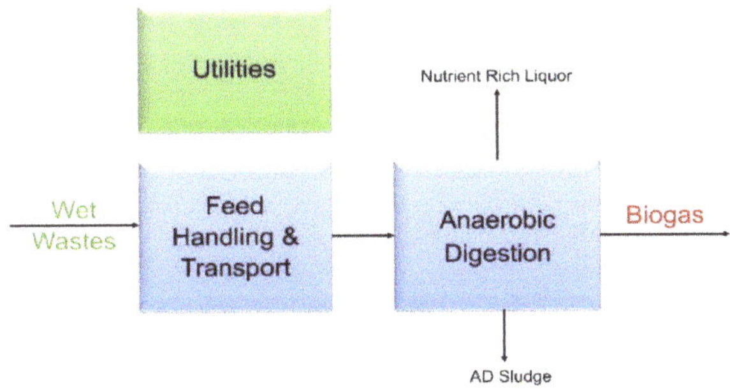

Figure 3. Process Flow Diagram for Modeling AD using Aspen Plus.

The design of an AD system can consist of a single stage or of multiple stages with the processing steps of hydrolysis, acidogenesis, acetogenesis, and methanogenesis, separated or combined. There are two types of AD digesters. Mesophilic digesters operate at a lower temperature as compared to thermophilic digesters. The digester performance depends greatly on reactor configuration and total vs. or total percentage COD in the biosolids. The loading rate can be calculated using Equation (1) [77], while sizing of the digesters is based on the solids retention time and the loading rate of the vs. into the system.

$$Loading\ rate\left(\frac{mgCOD}{m^3 * day}\right) = \frac{organic\ mater\left(\frac{mgCOD}{m^3}\right) * Flow\ rate\left(\frac{m^3}{day}\right)}{operating\ volume(m^3)} \tag{1}$$

In addition, the stoichiometric reactions modeled in the AD reactor are shown in Equations (2)–(4). The production of biogas from Aspen simulation is based on the fractional conversion of carbohydrates, proteins, lipids, and other components based on defined COD reductions in Table 1.

$$Carbohydrates + H_2O \rightarrow 3\ CH_4 + 3\ CO_2 \tag{2}$$

$$Proteins + 0.591\ H_2O \rightarrow 0.028\ AD\ Sludge + 0.4375\ CH_4 + 0.262\ NH_3 + 0.01\ H_2S + 0.4225\ CO_2 \tag{3}$$

$$Lipids + 23.6\ H_2O + 1.452\ NH_3 \rightarrow 1.452\ AD\ Sludge + 36.37\ CH_4 + 13.37\ CO_2 \tag{4}$$

We utilize the values of VS. loading rate from the literature to maximize biogas production. We use vs. loading of 0.95 kg VS/m^3 AD/day for wastewater sludge [56], while a higher value of 4.7 kg VS/m^3 AD/day is used for food waste because of the high solids content. For swine manure, the vs. loading rate is 1.5 kg VS/m^3 AD/day [70], while vs. loading for FOG is 3.5 kg VS/m^3 AD/day [78,79]. Then, the total volume needed for AD is calculated based on both the hydraulic retention time and the total inlet feed flow rate. An in-house excel model is created to import mass and energy data from the process model which are then used to size and cost the AD reactor.

In addition to the AD reactor, we consider the costs of additional capital equipment including biogas blowers, feed screws, coolers, pumps, holding tanks, mixers, and heat exchangers to include other components of the AD system. We follow a similar methodology as described in the previous section to determine the annualized capital and operating costs considering a straight 30-year depreciation to estimate the biogas production costs in USD/GJ.

3.3.3. Cost Data from CapdetWorks

Like Aspen Plus applied for chemical processes, CapdetWorks is a wastewater industry standard that has been used for various analyses by the U.S. Environmental Protection Agency (EPA) [80,81]. It is an effective software that has a built-in "what if" scenario (unlike Aspen) that performs sensitivity analyses much faster and more flexibly while providing accurate results. CapdetWorks provides two ways to define inlet streams—one is to use the defined wastewater inlet stream from the model, and the other is to use a defined biosolids feed stream (typically applied to food, manure wastes). When using the wastewater sludge, the stream has a low solids and COD content; thus, the process undergoes several dewatering steps to reach the AD unit. Although we model the full wastewater facility with dewatering steps in CapdetWorks (as shown in Figure 4), only the cost of the AD unit is considered as part of this analysis. The food waste and swine manure are fed directly into the AD, as shown in Figure 5.

Though CapdetWorks requires few variable inputs, it is difficult to obtain cost, biogas yields, and sizing values for FOGs because of its limitations to vary COD/VS values. Thus, we do not model FOG in this simulation software. To be consistent with other models, we consider additional piping (4.5% of ISBL), project contingency (10% of TDC), other costs (10% of TDC), and working capital (5% of FCI), in the CapdetWorks economic analysis. The capital costs are depreciated straight-line in a 30-year period, while the amortized operating costs are directly obtained from the model. It is important to note that Aspen Plus simulation software has the ability to perform complex calculations for batch and continuous operations for any type of chemical facility while CapdetWorks is a wastewater treatment industry standard software which can also be used for simulation of biosolids processes.

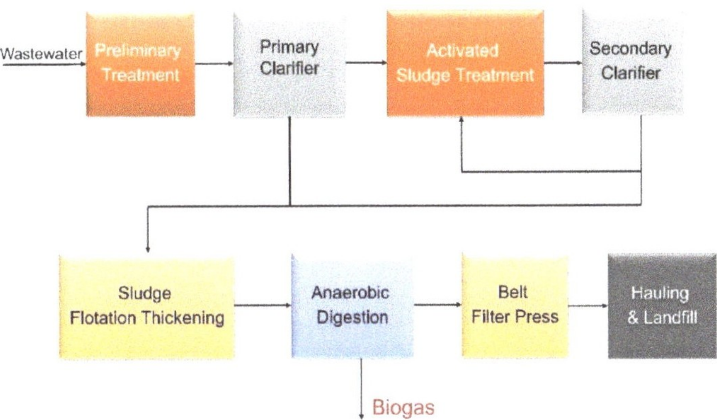

Figure 4. Process Flow Diagram of a WWT Facility as Modeled in CapdetWorks.

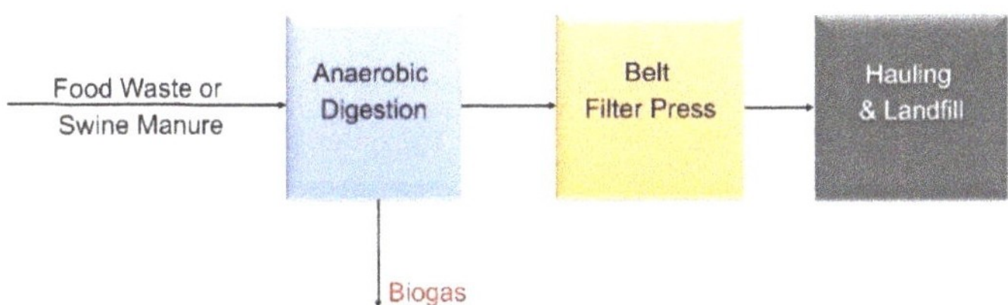

Figure 5. Process Flow Diagram of an AD Facility Converting Food Waste or Manure as Modeled in CapdetWorks.

4. Results

4.1. Biogas Composition and Energy Yields

Our results suggest that the biogas composition for wastewater sludge and food waste are 60% and 56% methane, respectively, while the methane content of the biogas is 53% for swine manure and 70% for FOG. The deviation of the biogas composition is due to compositional distribution of lipids, carbohydrates, proteins, and other fermentable components in the feed. Although swine manure has low CH_4 content at 53%, which is most likely due to the higher amount of unfermentable components such as lignin and ash when compared with other wastes, there are several literatures that show CH_4 content in swine manure to be typically around 65–75%, mainly due to several removal practices followed by the facilities [15,82]. Moreover, co-digesting swine manure with other wastes such as winery wastewater and vegetable processing wastes could increase CH_4 content to 64% and 69%, respectively, depending on operating conditions [83,84]. Li et al. [85] also showed methane content in excess of 62% when co-digesting swine manure with cow dung in dry AD process. FOG, on the other hand, produces a biogas product with 70% CH_4, likely due to the high lipids content and zero ash and lignin in the feed. Co-digestion of FOG with other wet wastes (sludge or food waste) would likely reduce the CH_4 content back to ~60% or lower depending on the co-digestion ratio and COD digestibility.

Figure 6 shows the theoretical biogas energy yields for different wet wastes as compared to range of the literature values.

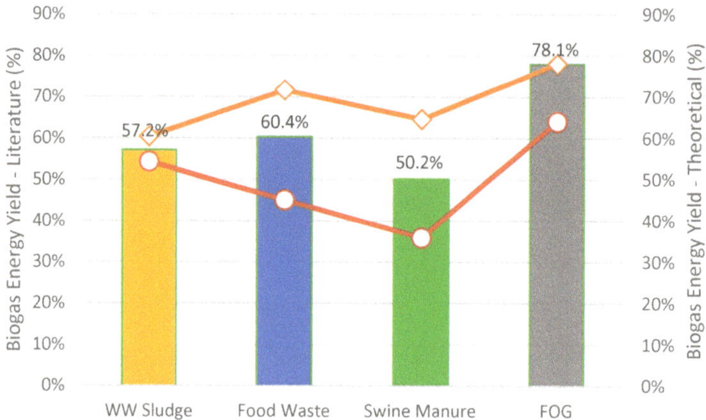

Figure 6. Biogas Energy Yield (%) from Wet Waste Feedstocks. The bar chart shows the theoretical values of energy yield from different wastes while the line chart shows the range of energy yields from the literature.

As mentioned previously, the percentage energy yields shown in the figure are estimated by the energy content of biogas divided by the feedstock energy content to the process. For example, based on the biogas yield of 0.65 m^3/kg TS (646 m^3/t TS, as reported in Table 1) of food waste [86], energy content of 24.2 MJ/kg TS (20.8 MMBtu/dry ton) for food waste [87], and calorific value of 22.6 MJ/m^3 (21,422 Btu/m^3 at 60% CH$_4$) for biogas, the theoretical biogas energy yield is estimated to be 60.4%. This methodology is utilized for each feedstock and the numbers are compared with those obtained from the literature to estimate the range. As shown in Figure 6, the energy yield for wastewater sludge ranges from 54–60%, while it ranges from 45–72% for food waste. Likewise, the energy yield for swine manure ranges from 36–65%, while for FOG, it is 64–78%, with fat as the source component. Thus, the energy yield for FOG is highest at 78.1% and is lowest for swine manure at 36%, among all the wastes. This is because of the lower biogas production from swine manure as compared to other wastes, due to the high quantity of non-fermentable components (e.g., ash) in the feed.

It should be noted that the biogas energy yields from the literature for several wastes are estimated to be higher than theoretical values because of high methane production through pretreatment technologies utilized prior to AD, use of inoculum to increase fermentable composition, or variation of operation parameters (e.g., pH, temperature, loading rate, HRT, etc.,) to boost the productivity. In contrast, the theoretical energy yields are based on biogas composition from fixed COD reduction and without considering any pretreatment technologies and additional operational changes.

4.2. Biogas Economics

To provide further insights into the economic potential of producing biogas from wet wastes via the AD process, we estimate the costs of AD units from the literature, Aspen Plus, and CapdetWorks modeling. As mentioned previously, the cost estimates include direct (e.g., equipment costs) and indirect costs (e.g., project contingency, additional piping) along with the working capital, considering a 30-year straight line capital depreciation.

Figure 7a–d shows the biogas production costs from several literature data sources [88–91], Aspen process model [76], and CapdetWorks model (except FOG) for facilities using wastewater sludge (1–300 MGD), food waste (1–250 wet tons/day), swine manure (1–250 wet tons/day), and FOG (1–200 wet tons/day), respectively. As shown from the figure, the cost for biogas production from wastewater sludge decreases with increasing scale, following economies of scale. The cost numbers estimated from Aspen and CapdetWorks simulations are in the middle of the plotted literature results [88–91], showing good agreement of simulated values with the literature. However, the cost numbers are on the optimistic side considering the recent agreement with literature values at plant scales lower than 10 MGD except for Misra [91]. But the cost numbers are quite off for a couple of the literature works at scales greater than 10 MGD. This is mainly due to varying sludge composition across different WWT facilities, some of which may also include pretreatment technologies to make sludge more attractive for the AD process. In contrast, the modeling aspects only consider a fixed composition of wastewater sludge without considering any pretreatment, which causes the disparity in cost numbers. In addition, the cost advantages due to increased output for large-scale facilities helps reduce the cost of biogas to around ~1 USD/GJ, as shown in Misra [91].

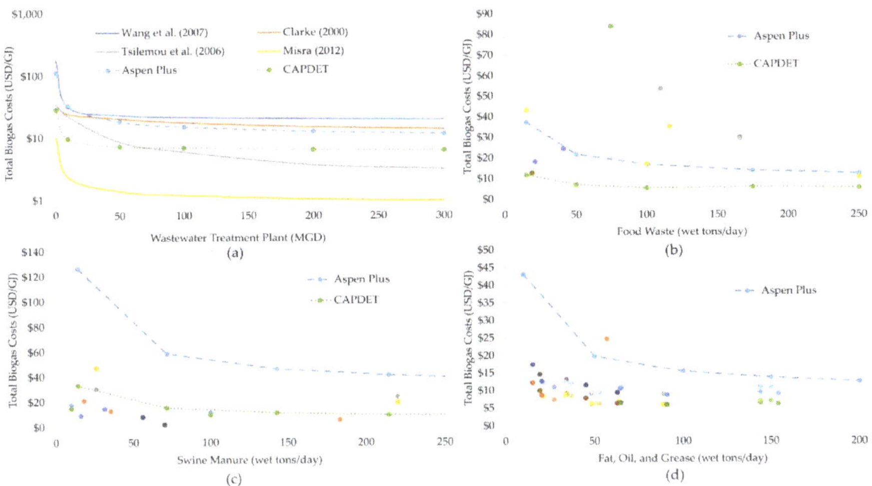

Figure 7. Total Biogas Production Costs in USD/GJ for AD Processing (**a**) Wastewater Treatment Sludge, (**b**) Food Waste, (**c**) Swine Manure, and (**d**) Fat, Oil, and Grease. Please note that that *y*-axis on Figure 7a is plotted on a log-scale.

For food waste, the literature data sources [77,92–95] have a wide range of cost values for AD utilizing food waste. This is mainly because the food waste utilized in literature has different sources, which includes food loss before or after meal preparation, waste after consumption, as well as food discarded in the process of manufacturing, distribution, retail, and food services, for which there is a wide variation in amount of VS and its conversion to biogas. In contrast, the costs of AD producing biogas as estimated from Aspen and CapdetWorks simulations follow a cost curve and are lower than literature values. A potential reason might be the high vs. loading rate assumed in the model for high solid content in the food waste. Moreover, the proximate analysis of different types of food (vegetable, meat, etc.) shows different physical properties that may lead to a wide range of AD costs and biogas yields.

Like wastewater sludge, the cost of biogas production from swine manure from the CapdetWorks simulation generally follows the dotted literature values indicating a very good agreement on what is being simulated in lieu of data found in the literature. The only exception is at lower plant scales (<50 wet tons/day), where simulation results presents a sharp increase in costs. In addition, the cost of biogas production from Aspen Plus simulation is higher compared to literature which may be because of the wide range of moisture content found in literature as opposed to fixed composition assumed in the simulations.

For FOG, the cost data from Aspen Plus are slightly higher than the literature at plant scale greater than 50 wet tons/day which has a wide range of values, while it is significantly higher for small facilities, mainly due to costs representing FOG co-digested or blended with other wastes. Additionally, the economies of scale play an important role when scaling the equipment cost as large-scale facilities have more cost-savings and competitive advantages as compared to small facilities with low production volumes. There are exceptions at large plant scales where the composition of co-digested waste could significantly alter the biogas production values and AD costs. It should be noted that the costs could deviate significantly from the actual FOG-to-biogas conversion cost for any plant scale depending on whether the waste primarily consisted of either FOG and when considering this as the sole source of feedstock. It should be noted that the values estimated from the simulation models are based on fixed composition of fermentable components (as described in Table 1) which may vary significantly from the literature values, producing wide variance in the cost results.

5. Discussion

The cost estimates obtained from Aspen Plus and CapdetWorks simulations are preliminary results based on predefined assumptions from the literature. Further technological advances in the AD process can enhance CH_4 content in biogas, which may potentially drive down the costs for AD to be economic at small plant scales. Some of the developments that have been tested include novel pretreatment operations (e.g., physical, chemical, and biological) to enhance biogas productivity and energy intensity [96]. Physical pretreatment includes mechanical and thermal energy disruption, while chemical pretreatment includes the use of chemical substrates to disintegrate waste for easy downstream operations. Biological pretreatment includes the application of microbial consortium or enzymes to enhance hydrolysis of the waste (mainly applied for biomass) for increasing CH_4 yield in the subsequent AD reactor. Another development is the co-digestion of wet waste with high organic feedstocks, which aims to balance the nutrient content and reduce the negative effects of toxic compounds on the process that ultimately increase biogas yield [97]. In addition, the optimization of operating parameters (e.g., improving COD reduction rates based on loading rate, temperature and pH control, etc.) would help enhance AD performance by keeping the system stable and controlled for high biogas selectivity. The configuration of AD reactors can also help alter operating parameters depending on the type of waste. Propositions have been made for novel digester systems including multi-phased digesters to enhance microbial growth in different chambers individually, use of membrane bioreactors, solid state digestion for waste >15% total solids, psychrophilic AD for cold temperature conditions, and integrated AD systems for producing multiple products [82,96]. A detailed analysis on these technologies around the economic and environmental standpoint should be carried out for better evaluating the tradeoffs between biogas yields and cost of adopting these technologies, which may ultimately help compete with natural gas prices.

Current Uses and Critical Issues Related to Biogas

Current uses. Biogas has a wide variety of applications across different industry sectors as mentioned previously. It can be directly used to produce heat and electricity via a CHP system, or it can be upgraded to remove water vapor, hydrogen sulfide, and CO_2 for use as a natural gas [98,99]. It can also be used as an engine fuel in internal combustion engines or fuel cells for production of mechanical work and/or electricity generation. Biogas can also be used as a fuel for agricultural pumps depending on the requirements [100], or can be directly upgraded to biofuels competing with biomass-based bioethanol and biodiesel production.

Additionally, biogas can be used to make valuable chemical products either through thermochemical or biological pathways. Thermochemical conversion typically begins by oxidizing the methane to form more reactive and, therefore, chemically malleable species. The biological conversion selectively produces specific compounds by taking advantage of the catalytic effect of methanotrophic or methylotrophic bacteria, by consuming methane as a carbon source. Methylotrophic and Methanotrophic bacteria are one class of bacteria, which is characterized by their ability to utilize a variety of different C-1 substrates including methane, methanol, methylated amines, halomethanes, and methylated compounds containing sulfur [101–104] as the carbon and energy sources [105–109].

There are three groups in methanotrophs, which are Alphaproteobacterial methanotrophs, Gammaproteobacterial methanotrophs, and Verrucomicrobial methanotrophs [110]. These organisms can be genetically engineered to produce a wide variety of chemicals of interest [111]. The main metabolism of methanotrophs is the methane oxidation via methanol to formaldehyde, which serves as an intermediate in catabolism and anabolism, breaking the rather stable C-H bond in methane under ambient conditions [107]. Formaldehyde or methanol is then oxidized to formate, which can either be further oxidized to CO_2 by a NAD-dependent formate dehydrogenase with the reducing power for methane metabolism [108,112], or serves as a key branch point to the serine pathway for carbon assimilation and catabolism [113,114], then into the tricarboxylic acid cycle.

Despite the potentially attractive cost of biogas, considerable technical challenges exist, specifically on overcoming the gas-liquid mass transport barrier when converting methane biologically. Currently, biological conversion of biogas is still in the early stage of research development as only the production of single cell protein and poly-hydroxy-butyrate from methane has been commercialized. Further targeted research can reveal other financially viable products, which may be of interest to industries.

Critical issues. Although biogas has a wide range of applications, there are several problems associated with the production of biogas. If the biogas from the AD process is not handled effectively, the emission of methane in the atmosphere could be hazardous and penalized heavily for GHG emissions. In addition, the overall positive impact of AD in terms of GHG emissions would be diminished if only a small percent of gas is emitted, as the global warming potential of methane is 23 times of CO_2 [14]. Also, the process utilizes a consortia of microbial population under a specific set of operating conditions, which if not managed properly, can lead to an unstable system and inefficient biogas production [20]. Moreover, the abundance of natural gas has pushed its prices to all-time lows, which makes it difficult for biogas produced from renewable sources to compete on a price basis.

The EPA has recognized the benefits of promoting net/low-carbon fuels derived from biogas. In the recent rulings, the EPA classified many sources of biogas from cellulosic feedstock for transportation fuels as part of the Renewable Fuel Standard (RFS) [115]. Furthermore, the use of biogas under the RFS can improve AD economics by allowing biogas (containing methane as an energy carrier) producers to generate Renewable Identification Numbers (RINs) [115,116]. The current cellulosic RIN credit is approximately USD5.7–USD8.6/GJ (USD6–USD9/MMBtu) [115]. Without these incentives, for producing fuels from biogas, it is harder to be economically sustainable.

The methane percentage in the biogas could be from 40–70% (with remaining of CO_2) and the market selling price of biogas varies from USD1.4–USD9.5/GJ (USD1.5–10/MMBtu) [117]. The exact composition of biogas produced varies by the composition of organics in the feed, including fats, proteins and carbohydrates, or carbon to O_2 ratios. This difference in chemical composition leads to the major classes of substrates in biomass having different expected yields and methane content. Lower methane content indicates lower energy yield from the traditional AD process concept, implying energy yield loss on unavoidable CO_2 production. Therefore, research efforts have been driven towards the production of higher value products (such as short chain organic acids and alcohols [118]) that could challenge traditional process of biogas production as a main driver of the AD process.

6. Conclusions

This paper summarizes the technical and economic perspectives of biogas production from wet wastes using detailed analysis of biogas energy yields to show the potential for waste utilization to satisfy growing energy demand. The energy yield can be used in predicting biogas production once the amount of waste (on either dry or wet basis) is known. Based on the total available waste resources and energy density, the total energy resource for biogas production via the AD process is highest for swine manure, followed by food waste, WWT sludge, and FOG.

This paper also provides a review on the economics of current WTE technology as compared to cost estimates obtained from simulation models, with consistent financial assumptions to ensure reasonable comparison across three sources. Cost estimates of biogas production from process simulations are in line with the data from the literature for all wet wastes except food waste, where some level of deviation is observed. This may be due to different sources of food waste (kitchen waste, winery waste, food waste prior and after usage, etc.) with varying compositions leading to a wide range of biogas yields and costs or may be due to variation of facility scales.

The cost of biogas production using AD has a wide range of values for each feedstock mainly due to economies of scale, and the cost variation would be larger if compared among different feedstock types. Moreover, it would be difficult for biogas to compete with the low prices of natural gas around the world. Thus, technological developments are required to increase methane content in biogas to be cost effective and energy competent with natural gas. Although several waste-to-biogas facilities are already in operation, additional research on process parameters such as maximizing the COD reductions for high biogas production rates or increasing methane content would help inspire future AD developments. This research provides recommendations on process challenges and economic potentials that would help the developers and investors make informed decisions prior to construction.

With methane being a main component in biogas, its emissions from the process can be hazardous and would lead to increased GHG emissions in the atmosphere. In addition, the second largest constituent of biogas is CO_2, which is unusable, diminishing its energy efficiency. Additional sustainability perspectives should be addressed next in comprehensive ways as we did for energy yields and cost aspects.

Author Contributions: Conceptualization, L.T.; Literature, A.H.B.; Chemical Process Modeling, A.H.B. and L.T.; Writing—Original draft preparation, A.H.B. and L.T.; Writing—review and editing, A.H.B. and L.T. All authors have read and agreed to the published version of the manuscript.

Funding: The funding for this work was provided by U.S. Department of Energy's BETO. The views expressed in the article do not necessarily represent the views of the DOE or the U.S. Government. The U.S. Government retains and the publisher, by accepting the article for publication, acknowledges that the U.S. Government retains a nonexclusive, paid-up, irrevocable, worldwide license to publish or reproduce the published form of this work, or allow others to do so, for U.S. Government purposes.

Acknowledgments: This work was authored by the National Renewable Energy Laboratory (NREL), operated by Alliance for Sustainable Energy, LLC, for the U.S. Department of Energy (DOE) under Contract No. DE-AC36-08GO28308. We appreciate all the editing help from our technical editor. We thank Beau Hoffman and Mark Philbrick of BETO for their generous review and suggestions.

Conflicts of Interest: The authors declare no conflict of interest.

References

1. Mazur, A. How does population growth contribute to rising energy consumption in America? *Popul. Environ.* **1994**, *15*, 371–378. [CrossRef]
2. Graham, Z. *Peak People: The Interrelationship between Population Growth and Energy Resources*; Energy Bulletin: Leederville, Australia, 2009.
3. Kataki, R.; Chutia, R.S.; Bordoloi, N.J.; Saikia, R.; Sut, D.; Narzari, R.; Gogoi, L.; Nikhil, G.N.; Sarkar, O.; Venkata Mohan, S. Biohydrogen Production Scenario for Asian Countries. In *Biohydrogen Production: Sustainability of Current Technology and Future Perspective*; Singh, A., Rathore, D., Eds.; Springer: New Delhi, India, 2017; pp. 207–235. [CrossRef]
4. Stern, D. The Role of Energy in Economic Growth. *Ecol. Econ. Rev.* **2011**, *1219*, 26–51.
5. Koberg, M.; Gedanken, A. Optimization of bio-diesel production from oils, cooking oils, microalgae, and castor and jatropha seeds: Probing various heating sources and catalysts. *Energy Environ. Sci.* **2012**, *5*, 7460–7469. [CrossRef]
6. BETO. *Biofuels and Bioproducts from Wet and Gaseous Waste Streams: Challenges and Opportunities*; Bioenergy Technologies Office (BETO): Washington, DC, USA, 2017.
7. Liu, A.; Ren, F.; Lin, W.Y.; Wang, J.-Y. A review of municipal solid waste environmental standards with a focus on incinerator residues. *Int. J. Sustain. Built Environ.* **2015**, *4*, 165–188. [CrossRef]
8. Rushton, L. Health hazards and waste management. *Br. Med. Bull.* **2003**, *68*, 183–197. [CrossRef]
9. Capson-Tojo, G.; Rouez, M.; Crest, M.; Steyer, J.-P.; Delgenès, J.-P.; Escudié, R. Food waste valorization via anaerobic processes: A review. *Rev. Environ. Sci. Biotechnol.* **2016**, *15*, 499–547. [CrossRef]

10. Harrad, S.H.R.M. *The Health Effects of the Products of Waste Combustion*; Institute of Public and Environmental Health, University of Birmingham: Birmingham, UK, 1996.

11. WorldBank. Chapter 5—Waste Composition. In *Urban Development Series—Knowledge Papers*; World Bank Group: Washington, DC, USA, 2009.

12. Börjesson, M.; Ahlgren, E.O. Cost-effective biogas utilisation—A modelling assessment of gas infrastructural options in a regional energy system. *Energy* **2012**, *48*, 212–226. [CrossRef]

13. Tilche, A.; Galatola, M. The potential of bio-methane as bio-fuel/bio-energy for reducing greenhouse gas emissions: A qualitative assessment for Europe in a life cycle perspective. *Water Sci. Technol. J. Int. Assoc. Water Pollut. Res.* **2008**, *57*, 1683–1692. [CrossRef]

14. Kleerebezem, R.; Joosse, B.; Rozendal, R.; Van Loosdrecht, M.C.M. Anaerobic digestion without biogas? *Rev. Environ. Sci. Bio/Technol.* **2015**, *14*, 787–801. [CrossRef]

15. Chiumenti, R.; Chiumenti, A.; da Borso, F.; Limina, S. *Anaerobic Digestion of Swine Manure in Conventional and Hybrid Pilot Scale Plants: Performance and Gaseous Emissions Reduction*; American Society of Agricultural and Biological Engineers: St. Joseph, MI, USA, 2009; Volume 4.

16. Global Methane Initiative. *Overview of Anaerobic Digestion for Municipal Solid Waste*; Global Methane Initiative: Washington, DC, USA, 2016.

17. Chen, L.N.H. *Anaerobic Digestion Basics*; University of Idaho Extension: Moscow, ID, USA, 2014.

18. Water Environment Federation. *Anaerobic Digestion Fundamentals—Fact Sheet*; Water Environment Federation: Alexandria, VA, USA, 2017.

19. Weiland, P. Biogas production: Current state and perspectives. *Appl. Microbiol. Biotechnol.* **2010**, *85*, 849–860. [CrossRef]

20. Mao, C.; Feng, Y.; Wang, X.; Ren, G. Review on research achievements of biogas from anaerobic digestion. *Renew. Sustain. Energy Rev.* **2015**, *45*, 540–555. [CrossRef]

21. Jain, M.K.; Singh, R.; Tauro, P. Anaerobic digestion of cattle and sheep wastes. *Agric. Wastes* **1981**, *3*, 65–73. [CrossRef]

22. Bowen, E.J.; Dolfing, J.; Davenport, R.J.; Read, F.L.; Curtis, T.P. Low-temperature limitation of bioreactor sludge in anaerobic treatment of domestic wastewater. *Water Sci. Technol.* **2014**, *69*, 1004–1013. [CrossRef] [PubMed]

23. Mtui, G.Y.S. Trends in Industrial and Environmental Biotechnology Research in Tanzania. *Afr. J. Biotechnol.* **2007**, *6*, 2860–2867.

24. Singh, S.K.; Kadi, S.; Prashanth, B.; Nayak, S.K. Factors Affecting Anaerobic Digestion of Organic Waste. *Int. J. Eng. Res. Mech. Civ. Eng.* **2018**, *3*.

25. Rincón, B.; Borja, R.; González, J.M.; Portillo, M.C.; Sáiz-Jiménez, C. Influence of organic loading rate and hydraulic retention time on the performance, stability and microbial communities of one-stage anaerobic digestion of two-phase olive mill solid residue. *Biochem. Eng. J.* **2008**, *40*, 253–261. [CrossRef]

26. Kougias, P.; Tiwari, V.; Barshes, N.R.; Bechara, C.F.; Lowery, B.; Pisimisis, G.; Berger, D.H. Modeling anesthetic times. Predictors and implications for short-term outcomes. *J. Surg. Res.* **2013**, *180*, 1–7. [CrossRef]

27. Kim, J.; Park, C.; Kim, T.H.; Lee, M.; Kim, S.; Kim, S.W.; Lee, J. Effects of various pretreatments for enhanced anaerobic digestion with waste activated sludge. *J. Biosci. Bioeng.* **2003**, *95*, 271–275. [CrossRef]

28. Lee, D.H.; Behera, S.K.; Kim, J.W.; Park, H.S. Methane production potential of leachate generated from Korean food waste recycling facilities: A lab-scale study. *Waste Manag.* **2009**, *29*, 876–882. [CrossRef]

29. Turovskiy, I.S.; Mathai, P.K. *Wastewater Sludge Processing*; John Wiley & Sons: Hoboken, NJ, USA, 2006; p. 354.

30. Seiple, T.E.; Coleman, A.M.; Skaggs, R.L. Municipal wastewater sludge as a sustainable bioresource in the United States. *J. Environ. Manag.* **2017**, *197*, 673–680. [CrossRef]

31. Manara, P.; Zabaniotou, A. Towards sewage sludge based biofuels via thermochemical conversion—A review. *Renew. Sustain. Energy Rev.* **2012**, *16*, 2566–2582. [CrossRef]

32. Ebner, J.H.; Labatut, R.A.; Lodge, J.S.; Williamson, A.A.; Trabold, T.A. Anaerobic co-digestion of commercial food waste and dairy manure: Characterizing biochemical parameters and synergistic effects. *Waste Manag.* **2016**, *52*, 286–294. [CrossRef] [PubMed]

33. Nazlina, H.; Aini, A.N.; Ismail, F.; Yusof, M.; Hassan, M. Effect of different temperature, initial pH and substrate composition on biohydrogen production from food waste in batch fermentation. *Asian J. Biotechnol.* **2009**, *1*, 42–50.

34. Zhang, C.; Su, H.; Baeyens, J.; Tan, T. Reviewing the anaerobic digestion of food waste for biogas production. *Renew. Sustain. Energy Rev.* **2014**, *38*, 383–392. [CrossRef]

35. Wang, K.; Yin, J.; Shen, D.; Li, N. Anaerobic digestion of food waste for volatile fatty acids (VFAs) production with different types of inoculum: Effect of pH. *Bioresour. Technol* **2014**, *161*, 395–401. [CrossRef]

36. Uçkun Kiran, E.; Trzcinski, A.P.; Ng, W.J.; Liu, Y. Bioconversion of food waste to energy: A review. *Fuel* **2014**, *134*, 389–399. [CrossRef]

37. Li, Y.; Jin, Y.; Li, H.; Borrion, A.; Yu, Z.; Li, J. Kinetic studies on organic degradation and its impacts on improving methane production during anaerobic digestion of food waste. *Appl. Energy* **2018**, *213*, 136–147. [CrossRef]

38. Lu, X.; Jin, W.; Xue, S.; Wang, X. Effects of waste sources on performance of anaerobic co-digestion of complex organic wastes: Taking food waste as an example. *Sci. Rep.* **2017**, *7*, 15702. [CrossRef]

39. ECN Database. *Database for Biomass and Waste*; Energy Research Center of the Netherlands: Petten, The Netherlands, 2012.

40. Turner, C. How to Compost Dow Dung Manure. Available online: http://compost-turner.net/composting-technologies/how-to-compost-cow-dung-manure.html (accessed on 9 July 2019).

41. Long, J.H.; Aziz, T.N.; De los Reyes, F.; Ducoste, J. Anaerobic co-digestion of fat, oil, and grease (FOG): A review of gas production and process limitations. *Process Saf. Environ. Prot.* **2012**, *90*, 231–245. [CrossRef]

42. EIA. *Methodology for Allocating Municipal Solid Waste*; Energy Information Administration: Washington, DC, USA, 2007.

43. Chartier, P.; Palz, W. *Energy from Biomass*; Springer: Dordrecht, The Netherlands, 2011.

44. Nasr, F.A. Treatment and Resuse of Sewage Sludge. *Environment* **1997**, *17*, 109–113.

45. Chua, K.H.; Cheah, W.L.; Tan, C.F.; Leong, Y.P. Harvesting biogas from wastewater sludge and food waste. *IOP Conf. Ser. Earth Environ. Sci.* **2013**, *16*. [CrossRef]

46. Kuo, J.; Dow, J. Biogas production from anaerobic digestion of food waste and relevant air quality implications. *J. Air Waste Manag. Assoc.* **2017**, *67*, 1000–1011. [CrossRef] [PubMed]

47. Nayono, S.E.; Gallert, C.; Winter, J. Co-digestion of press water and food waste in a biowaste digester for improvement of biogas production. *Bioresour. Technol.* **2010**, *101*, 6987–6993. [CrossRef] [PubMed]

48. Nguyen, D.D.; Chang, S.W.; Cha, J.H.; Jeong, S.Y.; Yoon, Y.S.; Lee, S.J.; Tran, M.C.; Ngo, H.H. Dry semi-continuous anaerobic digestion of food waste in the mesophilic and thermophilic modes: New aspects of sustainable management and energy recovery in South Korea. *Energy Convers. Manag.* **2017**, *135*, 445–452. [CrossRef]

49. Xia, Y.; Masse, D.I.; McAllister, T.A.; Beaulieu, C.; Ungerfeld, E. Anaerobic digestion of chicken feather with swine manure or slaughterhouse sludge for biogas production. *Waste Manag.* **2012**, *32*, 404–409. [CrossRef]

50. Cuetos, M.J.; Fernandez, C.; Gomez, X.; Moran, A. Anaerobic co-digestion of swine manure with energy crop residues. *Biotechnol. Bioprocess Eng.* **2011**, *16*, 1044–1052. [CrossRef]

51. Chae, K.J.; Yim, S.K.; Choi, K.H.; Park, W.K.; Lim, D.K. *Anaerobic Digestion of Swine Manure: Sung-Hwan Farm-Scale Biogas Plant in Korea*; Division of Agriculture Environment and Ecology, National Institute of Agricultural Science and Technology: Kyungki, Korea, 2002; pp. 564–571. Available online: https://pdfs.semanticscholar.org/b51b/15d425e388daeccb6a5263b099ee8fad652a.pdf (accessed on 5 September 2019).

52. Lobato, A.; Cuetos, M.; Gómez, X.; Morán, A. Improvement of biogas production by co-digestion of swine manure and residual glycerine. *Biofuels* **2014**, *1*. [CrossRef]

53. Gonzalez-Fernandez, C.; Leon-Cofreces, C.; Garcia-Encina, P.A. Different pretreatments for increasing the anaerobic biodegradability in swine manure—ScienceDirect. *Bioresour. Technol.* **2008**, *99*, 8710–8714. [CrossRef]

54. Martinez, E.J.; Redondas, V.; Fierro, J.; Gomez, X.; Moran, A. Anaerobic Digestion of High Lipid Content Wastes: FOG Co-digestion and Milk Processing FAT Digestion. *J. Residuals Sci. Technol.* **2011**, *8*, 53–60.

55. Skaggs, R.L.; Coleman, A.M.; Seiple, T.E.; Milbrandt, A.R. Waste-to-Energy biofuel production potential for selected feedstocks in the conterminous United States. *Renew. Sustain. Energy Rev.* **2018**, *82*, 2640–2651. [CrossRef]

56. USEPA. *Inventory of U.S. Greenhouse Gas Emissions and Sinks: 1990–2012*; US Environmental Protection Agency: Washington, DC, USA, 2014.

57. Zhang, Y.; Wang, X.C.; Cheng, Z.; Li, Y.; Tang, J. Effect of fermentation liquid from food waste as a carbon source for enhancing denitrification in wastewater treatment. *Chemosphere* **2016**, *144*, 689–696. [CrossRef]

58. Nijaguna, B.T. *Biogas Technology*; New Age International: Kochi, India, 2007.

59. Vranitzky, R.; Lahnsteiner, D.J. Sewage Sludge Disintegration Using Ozone—A Method of Enhancing the Anaerobic Stabilization of Sewage Sludge. 2003. Available online: https://www.semanticscholar.org/paper/SEWAGE-SLUDGE-DISINTEGRATION-USING-OZONE-A-METHOD-Vranitzky/50db49aefe72072c3b6584384efdc92d6d698162 (accessed on 5 September 2019).

60. Ligero, P.; Soto, M. Sludge granulation during anaerobic treatment of pre-hydrolysed domestic wastewater. *Water SA* **2002**, *28*, 307–311. [CrossRef]

61. Zitomer, D. *Myths and Misconceptions of Digesters*; Midwest Manure Summit, University of Wisconsin: Green Bay, Wisconsin, 2013. Available online: https://fyi.extension.wisc.edu/midwestmanure/files/2013/03/Myths-Minsconceptions-of-Digesters-Zitomer.pdf (accessed on 5 September 2019).

62. Negre, J.; Jonsson, L. *Sludge Treatment in an Anaerobic Bioreactor with External Membranes*; Stockholm Vatten: Bromma, Sweden, 2010.

63. Bachmann, N. *Sustainable Biogas Production in Municipal Wastewater Treatment Plants*; European Commission: Brussels, Belgium, 2015.

64. Achinas, S.; Achinas, V.; Euverink, G.J.W. A Technological Overview of Biogas Production from Biowaste. *Engineering* **2017**, *3*, 299–307. [CrossRef]

65. Gray, D.; Suto, P.; Peck, C. Anaerobic Digestion of Food Waste. 2008. Available online: https://archive.epa.gov/region9/organics/web/pdf/ebmudfinalreport.pdf (accessed on 10 October 2019).

66. Paritosh, K.; Kushwaha, S.K.; Yadav, M.; Pareek, N.; Chawade, A.; Vivekanand, V. Food Waste to Energy: An Overview of Sustainable Approaches for Food Waste Management and Nutrient Recycling. *BioMed Res. Int.* **2017**, *2017*. [CrossRef]

67. ECN. Phyllis2—Pig Manure (#1715). Available online: https://www.ecn.nl/phyllis2/Biomass/View/1715 (accessed on 10 October 2019).

68. Ministry of Agriculture, Natural Resources and Environment. *Database on Waste Disposal Permits*; Ministry of Agriculture, Natural Resources and Environment: Nicosia, Cyprus, 2011.

69. Biogas Yields and Feedstock Productivity—Anaerobic Digestion. Available online: http://www.biogas-info.co.uk/about/feedstocks/ (accessed on 19 October 2019).

70. Theofanous, E.; Kythreotou, N.; Panayiotou, G.; Florides, G.; Vyrides, I. Energy production from piggery waste using anaerobic digestion: Current status and potential in Cyprus. *Renew. Energy* **2014**, *71*, 263–270. [CrossRef]

71. Brenan, J.; Pierce, C.; Hickey, D.R.O.S.M. *Dairy Co-Digestion Using and Anaerobic Digester Research Project*; California Energy Commission: Sacramento, CA, USA, 2016.

72. Kabouris, J.C.; Tezel, U.; Pavlostathis, S.G.; Engelmann, M.; Dulaney, J.; Gillette, R.A.; Todd, A.C. Methane recovery from the anaerobic codigestion of municipal sludge and FOG. *Bioresour. Technol.* **2009**, *100*, 3701–3705. [CrossRef]

73. Nas, M.N.; Mutlu, N.; Read, P.E. Random amplified polymorphic DNA (RAPD) analysis of long-term cultured hybrid hazelnut. *Hortscience* **2004**, *39*, 1079–1082. [CrossRef]

74. SRI Consulting. U.S. Producer Price Indexes—Chemicals and Allied Products/Industrial Inorganic Chemicals Index. In *Chemical Economics Handbook*; IHS Markit: London, UK, 2008.

75. Bureau of Labor Statistics Data Website. *National Employment, Hours, and Earnings Catalog, Industry: Chemicals and Allied Products, 1980–2009*; Bureau of Labor Statistics: Washington, DC, USA, 2009.

76. *AspenPlus^{TM}*; Release 7.2, Aspen Technology Inc.: Cambridge, MA, USA, 2007.

77. Arsova, L. *Anaerobic Digestion of Food Waste: Current Status, Problems and an Alternative Product*; Columbia University: New York, NY, USA, 2010.

78. Davis, R.; Kinchin, C.; Markham, J.; Tan, E.; Laurens, L.M.L.; Sexton, D.; Knorr, D.; Schoen, P.; Lukas, J. *Process Design and Economics for the Conversion of Algal Biomass to Biofuels: Algal Biomass Fractionation to Lipid- and Carbohydrate-Derived Fuel Products*; National Renewable Energy Laboratory (NREL): Golden, CO, USA, 2014.

79. Birdsall Services Group. Anaerobic digestion of fats oils and grease to generate biogas biopower feasibility study. Available online: https://rucore.libraries.rutgers.edu/rutgers-lib/46247/PDF/1/play/ (accessed on 9 December 2019).

80. Harris, R.W.; Cullinane, M.J., Jr.; Sun, P.T. *Process Design and Cost Estimating Algorithms for the Computer Assisted Procedure for Design and Evaluation of Wastewater Treatment Systems (CAPDET)*; Army Engineer Waterways Experiment Station: Vicksburg, MS, USA, 1982.

81. Rossman, L.A. *Computer-Aided Synthesis of Wastewater Treatment and Sludge Disposal Systems*; Municipal Environmental Research Laboratory, Office of Research and Development, US Environmental Protection Agency: Washington, DC, USA, 1980; Volume 79.

82. Chiumenti, A.; Chiumenti, R.; da Borso, F.; Limina, S. *Comparison Between Dry and Wet Fermentation of Biomasses as Result of the Monitoring of Full Scale Plants*; American Society of Agricultural and Biological Engineers: St. Joseph, MI, USA, 2012; Volume 5.

83. Riaño, B.; Molinuevo, B.; García-González, M.C. Potential for methane production from anaerobic co-digestion of swine manure with winery wastewater. *Bioresour. Technol.* **2011**, *102*, 4131–4136. [CrossRef]

84. Molinuevo-Salces, B.; González-Fernández, C.; Gómez, X.; García-González, M.C.; Morán, A. Vegetable processing wastes addition to improve swine manure anaerobic digestion: Evaluation in terms of methane yield and SEM characterization. *Appl. Energy* **2012**, *91*, 36–42. [CrossRef]

85. Li, J.; Jha, A.K.; Bajracharya, T.R. Dry Anaerobic Co-digestion of Cow Dung with Pig Manure for Methane Production. *Appl. Biochem. Biotechnol.* **2014**, *173*, 1537–1552. [CrossRef]

86. East Bay Municipal Utility District. *Anaerobic Digestion of Food Waste*; EPA Region 9: San Francisco, CA, USA, 2008.

87. EIA. *Methodology for Allocating Municipal Solid Waste to Biogenic and Non-Biogenic Energy*; Energy Information Administration: Washington, DC, USA, 2007.

88. Wang, L.K.; Shammas, N.K.; Hung, Y.-T. *Biosolids Treatment Processes*; Humana Press: Totowa, NJ, USA, 2007; Volume 6.

89. Clarke, W.P. Cost-benefit analysis of introducing technology to rapidly degrade municipal solid waste. *Waste Manag. Res.* **2000**, 18. [CrossRef]

90. Tsilemou, K.; Panagiotakopoulos, D. Approximate cost functions for solid waste treatment facilities. *Waste Manag. Res.* **2006**, *24*, 310–322. [CrossRef] [PubMed]

91. Misra, K.B. *Clean Production: Environmental and Economic Perspectives*; Springer: Berlin/Heidelberg, Germany, 2012.

92. Moriarty, K. *Feasibility Study of Anaerobic Digestion of Food Waste in St. Bernard, Louisiana. A Study Prepared in Partnership with the Environmental Protection Agency for the RE-Powering America's Land Initiative: Siting Renewable Energy on Potentially Contaminated Land and Mine Sites*; National Renewable Energy Lab (NREL): Golden, CO, USA, 2013.

93. ILSR. *Update on Anaerobic Digester Projects Using Food Waste in North America*; Institute for Local Self-Reliance (ILSR): Washington, DC, USA, 2010.

94. Davis, R.C. *Anaerobic Digestion—Pathways for using Waste as Energy in Urban Settings*; The University of British Columbia: Vancouver, BC, Canada, 2014.

95. CalRecycle. *Digesting Urban Residuals Forum*; CalRecycle: City of San Jose, CA, USA, 2012.

96. Zhang, Q.; Hu, J.; Lee, D.-J. Biogas from anaerobic digestion processes: Research updates. *Renew. Energy* **2016**, *98*, 108–119. [CrossRef]

97. Colantoni, A.; Carlini, M.; Mosconi, E.; Castellucci, S.; Villarini, M. An Economical Evaluation of Anaerobic Digestion Plants Fed with Organic Agro-Industrial Waste. *Energies* **2017**, *10*, 1165. [CrossRef]

98. Andriani, D.; Wresta, A.; Atmaja, T.D.; Saepudin, A. A review on optimization production and upgrading biogas through CO2 removal using various techniques. *Appl. Biochem. Biotechnol.* **2014**, *172*, 1909–1928. [CrossRef] [PubMed]

99. Niesner, J.; Jecha, D.; Stehlík, P. Biogas Upgrading Technologies: State of Art Review in European Region. *Chem. Eng. Trans.* **2013**, *35*, 517–522.

100. Krich, K.; Augenstein, D.; Batmale, J.P.; Benemann, J.; Rutledge, B.; Salour, D. *Biomethane from Dairy Waste—A Sourcebook for the Production and Use of Renewable Natural Gas in California*; Western United Dairymen: Modesto, CA, USA, 2005.

101. Anthony, C. *The Biochemistry of Methylotrophs*; Academic Press: London, UK, 1982.

102. Anthony, C. Bacterial oxidation of methane and methanol. *Adv. Microb. Physiol.* **1986**, *27*, 113–210.

103. Dijkhuizen, L.; Levering, P.; De Vries, G. *The Physiology and Biochemistry of Aerobic Methanol-Utilizing Gram-Negative and Gram-Positive Bacteria*; Springer: Boston, MA, USA, 1992.

104. Hanson, R.; Tsien, H.; Tsuji, K.; Brusseau, G.; Wackett, L. Biodegradation of low-molecular-weight halogenated hydrocarbons by methanotrophic bacteria. *FEMS Microbiol. Lett.* **1990**, *87*, 273–278. [CrossRef]

105. Buswell, A.; Sollo , F., Jr. The mechanism of the methane fermentation. *J. Am. Chem. Soc.* **1948**, *70*, 1778–1780. [CrossRef]

106. Hamer, G.; Hedén, C.G.; Carenberg, C.O. Methane as a carbon substrate for the production of microbial cells. *Biotechnol. Bioeng.* **1967**, *9*, 499–514. [CrossRef]

107. Jiang, H.; Chen, Y.; Jiang, P.; Zhang, C.; Smith, T.J.; Murrell, J.C.; Xing, X.-H. Methanotrophs: Multifunctional bacteria with promising applications in environmental bioengineering. *Biochem. Eng. J.* **2010**, *49*, 277–288. [CrossRef]

108. Hanson, R.S.; Hanson, T.E. Methanotrophic bacteria. *Microbiol. Rev.* **1996**, *60*, 439–471. [CrossRef] [PubMed]

109. Schrader, J.; Schilling, M.; Holtmann, D.; Sell, D.; Villela Filho, M.; Marx, A.; Vorholt, J.A. Methanol-based industrial biotechnology: Current status and future perspectives of methylotrophic bacteria. *Trends Biotechnol.* **2009**, *27*, 107–115. [CrossRef] [PubMed]

110. Op den Camp, H.J.; Islam, T.; Stott, M.B.; Harhangi, H.R.; Hynes, A.; Schouten, S.; Jetten, M.S.; Birkeland, N.K.; Pol, A.; Dunfield, P.F. Environmental, genomic and taxonomic perspectives on methanotrophic Verrucomicrobia. *Environ. Microbiol. Rep.* **2009**, *1*, 293–306. [CrossRef]

111. Henard, C.A.; Smith, H.; Dowe, N.; Kalyuzhnaya, M.G.; Pienkos, P.T.; Guarnieri, M.T. Bioconversion of methane to lactate by an obligate methanotrophic bacterium. *Sci. Rep.* **2016**, *6*, 21585. [CrossRef]

112. Anthony, C. How half a century of research was required to understand bacterial growth on C1 and C2 compounds; the story of the serine cycle and the ethylmalonyl-CoA pathway. *Sci. Prog.* **2011**, *94*, 109–137. [CrossRef]

113. Chistoserdova, L. Modularity of methylotrophy, revisited. *Environ. Microbiol.* **2011**, *13*, 2603–2622. [CrossRef]

114. Matsen, J.B.; Yang, S.; Stein, L.Y.; Beck, D.; Kalyuzhnaya, M.G. Global molecular analyses of methane metabolism in methanotrophic alphaproteobacterium, Methylosinus trichosporium OB3b. Part I: Transcriptomic study. *Front. Microbiol.* **2013**, *4*, 40. [CrossRef]

115. EPA Regulatory Announcement. *EPA Issues Final Rule for Renewable Fuel Standard (RFS) Pathways II and Modifications to the RFS Program, Ultra Low Sulfur Diesel Requirements, and E15 Misfueling Mitigation Requirements*; EPA: Washington, DC, USA, 2014.

116. USDA; EPA; DOE. *Biogas Opportunities Roadmap*; U.S. Department of Agriculture: Washington, DC, USA, 2014.

117. Murray, B.C.; Galik, C.S.; Vegh, T. *Biogas in the United States: An Assessment of Market Potential in a Carbon-Constrained Future*; Nicholas Institute for Environmental Policy Solutions: Durham, NC, USA, 2014.

118. Bhatt, A.H.; Ren, Z.; Tao, L. Value Proposition of Untapped Wet Wastes: Carboxylic Acid Production through Anaerobic Digestion. *iScience* **2020**, *23*, 101221. [CrossRef]

Article

Material Characterization and Substrate Suitability Assessment of Chicken Manure for Dry Batch Anaerobic Digestion Processes

Harald Wedwitschka [1,*], Daniela Gallegos Ibanez [1], Franziska Schäfer [1], Earl Jenson [2] and Michael Nelles [1,3]

[1]   Department Biochemical Conversion, DBFZ Deutsches Biomasseforschungszentrum gemeinnützige GmbH, Torgauer Straße 116, D-04347 Leipzig, Germany; Daniela.Ibanez@dbfz.de (D.G.I.); Franziska.Schaefer@dbfz.de (F.S.); Michael.Nelles@uni-rostock.de (M.N.)

[2]   Department Bio Industrial Services, Inno Tech Alberta, PO Bag 4000, HWY 16A and 75 Street, Vegreville, AB T9C 1T4, Canada; Earl.Jenson@InnoTechAlberta.ca

[3]   Department Waste and Resource Management, University of Rostock, Justus-von-Liebig Weg 6, D-18057 Rostock, Germany

*   Correspondence: Harald.Wedwitschka@dbfz.de; Tel.: +49-341-2434-562

Received: 31 July 2020; Accepted: 31 August 2020; Published: 7 September 2020

Abstract: Chicken manure is an agricultural residue material with a high biomass potential. The energetical utilization of this feedstock via anaerobic digestion is an interesting waste treatment option. One waste treatment technology most appropriate for the treatment of stackable (non-free-flowing) dry organic waste materials is the dry batch anaerobic digestion process. The aim of this study was to evaluate the substrate suitability of chicken manure from various sources as feedstock for percolation processes. Chicken manure samples from different housing forms were investigated for their chemical and physical material properties, such as feedstock composition, permeability under compaction and material compressibility. The permeability under compaction of chicken manure ranged from impermeable to sufficiently permeable depending on the type of chicken housing, manure age and bedding material used. Porous materials, such as straw and woodchips, were successfully tested as substrate additives with the ability to enhance material mixture properties to yield superior permeability and allow sufficient percolation. In dry anaerobic batch digestion trials at lab scale, the biogas generation of chicken manure with and without any structure material addition was investigated. Digestion trials were carried out without solid inoculum addition and secondary methanization of volatile components. The specific methane yield of dry chicken manure was measured and found to be 120 to 145 mL/g volatile solids (VS) and 70 to 75 mL/g fresh matter (FM), which represents approximately 70% of the methane potential based on fresh mass of common energy crops, such as corn silage.

Keywords: dry batch anaerobic digestion; chicken manure; percolation; permeability

---

1. Introduction

Due to increasing protein demand for human nutrition, poultry farming is one of the fastest developing sectors of animal husbandry. The FAO stated that, over the last fifty years, poultry was the livestock category with the highest head growth worldwide, with roughly a five-fold increase [?]. In Germany, the total number of chickens in the production cycle increased from 109 million animals in 2003, to approximately 158 million animals in 2016, and the number of chickens kept per farm increased from approximately 1217 to 3361 during this time span [?]. These growing livestock numbers lead to an increase of the process related waste material production. Nebel and Kühne [?] estimated the biomass potential of chicken manure from laying hens to be as much as 5 million t/a plus another 1.3 million t/a from chickens for egg production, pullets and broilers. Dry chicken manure is commonly utilized as agricultural fertilizer, which is rich in nitrogen, potassium, phosphate, calcium, magnesium and sulfur [?]. However, in regions with high livestock numbers, comprehensive manure management is required to reduce the risk of nitrate pollution of surface and ground water and gaseous ammonia (NH_3) emissions from manure storage and application. According to Haenel et al. [?] the NH_3 emissions of poultry breeding have a significant share (9%) of the overall NH_3 emissions from animal husbandry. More than half of these emissions are caused by manure handling

Chicken manure is an agricultural waste material, which is a particularly interesting substrate for biogas production [?]. Compared to other agricultural residue materials that are used as biogas production feedstock, dry chicken manure has a high energy potential based on fresh weight, with an average methane yield of 70–140 m^3/t. The average methane yields of cattle dung, cattle manure and pig manure are comparably lower with 33–36 m^3/t, 11–19 m^3/t and 12–21 m^3/t [?]. Since chicken manure is an agricultural residue material with a relatively high dry matter content only low production, storage and transportation costs are incurred. Valuable products from the anaerobic digestion process are biomethane, which can serve as a renewable fuel or as an energy source for thermal and electrical energy production, and digestate that can be used as compost, fertilizer or soil amendment. According to Bayrakdar et al. [?] nitrogen extraction from chicken manure digestate for the production of concentrated nitrogen fertilizer has the potential for value added production and reduction of greenhouse gas emissions. In current practice, chicken manure is utilized as a co-substrate by a number of agricultural wet digestion biogas plants. However, the high sand and nitrogen content of the feedstock can lead to technical and biological process problems at full scale, which is why the mono digestion of chicken manure in agricultural biogas plants has proven to be rather challenging [?].

The dry batch anaerobic digestion process is a waste treatment technology most appropriate for the treatment of stackable (non-free-flowing), fibrous and contaminant-containing dry feedstock materials. In contrast to conventional wet digestion systems, these systems rely purely on percolation of process fluid through the feedstock material within a garage type dry digester to facilitate the digestion process [?]. Process fluid is sprayed on top of the substrate and serves as a transport medium for heat, nutrients and material exchange and composes of water and effluent from the digestion process, which is sporadically recycled. Volatile components solubilized in the percolate are simultaneously digested in a separate digester (without or with minimal mixing) that services numerous garage type dry digesters. The feedstock materials are treated in batch operation, which means they are introduced once into the digester, are not actively mixed during the process, and are replaced after a given retention time by fresh substrate. The raw material is generally loaded into the garage type digester along with a fraction of previously digested material (inoculum), which helps to stabilize the process by preventing process acidification [?]. After 20 to 30 days of substrate retention time in the garage type digester, the digested feedstock is replaced with another mixture of fresh and digested material. A further special feature of the process is the discontinuous biogas formation resulting from the batch operation of the process. For this reason, dry batch fermentation plants consist of a series of box fermenters connected in series, which are operated at different batch start and finish times in a rotational cycle to ensure a somewhat constant and consistent biogas formation.

The advantage of this type of dry batch digestion process is a low susceptibility to contaminants, such as sand, stones, glass, plastics or metal, which can lead to an impairment of the stirrer and pump technology in a stirred tank system. Furthermore, the system is relatively unaffected by substances that tend to create sediment layers on the fermenter floor. Since the box system is opened, emptied and recharged after completion of a single fermentation cycle, larger sediment layers on the floor are avoided. In agricultural wet digestion systems large amounts of liquid are required i.e., liquid manure to adjust the water content and viscosity of feedstock mixtures with high dry matter contents. This results in the generation of high volumes of diluted digestate, which increases the costs of digestate storage and transport for agricultural use [?]. In the dry batch digestion process substrate materials can be treated in a stackable form, which results in an efficient utilization of the reactor volume and generation of less liquid digestate [?]. The production of a dry product with minimal wastewater generation makes the dry digestion process appear highly appealing for the treatment of dry chicken manure.

The dry batch anaerobic digestion process requires a feedstock material with sufficient permeable structure that enables percolation of liquids in the anaerobic stage to ensure the mentioned transportation function [? ?]. Insufficient material structure of the input material and insufficient structure stability during the digestion process can lead to a reduced biological degradation if the percolation of the substrate material is uneven or if the percolate accumulates within the digester, which often results in technical problems. Additionally, dry anaerobic batch digestion has to overcome several other challenges. Most dry batch digestion plants use stabilized digestate from previous batches as an inoculum source that needs to be recycled for process stabilization in the case of feedstock with high biodegradability and conversion rate. Consequently, with significant inoculum recycling, the plant capacity has to be larger, which leads to higher investment costs.

The aim of this study was to examine the feedstock suitability of chicken manure for dry batch digestion with regards to material composition and permeability under compaction. Hence, the novelty of this study surrounds oedometer tests that were conducted as a method for material characterization that guided conditioning practices to create an appropriate substrate material structure for dry anaerobic batch digestion of chicken manure at full scale. Furthermore, dry batch anaerobic digestion trials without solid inoculum addition and a secondary methanization digester were carried out with mixtures of chicken manure and different structure material contents. The aim of the digestion trials was to determine the digestibility of chicken manure without inoculation, and to evaluate the influence of structure material addition and material permeability on the specific methane yield. The ultimate goal is to increase the efficiency of percolation processes so that, on the one hand, new plants can enter directly into efficient operation (without long running-in times), and on the other hand, existing plants can optimize their operations and substrate utilization.

2. Materials and Methods

Chicken manure samples were tested for their material properties including total solids and volatile solids, nitrogen, protein, fat and fiber composition, permeability under compression and compressibility. Additionally, dry anaerobic batch digestion trials were carried out as a means to evaluate the digestion properties of chicken manure mixtures without inoculum addition and a secondary methanization digester.

2.1. Materials

Material characterization tests of chicken manure were carried out with samples collected during 2014 and 2020 from different regions of Saxonia in Germany. Wet sample materials were quickly cooled after sampling and stored at 5 °C before lab analysis. Structure materials such as beech wood chips, wheat straw and plastic carrier (Bioflow 40, RVT Process Equipment GmbH, Steinwiesen, Germany) were provided by the Deutsches Biomasseforschungszentrum gemeinnützige GmbH (DBFZ).

2.2. Analytical Methods

A multitude of material characterization tests were performed on chicken manure samples and substrate mixtures of chicken manure after structure material addition. Total solids (TS) and volatile solids (VS) were measured in accordance with DIN EN 12880 (2001) [?] and DIN EN 15935:2012-11 (2012) [?]. The pH-value was measured with a pH device 3310 (WTW Wissenschaftlich-Technische Werkstätten GmbH, Weilheim, Germany). Volatile organic acids, Weender feed analysis, ammonium nitrogen (NH$_4$-N) and total ammonia nitrogen (TAN) were determined as described in Liebetrau et al., (2016) [?]. Free ammonia (*FA*) concentrations were calculated according to Equation (1)

$$FA = 0.94412 * NH_4/(1 + 10((0.0925 + (2728,795/(t + 273.15)))) - pH))$$ (1)

where NH_4 is the total ammonium concentration, *pH* is the *pH*-value and *t* stands for temperature.

2.3. Measurement of the Compressibility and Permeability under Compaction

In order to determine the influence of the material compaction on substrate permeability and compaction, oedometer tests were conducted to simulate full scale conditions inside a percolation digester. A schematic illustration of the testing stand can be seen in Figure **??**. The method development is described in [?]. The height of the stacked substrate within the percolate digester, which can range between 2 and 3 m, results in material compaction and reduced permeability within the substrate heap. An oedometer test apparatus was used to determine the permeability of the sample material under similar material compaction by applying a defined force on the top of the sample, which simulates the compaction that occurs in a full-scale environment with several meters of material height.

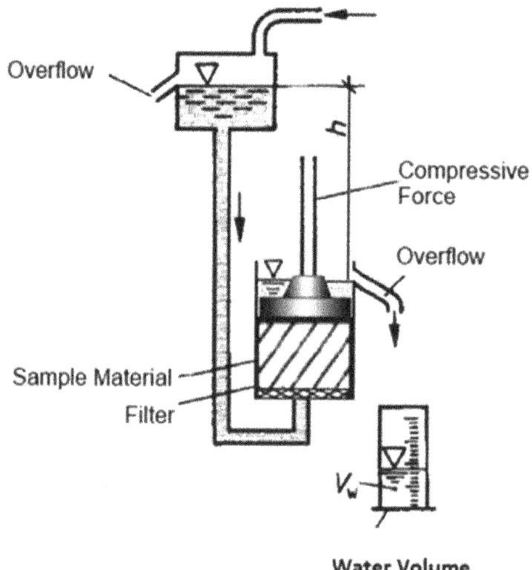

Figure 1. Schematic illustration of the oedometer testing appartus.

The sample material was placed in an oedometer with a 254 mm diameter and 240 mm height and compressive force was applied with a perforated compression plate with 247 mm diameter. The compression plate was connected to an air piston as seen in Figure **??** that supplied a constant but adjustable force. The hydraulic conductivity and the compressibility of the sample material were determined at loose compaction and low, medium, and high compression equivalent to different material heights encountered in full-scale dry fermentation processes. Hydraulic conductivity first requires measurement of the flow rate Q (m^3/s) through the sample, which is calculated from the volume of water Vw (L) passing through the sample material per unit of time t (s).

$$Q = \frac{V_w}{t}$$ (2)

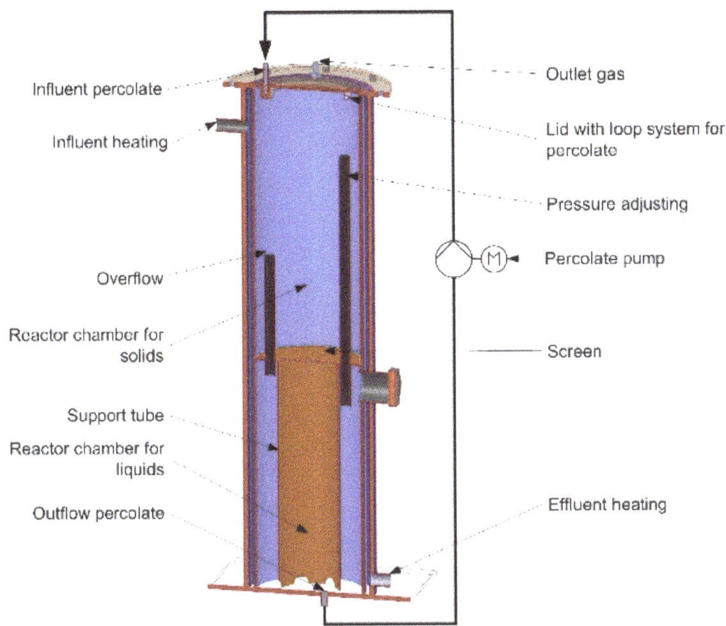

Figure 2. Schematic of percolate digester (dry digestion system) (Stur, Deutsches Biomasseforschungszentrum gGmbH (DBFZ)).

The hydraulic conductivity or permeability k_f (m/s) describes the ease with which the percolate can move through the pore spaces of the sample material. It is calculated by the quotient of the flow rate Q (m^3/s) multiplied by the material sample height l (m) divided by the material surface area in the oedometer A (m^2) multiplied by the hydrostatic height h (m).

$$k_f = \frac{Q * l}{A * h} \tag{3}$$

The simulated material heap height h_S (m) is calculated with the defined force applied by the perforated plate connected to the air piston and the oedometer diameter d (m), the density of the water saturated sample ρ (kg/m^3) and the mass m (kg) of the sample material describe the simulated material volume. The cumulative compaction or material compressibility is determined by dividing the material height before and after compaction.

$$h_s = \frac{4m}{\pi d^2 \rho} \tag{4}$$

The test setup was configured to measure the material permeability without compaction ($P1$) and with 1.5 m ($P2$) and 3.0 m ($P3$) simulated feedstock heap height. The material compressibility ($C2$) and ($C3$) was measured at the two mentioned compression levels.

2.4. Dry Anaerobic Batch Digestion Trials

Substrate material preparation—the material structure of the substrate was adjusted by intensive mixing of the dry chicken manure samples with structural material such as wood chips and wheat straw. After material conditioning the percolate digesters were filled with app. 2.0 kg (10.0–12.0 L) of the substrate mixture to a heap height of app. 0.3 m.

Dry batch anaerobic digestion—in three test series, defined dry chicken manure and structure material mixtures were fermented in two percolate digesters. Each of the individual fermentation experiments were completed after a test period of approximately four weeks. Information about the type of structure material used and the percentage of its addition to dry chicken manure can be found in Table ??. A schematic representation of the dry digestion system is depicted in Figure ??.

Table 1. Dry batch anaerobic digestion trials and structure material addition.

	Designation	Structure Material	Structure Material Addition (w%)
Trial 1	C1	-	-
Trial 2	C2	-	-
Trial 3	S1	Straw	5
Trial 4	S2	Straw	10
Trial 5	W1	Woodchips	10
Trial 6	W2	Woodchips	10

The testing system consisted of two percolate digesters, each connected to a percolate pump for percolate recycling. The two double walled percolate digesters each had a substrate capacity of app. 13 L and were heated to 37 °C by a thermostatically controlled water bath. The hydrolysis and methanization occurred simultaneously in the percolate digesters. Each digester was filled with 2 kg of chicken manure with and without structure material addition. The material mixture compositions that were used are shown in Table ??. The hydrolysis process was implemented by percolation of 7.6 L of percolate with a percolation rate of 660 mL/h every 30 min for approx. 20 s. In order to generate sufficient biological activity, the percolate consisted of a mixture of tap water, percolate from a pilot scale dry batch anaerobic digestion plant and percolate from lab scale dry batch digestion trials. Organic acids formed by hydrolysis were dissolved in the percolate, collected in a buffer tank and continuously irrigated on top of the substrate heap in the percolate digester. The biogas production took place in the percolate digester and buffer tank within the base of the percolate digester (shown in Figure ??). Biogas quantity was continuously measured with a gas flow meter (Ritter TG 1, Ritter GmbH Germany) and the methane content was determined with an automated gas analysis device (AWITE Gas measurement device, Germany). The volatile fatty acid concentration and the ammonia nitrogen content was regularly recorded. Specific methane yield (SMY) was calculated in accordance with VDI guideline 4630, 2006 [?] and normalized to standard conditions (dry gas, 273.15 K, 1013.25 hPa). The SMY is based on the input mass of volatile solids of the substrate mix. In the case of the control and wood chip variant, only the chicken manure VS input was considered and in case of the straw and chicken manure mixture the total VS input was used for SMY calculation. SMY originated from the volatile fatty acid content in the percolate was calculated for each percolate digester separately under the assumption that 1 g degraded volatile fatty acids in the percolate equals approx. 350 mL CH_4. The volatile fatty acid concentration in the percolate was measured on a daily base in the first week of the test and 2 to 3 times per week after the first week.

2.5. Statistical Analyses

Pearson correlation analysis was performed to determine the strength of the relationship between physio-chemical characteristics (for instance material structure, percolation rate and total solids) and methane production after 35 days batch duration. Pearson's correlation coefficients were performed using SAS software (SAS version 10.0, SAS Institute, Inc., Cary, NC, USA).

3. Results and Discussion

3.1. Material Characteristics of Chicken Manure

The umbrella term chicken manure can be differentiated and includes chicken slurry, dry chicken excreta, dried chicken excreta, chicken manure, fresh chicken excreta. The chicken manure types can differ in the origin or bird purpose and by the chicken housing style i.e., broiler, laying hens, young animals. Chicken manure is the mixture of feces and urine excreted, which contains varying amounts of undigested feeding stuff, desquamated intestinal epithelium, residues of secretion, microorganism from the intestinal flora, metabolites excreted with the urine, as well as other components e.g., feather, egg leftovers, bedding material, grid material and soil. [?]

In this study, chicken manure was tested for the material properties of total solids (TS) and volatile solids (VS), nitrogen, protein, fat and fiber composition. The 42 tested samples were derived from various sources and varied in storage age, housing form and bedding material type. The material characterization tests were carried out with the aim of obtaining an overview of the material properties and material diversity. The results are summarized in Table ??. The water content of the analyzed samples based on fresh matter (FM) averaged approximately 48.3%FM, ranging from 28.6 to 77.0%FM. Dry samples with TS contents >75% were either long term stored manure samples or collected from young animal breeding. The volatile solid content varied between approx. 56.7–86.4%, which shows the high ash content of chicken manure with approximately 30%. Sand and grit materials contained in the chicken feed and bedding material remain as ash content in the manure. Anaerobic digestion of chicken manure in continuous stirred-tank reactor (CSTR) systems can cause several technical problems. In cases of sediment accumulation, the digester volume can be reduced or technical devices such as stirrers or pumps can be negatively affected. Sediment removal can require the removal of the digester roof and the use of heavy machinery. Since dry batch anaerobic processes operate discontinuously, they are not likewise affected by sediment accumulation. After a certain retention time the digested substrate material and sediments are replaced by new fresh substrate. Moving parts such as stirrers are not present [?]. However, sediments can cause problems in percolation systems too, if the substrate permeability is reduced or fine sand migration effects the pumping and percolation system.

Table 2. Material characteristics of chicken manure.

	TS	VS	NH$_4$-N	TKN	Raw Protein	Raw Fat	Raw Fibre
	$n = 50$	$n = 50$	$n = 44$	$n = 42$	$n = 42$	$n = 38$	$n = 44$
	%FM	%TS	%TS	%TS	%TS	%TS	%TS
Mean	48.3	69.5	0.8	4.7	21.6	3.6	14.0
Max	77.0	86.4	2.3	8.7	46.2	12.9	32.5
Min	28.6	56.7	0.1	1.5	1.5	0.1	1.7

n number of tested chicken manure samples obtained from various sources.

The Total Kjeldahl Nitrogen (TKN) content of 42 tested chicken manure samples from various sources was in a range of approx. 1.5 to 8.7%TS. On average the ammonium (NH$_4$-N) content was approx. 50% of total nitrogen. The calculated raw protein content of the 42 tested samples varied between 21.6% and 46.2%TS. The material characterization results confirm the findings of Nie et.al., 2015 [?] that chicken manure is a feedstock material with an comparably high nitrogen content, which can result in process instability of the anaerobic digestion process by free ammonia inhibition The raw fat content of 39 tested samples averaged approximately 3.6% of total solids, and indicated a comparably low fat concentration while the raw fiber content of 44 tested samples ranged from approximately 14.0 to 32.5% of total solids. Several samples contained straw stalks or remains of pelletized straw or saw dust, which is commonly used as bedding material. Generally, bedding material in manure like straw or wood saw dust consists of ligno-cellulose primarily. Massé et.al. [?] confirmed the biogas potential of ligno-cellulosic biomass such as wheat straw in dry digestion processes. Hendriks and Zeeman [?]

identified the crystallinity of cellulose, available surface area, and lignin content as limiting factors for the anaerobic decomposition of ligno-cellulosic biomass. In preliminary investigations [?] it was found that the fiber content can have positive effects on the feedstock permeability. The material characterization tests revealed the highly divers material composition of chicken manure and regularly feedstock testing is recommended prior anaerobic digestion.

3.2. Material Permeability of Chicken Manure

The permeability and compressibility characteristics of nine chicken manure samples with and without structure material addition were determined as a measure for the feedstock suitability for percolation processes. An oedometer test apparatus was used to simulate the material compression within the substrate heap as it occurs in full scale dry batch anaerobic digestion plants. Oedometer tests without compaction ($P1$) showed similar permeability properties for all tested chicken manure samples (see Table ??). When compressive force was applied on top of the samples to simulate material compaction in lower layers of the substrate heap, the testing method revealed very different permeability and compaction properties. In order to classify and interpret the oedometer test results, the permeability of maize silage as a reference material was tested, since it is a common feedstock used successfully in several German anaerobic dry digestion biogas plants with a demonstrated sufficient permeability. Chicken manure sample No.1 was tested as impermeable under compression forces that would be caused by 1.5 m substrate heap height. Sample No.2 was tested without structure material addition and found to be impermeable under compression forces, which would be caused by 3.0 m substrate heap height. However, the addition of plastic structure carriers to sample No.2 led to a 39% superior material permeability, compared to maize silage. Sample No.3 was tested as sufficiently permeable without structure material addition. The material was collected from young animal rearing and contained high amounts of bedding material and was comparably dry at 76.5%TS. Sample No.4 without structure material addition shows a 50% lower permeability compared to maize silage. In Figure ?? it is shown that the addition of structure building materials such as wood chips and straw to chicken manure sample No.4 leads to an improvement of the percolation properties. An addition of 10% mass woodchips or 5% mass straw to chicken manure results in permeability characteristics of the feedstock comparable to maize silage.

Table 3. Permeability of chicken manure with and without structure material addition.

Sample Material	Structure Material Addition [w%]	$P1$ [m/s]	$P2$ [m/s]	$C2$ [%]	$P3$ [m/s]	$C3$ [%]
MS	-	1.59×10^{-4}	1.08×10^{-4}	22.94	8.39×10^{-5}	31.76
1	-	1.91×10^{-4}	*n.p.*	*n.m.*	*n.p.*	*n.m.*
2	-	4.36×10^{-5}	9.58×10^{-6}	*n.m.*	*n.p.*	*n.m.*
2	5 (PSC)	2.14×10^{-4}	1.36×10^{-4}	4.47	1.38×10^{-4}	19.42
3	-	1.91×10^{-4}	1.38×10^{-4}	8.43	9.89×10^{-5}	18.54
4	-	1.02×10^{-4}	6.34×10^{-5}	3.28	4.32×10^{-5}	7.38
4	5 (WC)	1.43×10^{-4}	9.95×10^{-5}	7.53	6.87×10^{-5}	15.07
4	10 (WC)	1.49×10^{-4}	1.01×10^{-4}	13.70	8.02×10^{-5}	17.12
4	5 (St)	1.71×10^{-4}	1.10×10^{-4}	15.00	8.20×10^{-5}	19.38
4	10 (St)	1.80×10^{-4}	1.15×10^{-4}	14.29	9.39×10^{-5}	16.67

p permeability and *C* compressibility (without compaction ($P1$), 1.5 and 3.0 m simulated material height ($P2$ and $C2$, $P3$ and $C3$); MS Maize silage; PSC Plastic structure carriers; WC wood chips; St straw; Sample 1-from Laying hens; 2-Broiler fattening; 3-Young animals; 4-Laying hens.

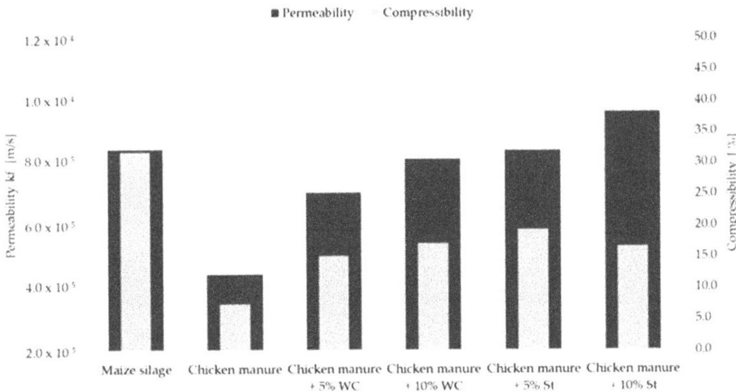

Figure 3. Permeability and compressibility characteristics of chicken manure with structure material addition relative to maize silage. WC, wood chips; St, straw.

The results shown in Table **??** indicate that in lower layers of the substrate pile, material compression increases as a result of increasing material weight and the associated pressure on top. Andre et al., 2015 [?] showed that substrate compaction leads to a reduction of the material permeability possibly due to a decrease in the material pore space. The correlation between material compression and permeability was also described by [?] for various agricultural residue materials. Some chicken manure samples from laying hens and broiler fattening were tested as impermeable when material heap height of 1–2 m was simulated. However, the results of this study confirm the findings by Hayes et al. [?] that structure materials can be used to increase the permeability of the feedstock mixture and thus improve the suitability of substrates with low material permeability for percolation processes. The correlations between physio-chemical parameters and permeability revealed that single physio-chemical characteristics did not significantly correlate with material permeability, indicating that the material permeability could be dependent on the storage age, housing form and bedding material type. The detailed results of Pearson's correlation coefficients are presented in supplementary Table S1. Chicken manure with a high proportion of dry fibrous bedding material could have a better material suitability for percolation processes compared to the sample materials, which were composed mainly of chicken manure alone.

The aim of this preliminary investigation was to determine the suitability of the material characteristics in terms of permeability under compacting forces and compactibility of chicken manure used in dry batch anaerobic digestion processes. In practice, there is a risk of technical problems if percolate cannot pass through the substrate material in the percolate digester. As a result, dead zones can occur in the lower part of the substrate material bed. In the worst case scenarios, large areas of the substrate heap are not sufficiently percolated, which can lead to deteriorated organic degradation or to the accumulation of percolate within the percolate digester [?]. However, further tests at full scale are required to determine if the oedometer testing apparatus results can be transferred completely to full scale. Further digestion trials should be conducted in order to determine if structure stability of the input material sustains during the digestion process and if the digestate properties can be manipulated by a targeted adjustment of the substrate properties by structure material addition. Positive results would be of interest for the handling and transportation of fermentation residues, since a very moist and structurally weak digestate may require more time for dewatering [?]. Improved dewaterability of structurally optimized chicken manure mixtures could contribute to an increased plant throughput in anaerobic dry digestion processes, since fermentation residues can only be removed from the percolator at the end of fermentation after the contained percolate has run off. Permeability and dewaterability tests of digestate samples were not part of the study, but preliminary tests at lab scale showed that structure material addition to the input material resulted in better percolate drainage [?].

3.3. Results of Dry Batch Anaerobic Digestion Trials

The aim of the digestion trials was to determine the biogas generation of chicken manure in a one step dry bath anaerobic digestion process without inoculum addition and a secondary methanization digester. A further aim was to determine the influence of structure material addition on the substrate methane yield. For this purpose, six dry batch anaerobic digestion trials were carried out with substrate mixtures of dry chicken manure and different additions of the structure materials straw and wood chips.

3.3.1. Process Conditions

The pH varied between 6.8 to 8.2 during the course of the digestion experiments (Figure ??a). During the first two weeks of the trial the pH-value decreased and reached the lowest values after approximately 7 days. During the second and third week the pH value increased and steadied until the end of the test. The graphs indicate that the pH-value was affected by the volatile fatty acid (VFA) and ammonium nitrogen concentration. Volatile fatty acid concentrations in the percolate increased during the test duration to a maximum of 32.8 g/L but decreased to the starting value of approx. 10 g/L in less than 30 days (Figure ??b). This could indicate that a secondary methanation step was not required to reduce the volatile components temporarily accumulated in the percolate. The buffer capacity of the percolate was sufficient to stabilise the pH-value above 6.8 even at peak VFA concentrations. The ammonium nitrogen (NH_4-N) concentrations ranged from 1.6 to a maximum of 7.2 g/L (Figure ??c) and free ammonia (FA) ranged from 0.1 to 1.2 g/L (Figure ??d). According to Hafner and Bisogni [?], free ammonia calculation for liquid digestate or percolate using the described formula can lead in some cases to an overestimation of the FA, as percolate has a rather complex material composition compared to water. Ammonium (NH_4+) and free ammonia (NH_3) are the predominant forms of inorganic nitrogen. Ammonium nitrogen (NH_4-N) is a degradation product of organic nitrogen such as proteins or urea during anaerobic digestion. In the digester liquid, NH_4-N is present as ammonium ions (NH_4+) and as free ammonia (NH3). Increasing pH or temperature, results in a higher percentage of NH_4-N present as free ammonia (NH_3). It has been shown that free ammonia is more toxic, as it can pass through the cell membrane, causing proton imbalance and potassium deficiency [?]. According to Ying et.al [?], Bujoczek et.al. [?] and Yenigün and Demirel [?], several research groups report that inhibitory NH_4-N concentration range between 1.5 to 5.0 g/L. However, significantly higher concentrations of up to 14 g/L are possible by microbial adaptation of the digestion process. Niu et al. [?] described a successfully recovery from ammonia inhibition of a CSTR system operated with chicken manure as substrate at TAN concentrations of 16 g/L after dilution and different washing steps. In the case of our study, an inhibition of the process biology can be expected, which possibly negatively effected the methane production.

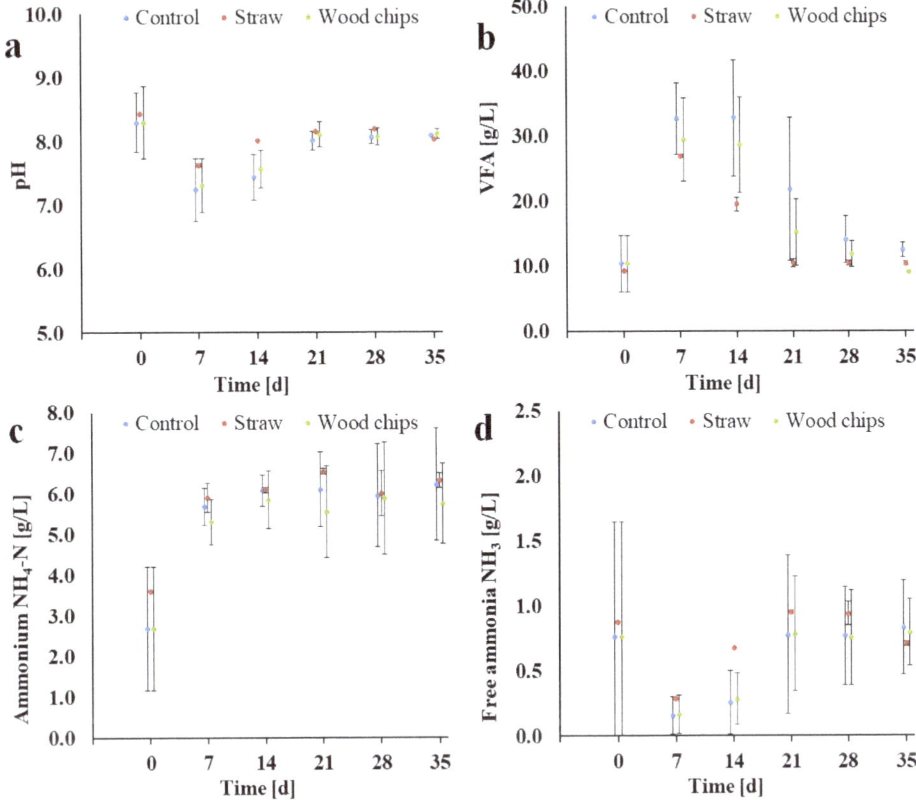

Figure 4. a–d. Process conditions during dry batch anaerobic digestion of chicken manure. pH value (**a**), volatile fatty acid concentration (**b**), ammonium concentration (**c**), free ammonia concentration (**d**).

3.3.2. Specific Methane Yield

In our study, chicken manure achieved an average SMY of 135 ± 20 mL/g VS or 75 ± 5 mL/g FM. According to FNR [?], the average biochemical methane potential of chicken manure is 280 mL/g VS or 90 mL/g FM. Several researchers investigated the SMY of chicken manure with or without bedding material in dry batch anaerobic processes. The SMY of chicken manure measured by Marchioro et.al. [?] was 74 mL/g VS and lower compared to the findings of this study. Other authors such as Bi et al. [?] or Wang et al. [?] published considerable higher methane yields of 280 mL/g VS and 437 mL/g VS respectively, partly due to ammonia extraction within the process. One possible reason for divergent results in various studies is the differences in physical and chemical composition of the manure, due to the method of rearing as well as the utilisation and kind of bedding material.

The methane concentration increased during the test duration to a maximum of approx. 68% as shown in Table S3.

In our study, chicken manure without structure material addition achieved SMYs of 127 ± 12 mL/g VS or 70 ± 6 mL/g FM (Table ??). The Figure ??a–c shows the cumulated daily biogas and methane production of the percolate digester and the calculated SMY of the VFA contained in the percolate. As depicted in Figure ??a, the cumulative methane production showed a low methane production during the first two weeks. Thereafter the methane production increased but the biogas production rate was not identical in the two parallel trials. The degradation of the organic content was possibly different due to an uneven percolate flow or distribution within the substrate material. However, the measured specific methane yield of the parallel trial was in a comparable range after 35 days. The biogas production did not reach a plateau phase at the end of the test. A longer test duration would have led to a higher methane production. However, retention times in dry batch processes in practice are commonly shorter. The digestion trials showed that dry batch anaerobic chicken manure digestion was possible, even without inoculum addition. However, an addition of inoculum to substrate material in the beginning could help to accelerate the initial methane production rate.

Table 4. Specific methane yield (SMY) of chicken manure with and without additives measured in dry batch anerobic digestion trials.

	SMY_1	SMY_2	SMY_{mean}	
	[mL/g VS]	**[mL/g VS]**	**[mL/g VS]**	**[mL/g FM]**
Control	136	119	127 ± 12	70 ± 6
Straw	138	132	135 ± 4	75 ± 2
Woodchips	175	111	143 ± 45	79 ± 25

The SMY of the straw variants was 6% higher compared to the control variant without structure material addition and resulted in 135 ± 4 mL/g VS or 75 ± 2 mL/g FM. Methane production occurred without a lag phase in the beginning. The methane concentration of the biogas exceeded 50% after a test duration of approx. 8 days. Data to the gas composition is summarized in Table S3. The control variant without structure material addition and the chicken manure mixture with wood chip addition showed a lower methane production rate and a methane concentration of 50%was reached after approx. 16 days test duration. As depicted in Figure ??b, a two-step gas production was observed in the straw variants with a first phase between the test start and day 20, and a second phase with a lower but steady methane production between day 20 and day 35. A possible reason could be the organic degradation of the additional straw content. Even though the added content of straw was different in both digesters (5% and 10% of the mass), the gas production was similar during the course of the trial. Possibly the straw addition to the chicken manure lead also to an increase of the pore space within the substrate mixture resulting in an even percolate distribution and yielded a comparable degradation of the organic content.

The digestion of the wood chip variants resulted in the highest SMY with 143 ± 45 mL/g VS or 79 ± 25 mL/g FM. Although the average methane generation was 11% higher compared to the control without structure material addition, the high standard deviation between the two digesters has to be considered. Until day 20, the methane production rate was similar to the control variant. Between day 21 and 35, the methane production increased but varied in the parallel digesters (see Figure ??c). The preliminary tests showed that wood chip addition to the chicken manure of 10% mass led to a very coarse substrate mixture. The material mixture possibly promotes channel effects, which results in percolate passing the substrate mix without an even distribution in the substrate heap, leading to only a partial organic degradation. Channel effects were observed in the material characterization tests when proportions of 10% mass wood chips where added to the substrate. Pearson's correlation showed a positive, but weak (r2 = 0.54), correlation between material permeability under compression (P3) and specific methane yield. Thus, chicken manure mixtures with wood chip addition showed best material permeability with 8.02×10^{-5} m/s and achieved the highest specific methane yield. The detailed results of results of Pearson's correlation coefficients are presented in supplementary Table S2.

It must be checked whether a very high use of structural substances is possible in full scale plant operation, because this reduces the plant throughput of the actual feedstock. Only very cost-effective structural materials, such as residual straw, were considered, since alternative structural materials, such as wood chips, are highly unlikely to be economically viable in practical application. A recovery of structural materials directly from the digestate is possible but challenging due to the unfavorable material properties that will impact the subsequent sieving. Under practical conditions, the use of screened residues as structure building material from chicken manure with bedding material appears to be more promising. In order to replace structural materials, which do not contribute to gas production in the mixture the material permeability of chicken manure can also be improved by mixing fibrous bedding material containing chicken manure with a high proportion of lignified biomass with moist, structurally weak chicken manure without bedding material, if both types of chicken manure are available simultaneously. A validation of the findings in full scale operations should be considered.

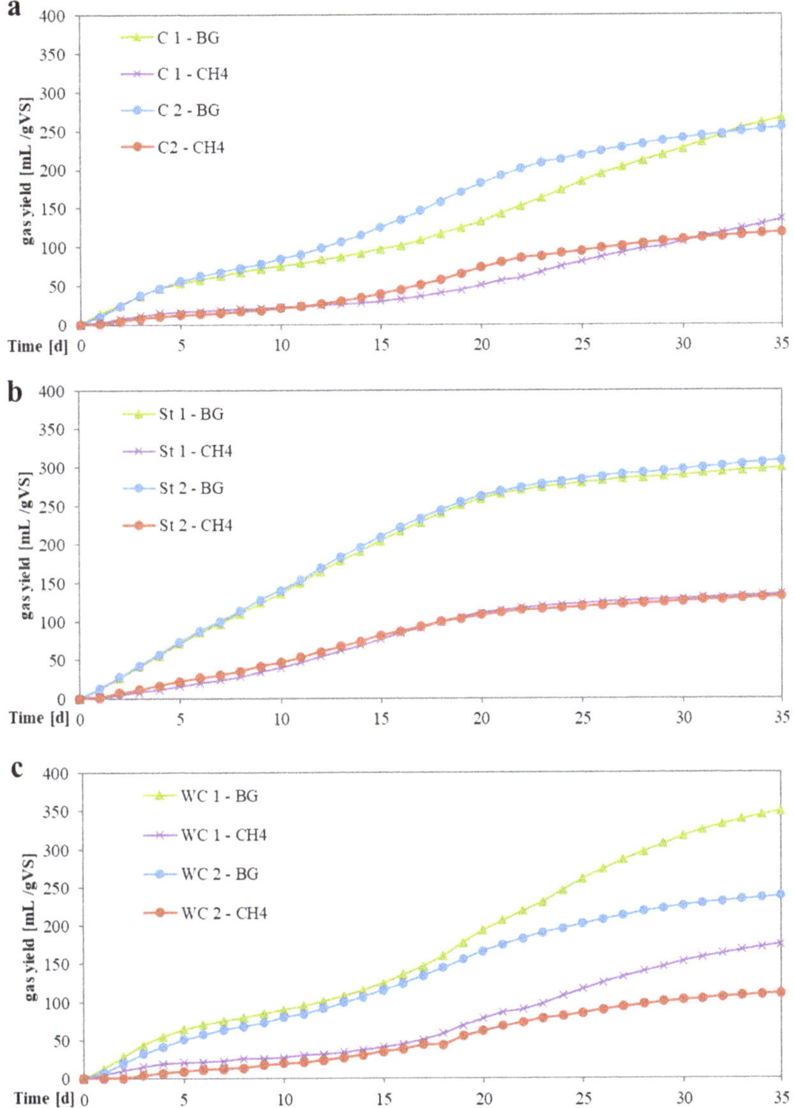

Figure 5. (a–c) Specific yield of biogas (BG) and methane (CH4) of chicken manure in dry batch anaerobic digestion. C, control variant (**a**); St, straw variant (**b**); WC, wood chip variant (**c**).

4. Conclusions

Material characterization tests of chicken manure were carried out to determine the feedstock suitability in terms of feedstock composition, permeability under compaction and compressibility for dry batch anaerobic digestion processes. Oedometer tests under compression showed that the permeability and structural stability of chicken manure can be improved through an initial addition of structural materials such as wood chips or wheat straw to the substrate mixture. Even originally impermeable chicken manure samples were suitable for percolation processes after an addition of 5% straw or wood chips (by mass). Channel effects with uneven percolate distribution were observed on a number of samples with 10% woodchips addition. Anaerobic dry batch digestion trials were carried out to investigate the biogas generation of dry chicken manure without inoculation and separate methanization step. The conducted digestion trials showed that dry batch anaerobic chicken manure digestion was possible without inoculum addition. However, the methane production showed a lag phase of approximately 2 weeks after the test start, and the measured methane yields were lower compared to published SMYs for wet digestion systems. In this study the obtained methane yield of chicken manure averaged 135 ± 20 mL/g VS or 75 ± 5 mL/g FM, which represents 70% of the gas potential based on fresh mass of common energy crops such as corn silage. The addition of straw and wood chips led to an increase of the specific methane yield by 6% and 11%, respectively.

Supplementary Materials: The following are available online at http://www.mdpi.com/2306-5354/7/3/106/s1, Table S1: Pearson's correlation coefficients of the physio-chemical charactieristics of chicken manure with material permeability, Table S2: Pearson's correlation coefficients of material permeability and compressibility with specific methane yield. Table S3: Biogas and methane production and gas composition the dry digestion of chicken manure with and without structure material addition.

Author Contributions: Conceptualization and methodology, H.W., F.S.; validation, H.W., D.G.I., F.S., E.J.; formal analysis, H.W., F.S., E.J.; investigation, H.W., F.S., D.G.I.; writing—original draft preparation, H.W.; writing—review and editing, H.W., E.J., F.S., M.N.; visualization, H.W.; supervision, F.S., H.W., M.N.; project administration, F.S.; funding acquisition, F.S. All authors have read and agreed to the published version of the manuscript.

Funding: This research project NovoHTK (funding code 03KB137A) was funded by the Federal Ministry of Economic Affairs and Energy (BMWi) under the program Energetische Biomassenutzung.

Acknowledgments: The authors would also like to thank for sample provision. The authors express their appreciation to colleagues from DBFZ-laboratory as well as to the student Weerachai Al Bon Jeng for his valuable contribution during the execution of sample material characterization and conduction of the dry batch anaerobic digestion trials.

Conflicts of Interest: The authors declare no conflict of interest. The funders had no role in the design of the study; in the collection, analyses, or interpretation of data; in the writing of the manuscript; or in the decision to publish the results.

References

1. FAO. *Nitrogen Inputs to Agricultural Soils from Livestock Manure. New Statistics*; Integrated crop management; Food and Agriculture Organization, FAO: Rome, Italy, 2018; Volume 24, p. 17. ISBN 978-92-5-130024-4.

2. BMEL-Statistik. Federal Ministry of Food and Agriculture—BMEL. November 2019. Available online: https://www.bmel-statistik.de/ernaehrung-fischerei/versorgungsbilanzen/eier/ (accessed on 14 July 2020).

3. Nebel, U.; Kühnel, M. *Survey of the Different Chicken Housing Systems and Accumulating Form of Manure/Slurry for the Derivative of a Standardised Form of Veterinary Drug Decomposition in Expositions Scenarios*; Federal Environment Agency Germany–Texte: Dessau-Roßlau, Germany, 2010; Volume 17, p. 74, ISSN 1862-4804.

4. Haenel, H.-D.; Rösemann, C.; Dämmgen, U.; Döring, U.; Wulf, S.; Eurich-Menden, B.; Freibauer, A.; Döhler, H.; Schreiner, C.; Osterburg, B. *Calculatons of Gaseous and Particulate Emissions from German Agriculture 1990–2016: Report on Methods and Data (RMD) 2018*; Johann Heinrich von Thünen-Institut: Braunschweig, Germany, 2018; p. 424. [CrossRef]

5. Rajagopal, R.; Massé, D. Start-up of dry anaerobic digestion system for processing solid poultry litter using adapted liquid inoculum. *Process Saf. Environ. Prot.* **2016**, *102*, 495–502. [CrossRef]

6. Federal Ministry of Food and Agriculture—FNR. Guide to Biogas from Production to Use. July 2012. Available online: https://mediathek.fnr.de/broschuren/fremdsprachige-publikationen/english-books/guide-to-biogas-from-production-to-use.html (accessed on 14 July 2020).

7. Bayrakdar, A.; Sürmeli, R.-Ö.; Çalli, B. Dry anaerobic digestion of chicken manure coupled with membrane separation of ammonia. *Bioresour. Technol.* **2017**, *244*, 816–823. [CrossRef] [PubMed]

8. Nie, H.; Jacobi, H.F.; Strach, K.; Xua, C.; Zhoua, H.; Liebetrau, J. Monofermentation of chicken manure: Ammonia inhibition and recirculation of the digestate. *Bioresour. Technol.* **2015**, *178*, 238–246. [CrossRef]

9. Luning, L.; Van Zundert, E.H.M.; Brinkmann, A.J.F. Comparison of dry and wet digestion for solid waste. *Water Sci. Technol.* **2003**, *48*, 15–20. [CrossRef] [PubMed]

10. Rocamora, I.; Wagland, S.T.; Villa, R.; Simpson, E.W.; Fernández, O.; Bajón-Fernández, Y. Dry anaerobic digestion of organic waste—A review of operational parameters and their impact on process performance. *Bioresour. Technol.* **2020**, *299*, 122681. [CrossRef]

11. Di Maria, F.; Barratta, M.; Biaconi, F.; Placidi, P.; Passeri, D. Solid anaerobic digestion batch with liquid digestate recirculation and wet anaerobic digestion of organic waste: Comparison of system performances and identification of microbial guilds. *Waste Manag.* **2017**, *59*, 172–180. [CrossRef]

12. Ge, X.; Xu, F.; Li, Y. Solid-state anaerobic digestion of lignocellulosic biomass: Recent progress and perspectives. *Bioresour. Technol.* **2016**, *205*, 238–249. [CrossRef]

13. Olivier, F.; Gourc, J.P. Hydro-mechanical behavior of Municipal Solid Waste subject to leachate recirculation in a large-scale compression reactor cell. *Waste Manag.* **2007**, *27*, 44–58. [CrossRef]

14. Cysneiros, D.; Banks, C.J.; Heaven, S.; Karatzas, K.-A.G. The role of phase separation and feed cycle length in leach beds coupled to methanogenic reactors for digestion of a solid substrate (Part 2): Hydrolysis, acidification and methanogenesis in a two-phase system. *Bioresour. Technol.* **2011**, *102*, 7393–7400. [CrossRef]

15. *DIN EN 12880. Characterization of Sludges—Determination of Dry Residue and Water Content*; DIN Deutsches Institut für Normung e. V.: Berlin, Germany, 2001.

16. *DIN EN 15935:2012-11. Sludge, Treated Biowaste, Soil and Waste—Determination of Loss on Ignition*; DIN Deutsches Institut für Normung e. V.: Berlin, Germany, 2012.

17. Liebetrau, J.; Pfeiffer, D.; Thrän, D. (Eds.) *Collection of Measurement Methods for Biogas—Methods to Determine Parameters for Analysis Purposes and Parameters That Describe Processes in the Biogas Sector*; Series of the Funding Programme "Biomass Energy Use"; Deutsches Biomasseforschungszentrum gemeinnützige GmbH: Leipzig, Germany, 2016; Volume 7, ISSN 2364-897X; Available online: https://www.energetische-biomassenutzung.de/fileadmin/user_upload/Downloads/Ver%C3%B6ffentlichungen/07_MMS_Biogas_en_web.pdf (accessed on 30 October 2019).

18. Wedwitschka, H.; Jenson, E.; Liebetrau, J. Feedstock Characterization and Suitability Assessment for Dry Anaerobic Batch Digestion. *Chem. Eng. Technol.* **2016**, *39*, 665–672. [CrossRef]

19. *VDI 4630: 2006. Fermentation of Organic Materials—Characterisation of the Substrate, Sampling, Collection of Material Data, Fermentation Tests*; Beuth Verlag GmbH: Berlin, Germany, 2006.

20. Massé, D.; Saady, N.M.C.; Gilbert, Y. Psychrophilic dry anaerobic digestion of cow feces and wheat straw—Feasibility studies. *Biomass Bioenergy* **2015**, *77*, 1–8. [CrossRef]

21. Hendriks, A.T.W.M.; Zeeman, G. Pretreatments to enhance the digestibility of lignocellulosic biomass. *Bioresour. Technol.* **2009**, *100*, 10–18. [CrossRef] [PubMed]

22. André, L.; Durante, M.; Pauss, A.; Lespinard, O.; Ribeiro, T.; Lamy, E. Quantifying physical structure changes and non-uniform water flow in cattle manure during dry anaerobic digestion process at lab scale—Implication for biogas production. *Bioresour. Technol.* **2015**, *192*, 660–669. [CrossRef] [PubMed]

23. Hayes, A.C.; Enongene, E.S.; Mervin, S.; Jenson, E. Techno-economic evaluation of a tandem dry batch, garage-style digestion-compost process for remote work camp environments. *Waste Manag.* **2016**, *58*, 70–80. [CrossRef]

24. Shewani, A.; Horgue, P.; Pommier, S.; Debenest, G.; Lefebvre, X.; Gandon, E.; Paul, E. Assessment of percolation through a solid leach bed in dry batch anaerobic digestion processes. *Bioresour. Technol.* **2015**, *178*, 209–216. [CrossRef]

25. Hafner, S.D.; Bisogni, J.J. Modeling of ammonia speciation in anaerobic digesters. *Water Res.* **2009**, *43*, 4105–4114. [CrossRef]

26. Ying, J.; Ewan, M.; Yue, Z.; Sonia, H.; Charles, B.; Philp, L. Ammonia inhibition and toxicity in anaerobic digestion: A critical review. *J. Water Process. Eng.* **2019**, *32*. [CrossRef]

27. Bujoczek, G.; Oleszkiewicz, J.; Sparling, R.; Cenkowski, S. High Solid Anaerobic Digestion of Chicken Manure. *J. Agric. Eng. Res.* **2000**, *76*, 51–60. [CrossRef]

28. Yenigün, O.; Demirel, B. Ammonia inhibition in anaerobic digestion—A review. *Process. Biochem.* **2013**, *48*, 901–911. [CrossRef]

29. Niu, Q.; Qiao, W.; Qiang, H.; Hojo, T.; Li, Y.Y. Mesophilic methane fermentation of chicken manure at a wide range of ammonia concentration: Stability, inhibition and recovery. *Bioresour. Technol.* **2013**, *137*, 358–367. [CrossRef]

30. Marchioro, V.; Steinmetz, R.L.R.; do Amaral, A.C.; Gaspareto, T.C.; Treichel, H.; Kunz, A. Poultry Litter Solid State Anaerobic Digestion: Effect of Digestate Recirculation Intervals and Substrate/Inoculum Ratios on Process Efficiency. *Frontiers* **2018**, *2*, 1–10. [CrossRef]

31. Bi, S.; Westerholm, M.; Qiao, W.; Xiong, L.; Mahdy, A.; Yin, D.; Song, Y.; Dong, R. Metabolic performance of anaerobic digestion of chicken manure under wet, high solid, and dry conditions. *Bioresour. Technol.* **2020**, *296*, 122342. [CrossRef] [PubMed]

32. Wang, X.; Gabauer, W.; Li, Z.; Ortner, M.; Fuchs, W. Improving exploitation of chicken manure via two-stage anaerobic digestion with an intermediate membrane contactor to extract ammonia. *Bioresour. Technol.* **2018**, *268*, 811–814. [CrossRef] [PubMed]

Article

Processing High-Solid and High-Ammonia Rich Manures in a Two-Stage (Liquid-Solid) Low-Temperature Anaerobic Digestion Process: *Start-Up and Operating Strategies*

Prativa Mahato [1,2], Bernard Goyette [1], Md. Saifur Rahaman [2] and Rajinikanth Rajagopal [1,*]

[1]   Sherbrooke Research and Development Center, Agriculture and Agri-Food Canada, 2000 College Street, Sherbrooke, QC J1M 0C8, Canada; p_mahato@encs.concordia.ca (P.M.); bernard.goyette@canada.ca (B.G.)
[2]   Department of Building, Civil and Environmental Engineering, Concordia University, Montreal, QC H3G 1M8, Canada; saifur.rahaman@concordia.ca
*   Correspondence: rajinikanth.rajagopal@canada.ca

Received: 1 June 2020; Accepted: 22 July 2020; Published: 25 July 2020

Abstract: Globally, livestock and poultry production leads to total emissions of 7.1 Gigatonnes of CO_2-equiv per year, representing 14.5% of all anthropogenic greenhouse gas emissions. Anaerobic digestion (AD) is one of the sustainable approaches to generate methane (CH_4) from manure, but the risk of ammonia inhibition in high-solids AD can limit the process. Our objective was to develop a two-stage (liquid–solid) AD biotechnology, treating chicken (CM) + dairy cow (DM) manure mixtures at 20 °C using adapted liquid inoculum that could make livestock farming more sustainable. The effect of organic loading rates (OLR), cycle length, and the mode of operation (particularly liquid inoculum recirculation-percolation mode) was evaluated in a two-stage closed-loop system. After the inoculum adaptation phase, aforementioned two-stage batch-mode AD operation was conducted for the co-digestion of CM + DM (Total Solids (TS): 48–51% and Total Kjeldahl Nitrogen (TKN): 13.5 g/L) at an OLR of 3.7–4.7 gVS/L.d. Two cycles of different cycle lengths (112-d and 78-d for cycles 1 and 2, respectively) were operated with a CM:DM mix ratio of 1:1 (*w/w*) based on a fresh weight basis. Specific methane yield (SMY) of 0.35 ± 0.11 L CH_4g/VS_{fed} was obtained with a CH_4 concentration of above 60% for both the cycles and Soluble Chemical Oxygen Demand (CODs) and volatile solid (VS) reductions up to 85% and 60%, respectively. For a comparison purpose, a similar batch-mode operation was conducted for mono-digestion of CM (TS: 65–73% and TKN: 21–23 g/L), which resulted in a SMY of 0.52 ± 0.13 L CH_4g/VS_{fed}. In terms of efficiency towards methane-rich biogas production and ammonia inhibitions, CM + DM co-digestion showed comparatively better quality methane and generated lower free ammonia than CM mono-digestion. Further study is underway to optimize the operating parameters for the co-digestion process and to overcome inhibitions and high energy demand, especially for cold countries.

Keywords: ammonia inhibition; chicken manure; dairy cow manure; high-solids anaerobic digestion; inoculum adaptation; volatile fatty acids

---

1. Introduction

In the last few decades, rapid growth in the population has been observed, which is further predicted to increase to 9.6 billion by 2050 [1]. In addition, the accelerated pace of urbanization and growing income is also noticed. Together, these factors pose severe challenges to the food and agriculture sectors. Along with the change in food habits, the elevation in manufactured agriculture products, mostly based on animal sources, the consumption of chicken meat, and egg production, has also increased by 50% and 36.5%, respectively, from 2000 to 2014 [2]. The demand for food is estimated to increase to 73% and 58% for meat and milk, respectively, by 2050. Consequently, this leads to mass production of livestock and, ultimately, a huge generation of manure. Manure causes emissions of greenhouse gases (GHG) [3] if not managed properly. Globally, the poultry-related emissions alone account for about 600 million tons of carbon dioxide (CO_2) equivalent per year [4], contributing to climate change and global warming. In order to manage the manure, one of the widely exercised solutions is the land application as it provides nutrition to the land. However, excessive land application of manure results in nutrition overloading in soil and water bodies, ending up in eutrophication [5]. Open land application of manure also contributes to methane (CH_4) emissions, which carries 23 times more global warming potential than CO_2 alone [6]. Another positive solution towards manure management can be composting as it reduces waste mass and produces valuable end products [7]; however, the huge loss of nitrogen (N) in the form of soluble nitrates is observed in composting, which eventually reduces the fertilizer value. Besides this, composting also causes odor nuisance and environmental side effects like air and water pollution, gases like NH_3, CH_4, and N_2O impacts air quality and, leaching and runoff due to precipitation causes high adverse effect on water pollution [8,9].

Anaerobic digestion (AD) is a sustainable approach to reduce the ill effects caused by improper processing of manure. In recent years, AD has received great attention due to its obvious advantage, i.e., reducing pollution, converting organic waste into high-quality biogas, which is useful in the form of heat and/or electricity [10]. Moreover, the generation of electricity through AD is a renewable process, thus reduces the cost of fossil fuels and their climatic side effects. Poultry litter is one of the highest biomethane potential organic substrates compared to dairy cow manure (DM). However, one of the major limitations of AD of chicken manure (CM) is the inhibition caused by the production of ammonia [11] due to which its potentiality cannot fully be exploited. CM is also high in solids (63 ± 10% Total Solids (TS)), which makes the process unsuitable in semi-liquid (10–15% TS) or wet (<10%) digesters as the dilution requirement would be 6–7 times than the normal practice to operate in these type of digesters. Similarly, high N in CM (Total Kjeldahl Nitrogen (TKN): 25–35 g/L) also demands huge dilution to avoid inhibitions during the AD process. Unfortunately, dilution by water requires comparatively high energy input, which makes the situation expensive and impractical to process the feedstocks rich in high-solids and high-ammonia. In this scenario, the co-digestion of CM with other crops or C-rich feedstocks could be a feasible method.

C/N ratio of CM ranges from 6.3 to 10 [12,13], and to operate the digester to its utmost condition, high carbon content is essential. On the other hand, the C/N ratio of DM is reported to be between 24 and 40 [12,14]. Therefore, co-digestion stabilizes the C:N ratio because of the composition of high lignocellulosic compounds in DM. Co-digestion also minimizes the risk of ammonia inhibitions and, in some cases, improves the methane content in the biogas. Co-digestion of manure also benefits in many ways, like the reduction of manpower in the segregation of waste to be processed. It avoids the separate storage, treatment, and handling of mixed waste [15]. Agriculture and Agri-Food Canada (AAFC) has successfully developed the AD biotechnology over the years to process poultry, swine, and cow manures operating at low temperatures. However, the potential of digesters to process the co-digestion of DM and CM at a TS > 50% using adapted liquid inoculum has not been studied. The positive results obtained from the study of the mono-digestion of CM has encouraged us to explore the possibilities of co-digesting CM + DM mix in an economical way.

This paper emphasized on the start-up and operating strategies for the development of low-temperature two-stage (liquid–solid) anaerobic co-digestion of CM + DM mixture using adapted inoculum. The primary objective of this study was to demonstrate the operational feasibility of two-stage process (i.e., liquid inoculum reservoir coupled with high-solid anaerobic digestion (HSAD) system), treating CM + DM at 20 ± 1 °C, and to encourage small-scale farmers to adopt this technology at low cost. An effort was made to develop the HSAD start-up protocol, using (i) acclimatized liquid inoculum since obtaining a huge quantity of solid inoculum to treat high-solids waste mix is practically not feasible at many farm locations; (ii) no mixing conditions, as mechanical mixers create complexity in full-scale operations. Besides, the scope of this study was also to assess the comparative performance of digesters co-digesting CM + DM and mono-digesting CM, especially in terms of methane concentration and free ammonia inhibition.

2. Materials and Methods

2.1. Feedstock and Inoculum

Fresh DM was obtained from the AAFC dairy farm located at our Sherbrooke Research and Development Center, whereas the fresh CM was sourced from a small-sized poultry farm located in Farnham (Quebec province). DM consisted of straw as bedding, whereas the bedding of CM composed of wood shavings. These bedding materials were used for the dairy cow/chicken productions by the farm itself. Hence, the manure used in this study was always contained in the bedding components. The manure was collected and stored in a cold room at 4 °C to prevent biological activity prior to feeding. For the feedstock characterization, manure was diluted and ground primarily to reduce the feed concentration for the analysis purpose and the homogenization of the solid samples, except for the TS and volatile solid (VS) analysis. The liquid inoculum used in the start-up phase was obtained from our ongoing laboratory-scale liquid sequencing batch AD, adapted to high-ammonia content chicken manure leachate. The summary of the feedstock and inoculum materials used is shown in Table 1.

Table 1. Summary of the materials used.

	Cycle 1	Cycle 2
Total weight of feedstock treated	7 kg (CM + DM)	4.7 kg (CM + DM)+ 4.7 kg (Dry inoculum from cycle 1)) = 9.4 kg
Quantity increment (%) per cycle	-	34% w/w
Mix ratio (CM:DM)	1:1	1:1
Volume of liquid inoculum	25 L	25 L
Solid substrate: liquid inoculum digester volumetric ratio	1:3.6	1:2.6
OLR (gVS/L.d) *	3.7	4.7

* OLR calculations were done based on the raw feedstock VS, and the formula used was OLR = VSi * (Q/V), where OLR: organic loading rate (g VS/L.d); VSi: VS of feedstock (CM + DM) in g/L; Q: quantity/flow rate of raw feedstock in kg or L/d; V: volume of the HSAD in L.

2.2. Experimental Setup of Two-Stage (Liquid–Solid) Anaerobic Digesters

The experimental arrangement consisted of two-stage (liquid–solid) anaerobic digesters (i.e., liquid inoculum reservoir coupled with HSAD system) for processing CM + DM mixture at 20 ± 1 °C. Two sets of digesters in duplicates with a total volumetric capacity of 40 L were operated in parallel. A set consisted of 2 digesters—one for liquid inoculum reservoir named "digester A", and the other for HSAD named "digester B". Digesters A and B were kept adjacent to each other, as shown in Figure 1. Therefore, the two sets of digesters were named as digester 1 (1A + 1B) and digester 2 (2A + 2B).

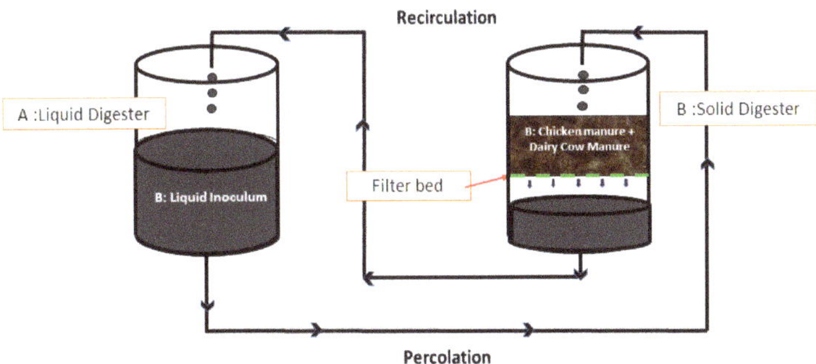

Figure 1. Schematic diagram of a single set of two-stage (liquid–solid) digesters.

The concept behind this coupled liquid–solid digesters arrangement was to enhance the digestion feasibility of the HSAD content, which was fed without any dry inoculum. Provisions made in such a way that a known volume of adapted liquid inoculum from 'digester A' was recirculated-percolated through the solid content in the HSAD (digester B), principally (i) to enhance mixing and, thus, waste-microbe interactions in 'digester B' and (ii) also to leach out a significant amount of Volatile Fatty Acids (VFA) and nitrogen from digesters 'B' to 'A'. By doing this, organic and VFA overloading in HSAD were minimized, but, at the same time, methane yield was increased since 'digester A' also contributed to producing biogas as it contained acclimatized microbes. This conception also aimed to increase the buffering capacity of the digesters by maintaining optimum pH and alkalinity in 'digester B'. Similarly, the liquid inoculum 'digester A', which was less in organic matter (Table 2), got fed and charged from 'digester B', aiding in additional methane production.

Table 2. Operating conditions of mono-digestion (CM) and co-digestion (CM + DM).

	CM (C1)	CM + DM (C1)	CM (C2)	CM + DM (C2)
Cycle length (retention time or treatment period)	70	112	85	78
Quantity of raw manure treated (kg)	5.4	7	6.5	4.7
Total volume of HSAD (L)	60	40	60	40
Total amount of solid material treated in HSAD (kg)	10	7	10.8	9.4
Total volume of liquid digester (L)	60	40	60	40
Active volume of liquid digester (L)	25			
Quantity and frequency of liquid inoculum percolated-recirculated	5L-thrice a week			
Mode of operation	Batch			
Temperature (°C)	20 ± 1			
OLR (gVS/L.d)	4.3	3.7	4.6	4.7

OLR = organic loading rate; CM = chicken manure; DM = dairy cow manure; C1 = cycle 1; C2 = cycle 2.

The digesters (A and B) were fit with the biogas pipeline to the tip tank for the release and quantification of the biogas produced. Digester A was connected with 3 additional pipelines; first one was connecting A and B; second, was linked to the pump for mixing. Mixing was done (just in digester A) every day for 5 min, mainly to homogenize the liquid content since it received leachate from digester B and also to release the space for air bubbles trapped in the anaerobic digesters. Similarly, the third one was connected to B, for recirculating the liquid inoculum from B to A. The first and the third pipe connections were responsible for percolation and recirculation of liquid inoculum. Five liters of inoculum from digester 'A' were recirculated to digester 'B' and then percolated back from digesters 'B' to 'A', thrice a week.

Embracing this set-up, altogether two batch feeding operations were conducted one after the other immediately; hence, they are named as "cycle 1" and "cycle 2", which represents retention time or treatment duration corresponding to each feeding. Cycle 1 was conducted for 112 days, while cycle 2 was conducted for 78 days only. The operation time or cycle length was mainly dependent upon the desirable methane concentration, methane yield, and VFAs accumulations. CM and DM were mixed in 1:1 (*w/w*) ratio for two reasons: (i) to operate the digester with TS of around 50% (instead of about 70% in CM); (ii) to maintain Total Chemical Oxygen Demand (COD_t)/TKN ratio in the range of 30. However, further study is underway in order to optimize several operating parameters, including COD_t/TKN ratio, as our prime aim is to operate at high ammonia levels. As presented in Table 1, a total of 7 kg of mixed manure was fed to the (HSAD) digester (cycle 1 operation). For cycle 2, about 4.7 kg of digested material resulted from cycle 1 was retained in the HSAD as a source of dry inoculum. This was done in order to reduce the start-up period by supplying adopted active microbes for the subsequent (batch) feeding. Our motive was to operate at short cycle length and to maintain a similar volumetric loading rate. Henceforth, about 4.7 kg of mixed (CM + DM) manure (refer Table 1) was mixed to the retained dry inoculum (i.e., 4.7 kg) and fed to the HSAD in order to have the substrate:dry inoculum ratio (*w/w*) close to 1:1. Once the stabilization occurred, substrate:dry inoculum ratio would be increased to accommodate more feedstocks for commercial benefits.

It is to be noted that the liquid inoculum used in cycle 1 was adapted to CM leachate with 5500 mg TKN/L. Since the adapted inoculum was not exposed to DM, longer retention time was required for cycle 1 operation to develop an acclimatized inoculum for cycle 2 operation. A volume of 25 L liquid inoculum was fed in the individual liquid digesters in both cycle 1 and cycle 2. The substrate to liquid inoculum digester volumetric ratio was maintained between 1:3.6 and 1:2.6 for cycles 1 and 2, respectively. The solid content of the mix was initiated with approximately 48% TS in cycle 1 and 51% TS in cycle 2.

A similar experimental set-up was used for CM mono-digestion. Two operational cycles (70-d for cycle 1 and 85-d for cycle 2) were conducted in order to have a performance comparison. Mono-digestion of CM was processed with the 65–73% TS, 4.3–4.6 gVS/L.d, and the co-digestion (CM + DM) was treated with 48–51% TS, 3.7–4.7 gVS/L.d. Two cycles of different cycle lengths, depending upon the consumption of VFAs, methane quality, and digester's stability factors, were carried out. The operating conditions of all the four processes (CM(C1), CM + DM (C1), CM(C2), and CM + DM(C2)) are shown in Table 2.

2.3. Analytical Methods

The bio-digesters were operated in a batch mode; therefore, the operational physio-chemical parameters were examined only for the liquid digesters on a weekly basis in order to assess the performance of the two-stage digesters. About 100 mL liquid samples were withdrawn from the liquid inoculum reservoir for the physiochemical analysis, whereas samples from the HSAD system was only taken twice *viz.* at the beginning and the end of operation since the HASD was not having weekly sampling provisions. Overall, 290 samples for biogas and 80 samples for physiochemical tests were taken during 190 days of the entire process of CM + DM. For CM alone, 240 gas samples and 50 samples were taken during 155 days of operation.

2.3.1. Biogas Analysis

The biogas production and its composition were checked for both A and B digesters on alternative days. The biogas samples were analyzed thrice a week (weekends not included from all the 4 bio-digesters (1A, 1B, 2A, and 2B)), and the volume of biogas was monitored every day using the wet tip gas meters. Methane concentration in the biogas was analyzed using a gas chromatograph (Micro GC 490, Agilent Technologies, Santa Clara, CA, USA) equipped with a thermal conductivity detector (TCD) and Helium gas as the carrier gas at a flow rate of 20 mL/min. The injector and oven temperatures were 110 °C and 180 °C, respectively.

2.3.2. Physiochemical Analysis

All the other samples were analyzed for the tests like pH, alkalinity, total solids (TS), volatile solids (VS), total COD (TCOD), soluble COD (CODs), TKN, ammonia nitrogen, and volatile fatty acids (VFAs). Along with this, TS and VS on a dry weight basis were determined following the guidelines given by the standard methods [16]. pH was determined by using pH Mettler Toledo AG 8603, SevenMulti (Schwerzenbach, Switzerland). Alkalinity was measured using Hach Lagne Sarl, Titralab AT1000 Series (Hach, Switzerland). COD was measured by using a closed reflux colorimetric method [16]. TKN and NH_3-N were analyzed using a 2460 Kjeltec Auto-Sampler System (FOSS, Sweden) following the macro-Kjeldahl method [16]. VFA was determined using a Perkin Elmer gas chromatograph, model Clarus 580 (Perkin Elmer, Shelton, CT, USA), mounted with a DB-FFAP high-resolution column, but before the evaluation of VFAs, samples were conditioned according to the procedures mentioned by Masse et al. (2003) [17]. Samples collected from digesters were first centrifuged at 41× *g* for 15 min and filtered through a 0.22 µm membrane before injected. The injection volume was 0.1 µL.

3. Results and Discussion

3.1. Characteristics of the Feedstock and the Inoculum

The characteristics of the inoculum and feedstocks are shown in Table 3. DM had low carbon in terms of CODt (~65% less) and nitrogen content in terms of TKN (~70% less) than CM, which complemented the DM to achieve a desirable nutrient content in the system for AD of CM + DM. The COD_t/TKN ratio of the CM + DM mixture in this study was around 30, which is considered as optimum value, as reported in [18]. However, for inoculum, this ratio was low in the range of 2–3, as it was acclimatized using high ammonia content wastes. The pH of CM, DM, or CM + DM mixture was always above 7.5, although high VFA concentrations of 11.6 g/L were detected for CM, mostly because of the high amount of alkalinity in the respective manures (Table 3). The biodegradability of CM, DM, and CM + DM mixture was generally higher (i.e., VS/TS = 86–89%).

Table 3. Characteristics of feedstock and inoculum.

Parameter	Cycle 1				Cycle 2			
	CM	DM	Inoculum	CM + DM	CM	DM	Inoculum	CM + DM
pH	8.68	7.58	7.86	8.2	8.88	8.13	8.37	8.1
COD_t (mg/L)	568,017	208,433	7121	405,534	565,885	188,341	5968	402,921
COD_s (mg/L)	114,768	44,852	4415	94,044	111,545	34,017	3915	96,944
Alkalinity (as mg/L $CaCO_3$)	33,282	13,932	13,313	12,649	30,486	11,126	9575	-
TS (%)	65	23.9	1.28	48	73	21.58	1.02	51
VS (%)	56	21.3	0.54	42	65	19.23	0.40	45
TKN (mg/L)	21,962	6749	3151	13,613	23,072	5194	2359	13,472
NH_3 (mg/L)	6070	1389	2732	3470	7229	1795	2117	-
TVFA (mg/L)	11,588	6973	24	10,582	10,914	6499	116	-
CODt/TKN	25.8	31	2	30	25	36	3	30

3.2. Influence of Operational Parameters in the Two-Stage AD Process Treating CM + DM Mixture

Two-sets of two-stage (liquid inoculum reservoir coupled with HSAD) AD digesters, treating CM + DM mixture, were operated for a total period of 190 days, in which cycle 1 was operated for 112 days (i.e., day 0–112), and then cycle 2 was done for 78 days (i.e., from day 113–190). Digester's performance was monitored by a wide range of several physicochemical parameters listed under Section 2.3.2, in order to develop a start-up solid-state AD protocol, using adapted liquid inoculum as a microbial source. Operational parameters, such as Organic Loading Rate (OLR), cycle length/treatment period, operating temperatures, recirculation-percolation rate and frequency, and the mode of operation, were controlled as they have a direct influence on the performance of the two-stage AD process. In addition to this, the effect of ammonia concentrations on the digester's performance was also given priority.

3.2.1. Performance of the Two-Stage AD at Different Cycles and OLRs: Biogas and Methane Production and Digester Buffering Indicators

The task of liquid inoculum reservoirs was not just limited to the dilution of solid digesters organic content or to supply active microbes but also played a vital role in providing signals of the ongoing metabolic activity in the HSAD. The indications from liquid digesters assisted in taking the required actions prior to the possible inhibitions that could occur in the system. Liquid digesters also participated in the generation of biogas in addition to the HSAD with a supply of new feed from each time the leachate was recirculated.

Figure 2a–c depict the biogas and methane production profiles and their yield along with the digester buffering indicators (pH and alkalinity). For cycle 1 (days 0–112) operation, the OLR was maintained at 3.7 g.VS/L.d, and the corresponding volumetric combined (liquid + HSAD) biogas production recorded was 9.4 ± 3.7 L/d (Figure 2a). Whereas for cycle 2 (days 113–190) operation, OLR was increased to about 4.7 g.VS/L.d, and the corresponding volumetric combined biogas production was observed to be 7.7 ± 1.8 L/d. The cumulative biogas volume was found to be stable in both cases. As far as the methane concentration in the biogas was concerned, during the cycle 1 operation, it took about 82 days, especially for the HSAD to reach 50% CH4, whereas, in cycle 2, it took only 42 days to attain the same value. Interestingly, methane content in the liquid inoculum reservoirs remained always higher for both the cycles (Figure 2b), which demonstrates that the process offered excellent quality of biogas, which remained fairly steady (70–75%) at the end of each cycle. High methane content also suggested that the methanogenic population in the liquid inoculum reservoirs was enhanced for this substrate (CM + DM mix). It is to be noted that a combined (liquid + HSAB) methane concentration at the end of each cycle had reached to about 70%.

Specific methane yield (SMY) is a parameter that quantifies the amount of methane generation per gram of the organic matter, such as VS or COD. The average SMY was reached to about 0.33 LCH₄/gVS at the end of cycle 1 operation (i.e., on day 112), whereas a similar result was obtained within 78 days in cycle 2 (Figure 2c). In addition to this, the degradation of organic matter in terms of CODt and VS was monitored. CODt and VS reductions of about 60% and 59%, respectively, were observed at the end of cycle 1. Whereas at the end of cycle 2, CODt and VS removal efficiencies were increased to about 76% and 62%, respectively, even at a shorter cycle length.

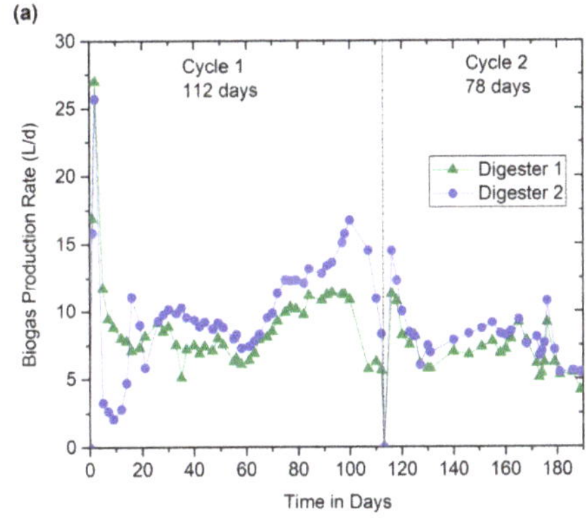

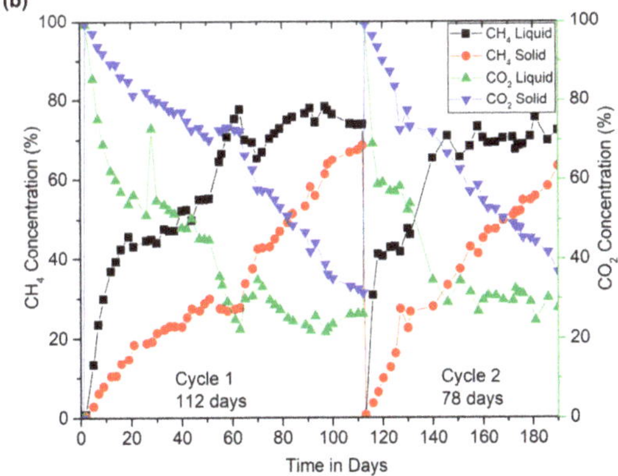

Figure 2. *Cont.*

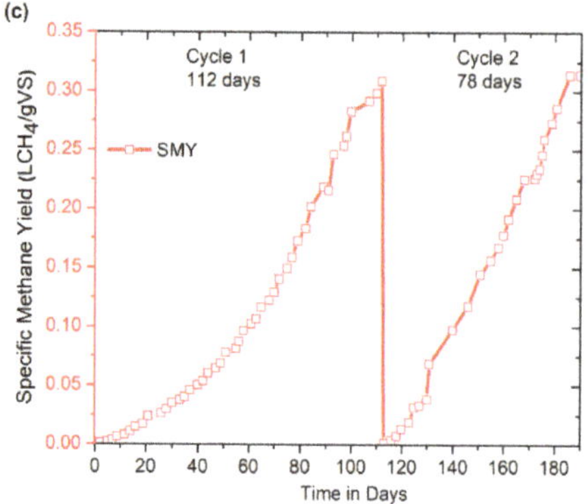

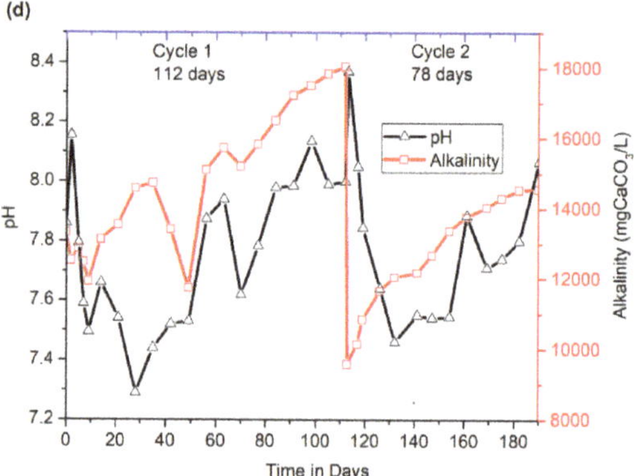

Figure 2. Performance of the liquid and solid digesters at different organic loading rates (OLRs) during 190 days of operation. (**a**) Biogas production rate; (**b**) Biogas composition; (**c**) Specific methane yield (SMY); (**d**) pH and alkalinity profiles.

From Figure 2d, at a pH range of 7.2–8.4, the alkalinity reached up to 18 g/L in cycle 1 operation and 14 g/L in cycle 2, respectively. A slight change in pH generally could affect the methanogenic activity in AD. However, abrupt changes in pH are balanced by sufficient alkalinity (buffering capacity). Generally, alkalinity generated in the AD system itself controls the system, which is assisted by high protein or nitrogen content in the manure. The levels of VFA remained low (total content below 900 mg/L) at the end of both, indicating high reactor stability, which was confirmed by the presence of more neutralized pH and higher alkalinity values within the digester. There was no sign of inhibition or nutrient deficiency at these operating conditions. The detailed results pertaining to the VFA concentrations are discussed in subsequent sections.

3.2.2. Performance Monitoring of Digesters: Correlation between VFAs, pH, and Methane Concentration

Figure 3 shows the correlation between pH and Total VFA (TVFA). VFAs are the intermediate products in the AD process, and their accumulation is advised to be avoided. The concentration of VFAs is one of the important parameters for the AD process as the increase in VFA indicates the initiation of the acidogenic phase; however, the rapid increase is a sign of inhibition of microorganisms responsible for methanogenesis. Fluctuations in VFA concentration change the pH with the change in hydrogen (H^+) ions released during the breakdown of organic matter. Maintaining optimal pH is a must for the survival of varieties of microorganisms playing a role in continuing the process without inhibition. The optimal pH range regarded is 6.8–7.2 for both acidogenic and methanogenic bacteria [19]. The pH range in this study was 7.2–8.4, with occasional fall and rise. The initial decrease in pH means the start of acidification, and a sudden increase indicates the termination of acidification at that point. The growth rate of methanogens is slower than the acidogens; therefore, methanogens require longer retention time than the acidogens, in order to consume the VFAs and produce methane-rich biogas. The extraction of VFAs is also possible by providing longer retention time and can be achieved with a batch mode of operation [20]. Low pH leads to the accumulation of acetic acid and hydrogen, which inhibits the degradation of propionic acid and ultimately accumulating VFAs. In the cycle 1 of this study, TVFA production went highest to 15 g/L (42-d), in which 10 g/L was acetic acid, and propionic acid was below 3 g/L. However, the case differed in the next cycle, which only generated a maximum of 4.5 g/L TVFAs (Figure 3). The reason behind comparatively low VFAs could be due to the amount of fatty acids, which declined rapidly due to an appreciable amount of methanogens generated from the previous cycle.

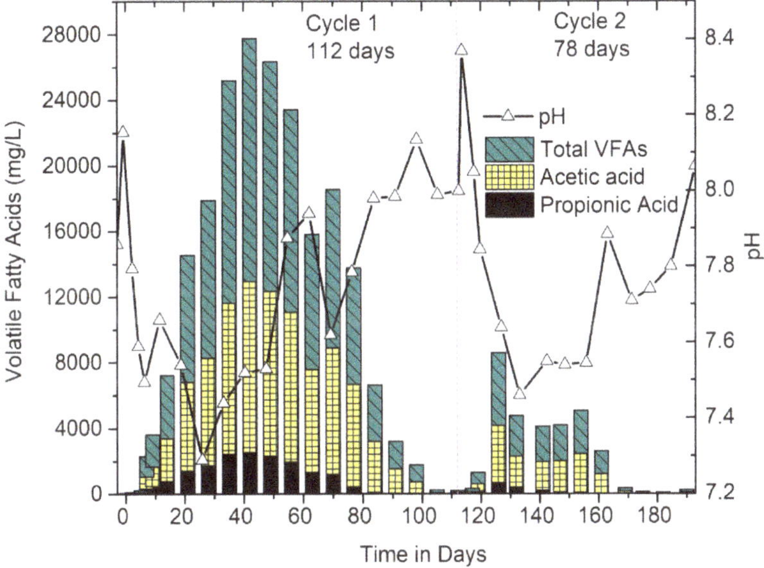

Figure 3. Correlation between volatile fatty acids (VFAs), pH, and methane concentration.

As shown in Figures 2b and 3, between days 51 and 63 of cycle 1, the methane quality was observed to be decreasing (29% on 51-d to 27% on 63-d) in the HSAD digester. Although the decrease was not significant, this could be due to the possibility of scarcity in methanogenic population; therefore, 10 L additional liquid inoculum was recirculated-percolated from liquid inoculum reservoir (digester A) to HSAD (digester B), which then increased the methane concentration due to the increase in their biomass activity. This also facilitated in leaching out the accumulated VFAs from HSAD to the liquid inoculum reservoir for further degradation. Henceforth, after day 65, a significant improvement in methane quality (approximately 45%) in HSAD and a rapid reduction in VFAs (8500 mg/L (66-d) to <200 mg/L (112-d)) in liquid inoculum reservoir were noticed (Figure 3). As far as the cycle 2 operation was concerned, comparatively less feed material was fed along with 50% (*w/w*) of the digested material (considered as dry inoculum) from the 1st cycle, which helped to shorten the cycle length with an enriched methane concentration over 50% with minimal VFA accumulations. However, as far as the OLR was concerned, due to the residual COD or VS accumulations from the digested material from cycle 1 operation, there was a slight increase compared to that of cycle 2. In this scenario, the better performance is mostly linked to the adapted microbial populations within the (liquid–solid) system interactions.

3.2.3. Performance Monitoring of Digesters: Ratio Limits

Monitoring the AD process requires proper selection of operational parameters depending upon its metabolic state. The parameters like pH, total alkalinity (TA), temperature, TVFA, and C/N ratio are important as they have a direct influence on the performance of the AD system. The proper understanding of these fundamental parameters and its implementation can exploit the AD to the fullest and avoid the inhibitions that can occur in certain conditions. One of the major factors, which expresses the stability of AD, is the ratio of TVFA/TA, which is reported to be less than 0.5 for high stability [21] and regarded optimal between 0.4 and 0.6 by [22], beyond which is indicated as overfed. Therefore, the profile of these parameters in the digesters during the operation is shown in Figure 4.

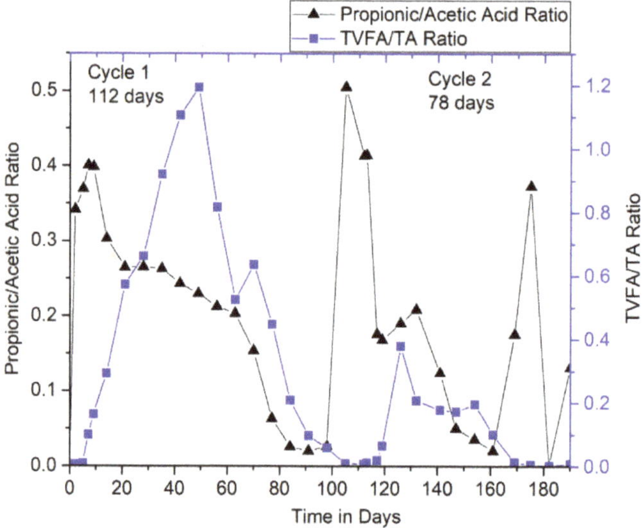

Figure 4. Evolution of TVFA/ TA and propionic acid/acetic acid ratios.

Ratio limits like TVFA/TA and propionic acid/acetic acid ratio are the key critical indicators of digester's crash. Studies suggest TVFA/TA to be less than 1 (preferably within a range of 0.1 and 0.6) [18,19] and propionic acid/acetic acid ratio to be less than 1.4 [23] for the high stability of the digester. In this study, the TVFA/TA ratio remained below 1 in the entire operation period, except for days between 40 and 50 of cycle 1 when it reached up to 1.2. This was an indication of inhibitions due to the lack of active microbes in HSAD. These results were in accordance with the methane concentration profiles (HSAD) (Figure 2b), and henceforth, about 10 L of liquid inoculum was recirculated-percolated to HSAD to overcome this situation. However, the propionic acid/acetic acid ratio reached only up to 0.5 throughout the operation (Figure 4). These indicators, demonstrating that the digesters were operating favorably without the risk of acid-buildup and better stability, attained with time.

3.2.4. Performance Monitoring of Digesters: Relationship between pH, Ammonia (TAN, FAN), and Temperature

The parameters in the AD process are inter-related; thus, an optimal pH and low temperature coupled with high alkalinity balanced the yield of free ammonia in this system. In Figure 5, FAN concentration was shown to be under 200 mg/L in cycle 1 and lower than 180 mg/L in cycle 2. TAN was 3.7 g/L at maximum in the 1st cycle; however, in the 2nd cycle, it was only 3.1 g/L. Generally, exceeding TAN results in the reduction of methane concentration and biogas production. This study, hence, justified the increase in methane-rich biogas with the reduction of TAN.

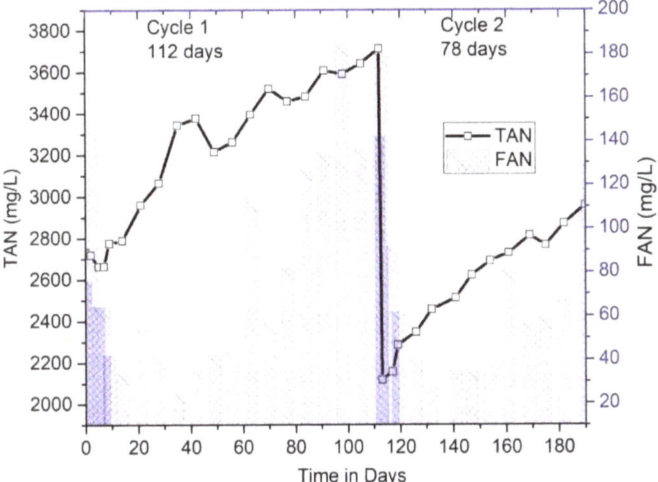

Figure 5. Evolution of total ammonia nitrogen (TAN) and free ammonia nitrogen (FAN) profiles.

Around 35–40% of the increase in total ammonia nitrogen (TAN) was observed from the commencement of the operation, and a considerable increment of free ammonia nitrogen (FAN) was also observed.

Ammonia has a significant role in supplying nutrients for microbial growth, maintaining buffering capacity (alkalinity) and stability of the digester. Ammonia is dependent on pH, temperature, alkalinity, and substrates. Ammonia exists in two forms: (i) free ammonia (NH_3) and (ii) ammonium (NH_4). Free ammonia is a gas and toxic, and ammonium is in ionized form, which is non-toxic salt. NH_3 or free ammonia nitrogen (FAN) and NH_4 together make total ammonia nitrogen (TAN). FAN takes part in the inhibitory actions in the AD process since the high concentration of free ammonia in the system ruptures the cell wall of the microbes, leading to cell lysis. Ammonia is mainly dependent upon temperature and pH mentioned by [24] in Equation (1).

$$FAN = TAN \left(1 + \frac{10^{-pH}}{10^{-\left(0.09018 + \frac{2729.92}{T(K)}\right)}}\right)^{-1} \tag{1}$$

where the temperature is in Kelvin (K); total ammonia nitrogen (TAN) and free ammonia nitrogen (FAN) are in mg/L.

Microorganisms, which are responsible for the entire AD process, are generally sensitive and survive at certain conditions. Similarly, the temperature is one of the important factors as the growth of the microbes is higher at a higher temperature. However, AD at a temperature > 50 °C is unstable and generates high FAN, especially while treating ammonia-rich wastes, which is an inhibitor for the process itself. FAN is directly proportional to the temperature; therefore, the generation of FAN is lesser at low-temperature conditions, contributing to the fewer chances of AD inhibition. For proper microbial growth at a lower temperature, substrate acclimation at low-temperature conditions is proven to be advantageous [10]. Therefore, in this study, liquid inoculum adapted to 20 ± 1 °C was utilized for the liquid digester in a two-stage operation, which led to the lower generation of FAN of up to 185 mg/L.

This study also reported the direct proportionality of FAN with different temperatures (for instance, 20 °C, 35 °C, and 55 °C), as shown in Figure 6, based on the formula provided in Equation (1) by extrapolating the concentration of FAN. This was done to derive a theoretical conclusion based on these calculations. Under an operating pH range of 7.2–8.4, FAN at 20 °C was a maximum of 185 mg/L. Extrapolating the results for FAN at different temperatures based on Equation (1) and operating pH range showed that at 35 °C (mesophilic), FAN could have reached up to 500 mg/L, and the same could have ascended to 1300 mg/L at 55 °C (thermophilic). Therefore, a theoretical extrapolation shows the concentration of FAN to be low and balanced at lower temperatures and inhibitory at higher temperatures.

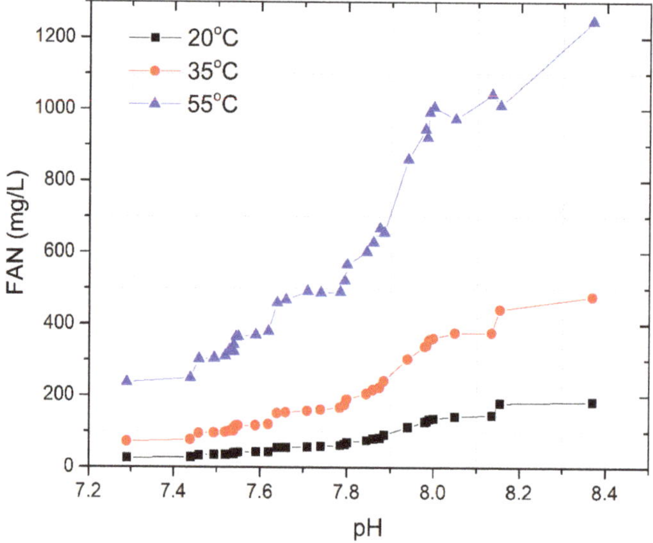

Figure 6. FAN at different temperatures under an operating pH range.

3.3. Comparative Study of Two-Stage (Liquid–Solid) AD of CM and Co-Digestion of CM + DM

The operating conditions of the two-stage (liquid–solid) AD of CM mono-digestion and CM + DM co-digestion are given in Table 2. The digesters were operated in a similar fashion in order to develop a start-up and operating strategies for these substrates. An attempt was made to compare digesters, treating these two substrates in terms of methane yield and its concentrations, and also the release of FAN during the AD processes in order to determine the inhibitory potential of ammonia in the respective feedstocks (Figure 7). The liquid inoculum used to start the digesters for both substrates were adapted to chicken manure leachate. Henceforth, the methane concentrations in cycle 1 of CM mono-digestion showed a quick start-up compared to that of CM + DM co-digestion.

In cycle 1 of both cases, the concentration of methane was approximately 58%. On the contrary to this, in the cycle 2 of CM + DM, on day 78, the CH_4 concentration was approximately 70%; however, the same for CM was around 60%, making a difference of up to 10% (Figure 7a). These results showed that co-digestion using DM had a positive effect in producing a comparatively better methane-rich biogas. However, methane yield or SMY obtained for the CM mono-digestion was 0.52 ± 0.13 L $CH_4g^{-1}VS_{fed}$, and for CM + DM co-digestion, it was 0.35 ± 0.11 L $CH_4g^{-1}VS_{fed}$ (detailed data not shown). Similarly, the volumetric biogas production was more in CM (13.6 ± 4 L/d) than CM + DM (7.7 ± 1.8 L/d); however, the quality of methane was observed to be better in the co-digestion process. Since the COD_t:TKN ratio was always higher than 25 for both CM and CM + DM mixture used in this study, better results for CM in terms of methane yield were observed as the CM has a better energy potential than DM.

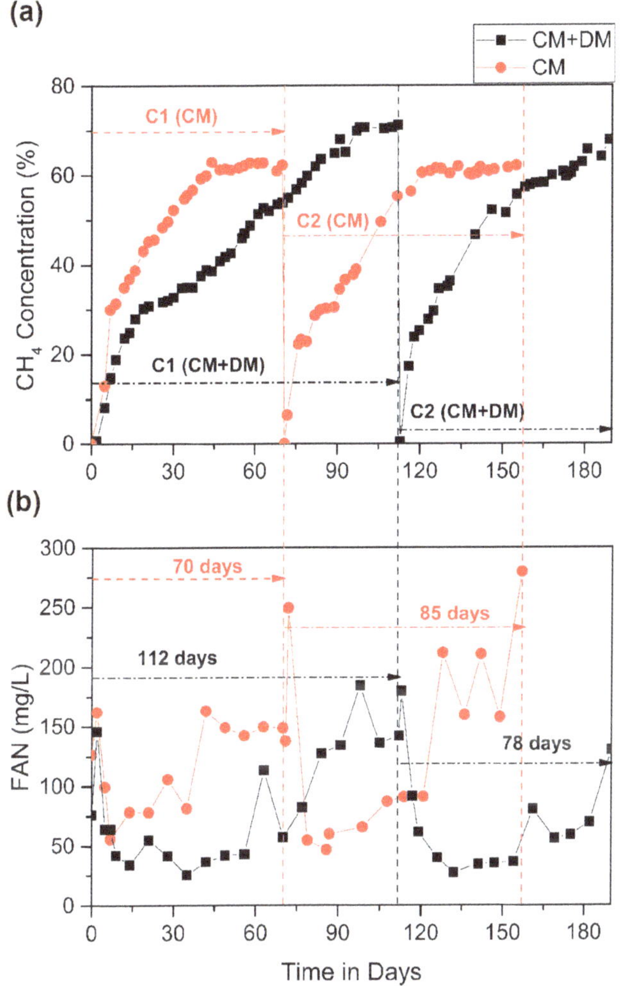

Figure 7. Comparative study of chicken manure (CM) mono-digestion and CM + dairy cow manure (DM) co-digestion: (**a**) Methane concentration profile; (**b**) FAN profile.

Furthermore, the release of FAN concentration for both the mono- and co-digestions was monitored to have a better perspective or forecasting of the ammonia inhibition (Figure 7b). It is evident from Figure 7b that FAN concentrations were comparatively lower in the co-digestion process than mono-digestion due to the dilution of higher ammonia content in CM by DM. Contribution to the generation of ammonia not only lies in the initial concentration of the feedstock but also during the biochemical process in AD. CM is high in nitrogen; hence, the initial concentration had a vital role in a higher concentration of FAN than that in CM + DM. On the 70th day of both cases, FAN was 150 mg/L in CM and 50 mg/L in CM + DM. This can be related to the high nitrogen content in CM mono-digestion than the co-digestion, which helped in the dilution of high ammonia. Although there was an increase in the FAN concentration for the mono-digestion of CM, no apparent inhibitions were reported for both the processes during the start-up phase. The probable reason was that the FAN levels were still in the tolerable range (always below 280 mg/L), and the VFA/alkalinity ratio always remained below 0.5 in both the cases, indicating that the digesters were operating favorably without the risk of acid-buildup. Thus, the presence of ammonia nitrogen did not inhibit the performance of the liquid inoculum reservoir, as well as HSAD, even at high OLRs. Even if the pH was not controlled in the bioreactors, there was no formation of foam or scum observed throughout the study. The mode of operation (process, temperature, percolation-recirculation rate, and frequency) and the appropriate choice of acclimatized inoculum at the start-up phase of the experiment allowed a high stabilization of CM + DM co-digestion, even at higher OLR (4.7 gVS/L.d) studied. Further study is underway to optimize the operating parameters, especially for the co-digestion process.

4. Conclusions

The proposed start-up study focused on two-stage (liquid inoculum reservoir coupled with HSAD) anaerobic digestion process using a closed-loop recirculation, and percolation mode operation was found efficient for the treatment of CM + DM at 20 ± 1 °C despite having a waste mix with high TKN (13.5–13.6 gN/L) and solid (TS: 48–51%) concentrations. Results showed that our system could generate a specific methane yield of 0.35 ± 0.11 L CH_4g/VS_{fed} at an OLR of 3.7–4.7 gVS/L.d. We also observed CH_4 concentrations above 60% with CODs and VS reduction by up to 85% and 60%, respectively. A comparative study was done using the same start-up protocol to perform the mono-digestion of CM (TKN: 21–23 g/L; TS: 65–73%). Although a better SMY (0.52 ± 0.13 L CH_4g$^{-1}VS_{fed}$) was obtained for mono-digestion of CM, co-digesting CM + DM showed a better methane quality and also generated comparatively lower FAN. However, no evident inhibitions due to ammonia or VFA accumulations were reported for both the processes during the start-up phase. Compared to the higher-temperature digestion process, more energy is expected to be available for farm uses, especially while treating high solids and ammonia-rich wastes.

Author Contributions: Conceptualization, R.R. and B.G.; methodology, R.R.; validation, M.S.R. and R.R.; investigation, R.R. and B.G.; resources, M.S.R.; writing—original draft preparation, R.R. and P.M.; writing—review and editing, R.R. and P.M.; supervision, R.R. and B.G.; funding acquisition, B.G. All authors have read and agreed to the published version of the manuscript.

Funding: Authors thank Agriculture and Agri-Food Canada (Project No. 2335) for providing financial support.

Conflicts of Interest: The authors declare no conflict of interest.

References

1. High Level Expert Forum—How to Feed the World in 2050. Available online: http://www.fao.org/fileadmin/templates/wsfs/docs/Issues_papers/HLEF2050_Global_Agriculture.pdf (accessed on 3 March 2020).
2. AVEC (Association of Poultry Processors and Poultry Trade in the EU Countries). Annual Report 2016. Available online: http://www.avec-poultry.eu/wp-content/uploads/2018/04/AVEC-2016-BAT.pdf (accessed on 8 June 2018).
3. Fatma, A.; Namba, N.; Kesseva, M.R.; Nishio, N.; Nakashimada, Y. Enhancement of methane production from co-digestion of chicken manure with agricultural wastes. *Bioresour. Technol.* **2014**, *159*, 80–87.
4. MacLeod, M.; Gerber, P.; Mottet, A. *Greenhouse Gas Emissions from Pig and Chicken Supply Chains*; Food and Agriculture Organization: Rome, Italy, 2012.
5. Werner, F.; Wang, X.; Gabauer, W.; Ortner, M.; Li, Z. Tackling ammonia inhibition for efficient biogas production from chicken manure: Status and technical trends in Europe and China. *Renew. Sustain. Energy Rev.* **2018**, *97*, 186–199.
6. Intergovernmental Panel on Climate Change (IPCC). *Climate Change 2001: The Scientific Basis*; Cambridge University Press: Cambridge, UK, 2001.
7. Sun, X.; Lu, P.; Jiang, T.; Schuchardt, F.; Li, G. Influence of bulking agents on CH_4, N_2O, and NH_3 emissions during rapid composting of pig manure from the Chinese Ganqinfen system. *J. Zhejiang Univ. Sci. B* **2014**, *15*, 353–364. [CrossRef] [PubMed]
8. Lynch, D.; Henihan, A.M.; Bowen, B.; Lynch, D.; McDonnell, K.; Kwapinski, W. Utilisation of poultry litter as an energy feedstock. *Biomass Bioenergy* **2013**, *49*, 197–204. [CrossRef]

9. Kelleher, B.P.; Leahy, J.J.; Henihan, A.M.; O'Dwyer, T.F.; Sutton, D.; Leahy, M.J. Advances in poultry litter disposal technology—A review. *Bioresour. Technol.* **2002**, *83*, 27–36. [CrossRef]

10. Rajinikanth, R.; Massé, D.I.; Singh, G. A critical review on inhibition of anaerobic digestion process by excess ammonia. *Bioresour. Technol.* **2013**, *143*, 632–641.

11. Suleyman, K.Y.S.; Kocak, E. Anaerobic digestion technology in poultry and livestock waste treatment—A literature review. *Waste Manag. Res.* **2009**, *27*, 3–18.

12. Matheri, A.N.; Ndiweni, S.N.; Belaid, M.; Muzenda, E.; Hubert, R. Optimising biogas production from anaerobic co-digestion of chicken manure and organic fraction of municipal solid waste. *Renew. Sustain. Energy Rev.* **2017**, *80*, 756–764. [CrossRef]

13. Singh, K.; Lee, K.; Worley, J.; Risse, L.M.; Das, K.C. Anaerobic digestion of poultry litter: A review. *Appl. Eng. Agric.* **2010**, *26*, 677–688. [CrossRef]

14. Siddique, M.N.I.; Munaim, M.S.A.; Wahid, Z.A. Mesophilic and thermophilic biomethane production by co-digesting pretreated petrochemical wastewater with beef and dairy cattle manure. *J. Ind. Eng. Chem.* **2014**, *20*, 331–337. [CrossRef]

15. Li, R.; Chen, S.; Li, X.; Lar, J.S.; He, Y.; Zhu, B. Anaerobic co-digestion of kitchen waste with cattle manure for biogas production. *Energy Fuels* **2009**, *23*, 2225–2228. [CrossRef]

16. Eaton, A.D.; Clesceri, L.S.; Rice, E.W.; Greenberg, A.E.; Franson, M.A.H.A. *APHA: Standard Methods for the Examination of Water and Wastewater*; Centennial Edition; APHA, AWWA, WEF: Washington, DC, USA, 2005.

17. Massé, D.I.; Masse, L.; Croteau, F. The effect of temperature fluctuations on psychrophilic anaerobic sequencing batch reactors treating swine manure. *Bioresour. Technol.* **2003**, *89*, 57–62. [CrossRef]

18. Esposito, G.; Frunzo, L.; Giordano, A.; Liotta, F.; Panico, A.; Pirozzi, F. Anaerobic co-digestion of organic wastes. *Rev. Environ. Sci. Bio Technol.* **2012**, *11*, 325–341. [CrossRef]

19. Khanal, S.K. *Anaerobic Biotechnology for Bioenergy Production: Principles and Applications*; John Wiley & Sons: Hoboken, NJ, USA, 2011.

20. Li, H.L.; Guo, X.L.; Cao, F.F.; Wang, Y. Process evolution of dry anaerobic co-digestion of cattle manure with kitchen waste. *Chem. Biochem. Eng. Q.* **2014**, *28*, 61–166.

21. Rajagopal, R.; Ramirez, I.; Steyer, J.P.; Mehrotra, I.; Kumar, P.; Escudie, R.; Torrijos, M. Experimental and modeling investigations of a hybrid upflow anaerobic sludge-filter bed (UASFB) reactor. *Water Sci. Technol.* **2008**, *58*, 109–117.

22. Brown, D.; Li, Y. Solid state anaerobic co-digestion of yard waste and food waste for biogas production. *Bioresour. Technol.* **2013**, *127*, 275–280. [CrossRef] [PubMed]

23. Ehimen, E.A.; Sun, Z.F.; Carrington, C.G.; Birch, E.J.; Eaton-Rye, J.J. Anaerobic digestion of microalgae residues resulting from the biodiesel production process. *Appl. Energy* **2011**, *88*, 3454–3463. [CrossRef]

24. Hansen, K.H.; Angelidaki, I.; Ahring, B.K. Anaerobic digestion of swine manure: Inhibition by ammonia. *Water Res.* **1998**, *32*, 5–12. [CrossRef]

Article

Influence of Enzyme Additives on the Rheological Properties of Digester Slurry and on Biomethane Yield

Liane Müller [1], Nils Engler [1,*], Kay Rostalsky [2], Ulf Müller [1], Christian Krebs [1] and Sandra Hinz [3]

[1] DBFZ Deutsches Biomasseforschungszentrum gemeinnützige GmbH, Biochemical Conversion Department, Torgauer Straße 116, 04347 Leipzig, Germany; liane.mueller@dbfz.de (L.M.); ulf.mueller@dbfz.de (U.M.); christian.krebs@dbfz.de (C.K.)

[2] Repowering Technik Ost GmbH, Reideburger Str. 43, 06116 Halle (Saale), Germany; k.rostalsky@repowering-technik-ost.de

[3] DUPONT/ Genencor International BV, Archimedesweg 30, 2333 CN Leiden, The Netherlands; sandra.hinz@dupont.com

* Correspondence: nils.engler@dbfz.de; Tel.: +49-341-2434389

Received: 8 April 2020; Accepted: 2 June 2020; Published: 4 June 2020

Abstract: The use of enzyme additives in anaerobic digestion facilities has increased in recent years. According to the manufacturers, these additives should increase or accelerate the biogas yield and reduce the viscosity of the digester slurry. Such effects were confirmed under laboratory conditions. However, it has not yet been possible to quantify these effects in practice, partly because valid measurements on large-scale plants are expensive and challenging. In this research, a new enzyme product was tested under full-scale conditions. Two digesters were operated at identic process parameters—one digester was treated with an enzyme additive and a second digester was used as reference. A pipe viscometer was designed, constructed and calibrated and the rheological properties of the digester slurry were measured. Non-Newtonian flow behavior was modelled by using the Ostwald–de Baer law. Additionally, the specific biomethane yield of the feedstock was monitored to assess the influence of the enzyme additive on the substrate degradation efficiency. The viscosity measurements revealed a clear effect of the added enzyme product. The consistency factor K was significantly reduced after the enzyme application. There was no observable effect of enzyme application on the substrate degradation efficiency or specific biomethane yield.

Keywords: anaerobic digestion; enzyme application; rheology of digestate

---

1. Introduction

Anaerobic digestion (AD) is an established technology to gain energy from biogenic materials, agricultural residues and biowaste. By end of 2016, there were approximately 8700 biogas plants in operation in Germany [1], and the total number in Europe exceeds 18,200 [2].

The majority of biogas digesters currently operating in Europe are designed as continuously stirred tank reactors (CSTR). This reactor type requires more or less permanent homogenization and is therefore equipped with agitators. A broad variety of mixer types and mixing techniques have been developed and implemented, depending on the reactor size, feedstock and operation procedure. Nevertheless, mixing is still an issue and technical problems with mixers and stirrers came second in a survey of technical problems at biogas facilities in Germany [3,4].

Studies [5,6] have shown that the electricity required for agitation accounts for approximately 30–50% of the total electricity consumption of a typical biogas plant. This is equivalent to 5–9% of the electricity produced. Optimizing the efficiency of stirring and the minimizing energy demand of stirrers is therefore an important issue for the further optimization of the efficiency of biogas digesters.

Additionally, a correct determination of hydraulic retention time (HRT), an important process parameter, assumes that the digester volume is used at its full capacity. Incomplete homogenization due to improper mixing leads to dead zones inside the digester and hence to a reduced working volume and, as a consequence, to a reduced HRT. Therefore, proper stirring is fundamental for this type of reactor in order to achieve high biogas and methane yields as well as process stability [7].

A number of additives to optimize the anaerobic digestion process have been developed in recent years. Amongst them are enzyme mixtures explicitly designed to improve the biogas yield and affect the slurry rheology in a positive way. Producers claim that these additives can improve the methane yield and substrate degradation efficiency and lead to a reduced energy demand for mixing and pumping. Reliable data on the effectiveness of enzyme application in full-scale biogas digesters are however scarce, not least because measurements under practical conditions are very difficult.

Due to the non-Newtonian rheology of the digester slurry, a simple determination of the viscosity is not possible and more sophisticated approaches are necessary. In particular, the equations developed by Ostwald/de Waele and Herschel–Bulkley are commonly used to describe the rheological characteristics of shear-thinning fluids, such as sewage sludge or digester slurry [8]. Although the Ostwald–De Waele model is not meaningful from a physical point of view, (apparent viscosity goes towards zero for very high shear rates), it can provide a good agreement with practice for limited and defined ranges of validity. Moreover, the coefficients required for this model can be well determined experimentally. The Ostwald/de Waele approach is therefore widely used [9,10].

Research on the rheology of digester slurry has gained more attention in recent years. Many authors have worked on the characterization of AD digester slurry. The works of Baudez et al. [11,12] focused on sewage sludge and revealed that for different flow regimes (which means different shear rates), different models are suitable to describe the rheology. Other authors studied the influence of origin of digester slurry on rheological parameters. According to these findings, the rheology of digester slurry strongly depends on its origin and several process parameters, such as total solids content (TS), particle size and particle size distribution [13]. Furthermore, the type and configuration of the measurement device are of great importance because digester slurry is a non-Newtonian three-phase fluid containing liquid, gas bubbles and solid particles, partly in the form of longer fibres. Rotational viscometers in a classical cup-and-bob configuration have been used by some authors [11]. These devices turned out to be less suitable when the slurry contains larger particles or longer fibre structures. However, the removal of the interfering particles alters the sample in a way that is not permissible [14]. Agitator viscometers have been used by other authors facing more or less the same problems [15]. Misleading or inconsistent results are possible due to inappropriate measurement and experimental setup as well as interpretation of data [16]. There is still a lack of reliable, validated and practically applicable methods to determine the rheological properties of digester slurry.

The effect of enzyme supplementation on the AD process is controversially discussed. The role of enzymes in ruminal digestion processes is quite clear, and a significant fibre disintegration effect has been scientifically proven [17]. However, the process conditions—in particular the pH and microbial community—in the bovine digestive tract are very different from those in biogas digesters. Investigations into the stability of enzymes in AD facilities showed a very fast degradation and implied a neglectable effect [18]. An improved degradation and solubilization of lignocellulose material was observed; this, however, did not lead to an increase in the specific methane yield [19]. Other authors [20–22] have reported a significant increase in the methane yield of energy crops by enzyme application in lab digesters. Although the effect of enzyme application on the substrate degradation efficiency is still matter of discussion, plant operators frequently confirm positive overall effects, like the improved flow behaviur of the digester slurry, reduced floating layers and reduced sedimentation. However, they have been without scientific evidence until now.

This work aims at obtaining reliable measurement results on the influence of enzyme application on the biogas yield and viscosity of digester slurry under the conditions of AD plant operation in practice.

2. Materials and Methods

2.1. Anaerobic Digestion Tests

The experiments were carried out at a full-scale biogas research plant situated at Deutsches Biomasseforschungszentrum gemeinnützige GmbH (DBFZ) in Leipzig. The research biogas plant comprises two continuously stirred tank reactors (CSTR) with a gross volume of 208 m^3 each. They are equipped with a common feeding system that allows adding solid feedstock via a solid feeder, as well as liquid feedstock from a separate storage tank of 174 m^3. The standard operation mode is at a mesophilic temperature (39 °C). The plant is equipped with extensive measuring technology for all the relevant process parameters, such as feedstock mass flow, process temperature, gas flow and gas composition. A schematic plant design is shown in Figure 1.

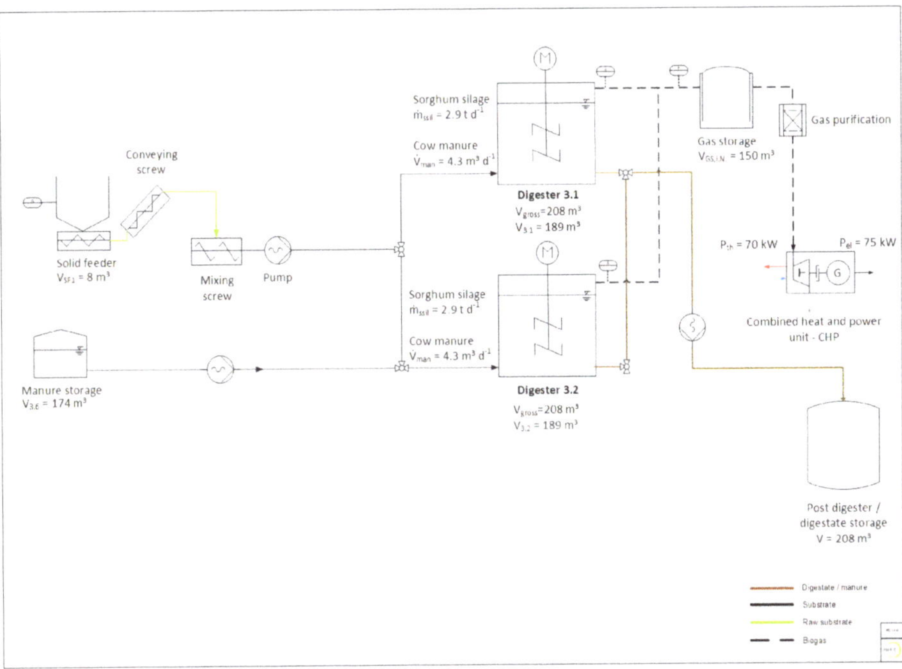

Figure 1. Scheme of the research biogas plant at Deutsches Biomasseforschungszentrum (DBFZ).

In order to assess the influence of enzyme application, both digesters of the research plant were operated in parallel mode. The operating volume was 189 m^3 for each digester during the experiment. The feedstock was a mixture of sorghum silage and cattle manure (40%/60% mass, based on fresh matter). The cattle manure originated from a cattle farm located near Leipzig. Ten batches of cattle manure were delivered during the trial period. Due to unforeseen shortages, it was necessary to change the supplier for the sorghum silage during the trial period. The silages with different origins are identified below as Sorghum silage I and Sorghum silage II. Four batches of Sorghum silage I and five batches of Sorghum silage II were used for the experiment.

Digester 3.1 received the feedstock mix described above and additionally the enzyme additive described in Section 2.2. Digester 3.2 was used as a reference and therefore received the same substrate mixture without the enzyme additive. The organic loading rate was maintained at 5.0 kg$_{VS}$/(m^3·d)$^{-1}$ for both digesters, resulting in an HRT of 25 days on average. After a pre-period of 35 days (1.5× HRT), the enzyme application started in digester 3.1 and was maintained over 97 days (approx. 4× HRT).

Subsequently, a phase-out period of 28 d (1× HRT) followed. All the described process parameters were monitored over the entire trial period. The feedstock mass flow, process temperature, gas flow and gas composition were controlled online. The process parameters total solids (TS), volatile solids (VS), pH, volatile fatty acids (VFA), alkalinity and ammonia were controlled in order to evaluate the process stability.

2.2. Enzyme Product

The enzyme product used in this study is a formulation developed within the joint research project *DEMETER*—demonstrating more efficient enzyme production to increase biogas yields [23]. This research project aimed to develop an efficient enzyme production process and demonstrate effects of enzyme application in a full-scale operation. The enzyme product contains cellulases and xylanases, which all originate from the strain Myceliophthora thermophila C1. The dosage carried out according to the manufacturer's specifications was 0.45 g per kg_{VS} of substrate.

2.3. Analytic Methods

The feedstock and digester slurry were regularly analyzed. The total solids (TS) and volatile solids (VS) were analyzed according to the European standard DIN EN 15934 and DIN EN 15935. The Weender method was used to determine the crude fat (XL), crude protein (XP) and crude fibre (XF). The process stability parameters were measured two times per week. The VFA and alkalinity were determined through titration using an automatic titration system, Mettler Toledo Rondo 60/T90 (Mettler-Toledo gmbH Gießen, Germany). The ammonia nitrogen was analyzed in the liquid phase of the digester slurry by the photometric method (Spectrophotometer DR 3900, Hach Lange GmbH, Germany). All the methods mentioned above are described in detail in a collection of measurement methods for biogas [24]. The specific methane potential was measured in batch tests in lab scale (1 L) using the Automatic Methane Potential Test System (AMPTS) (Bioprocess control, Sweden). Tests were carried out at a mesophilic temperature (38 °C). The test protocol followed the German guideline VDI 4630, which implements the recommendations of a European expert commission [25]. The feedstock analyses are shown in Table 1.

Table 1. Feedstock analyses of the substrates used.

Substrate	TS	VS	Crude Fat	Crude Protein	Crude Fibre	CH_4-Potential
	%	%TS	$g·kg^{-1}$ TS	$g·kg^{-1}$ TS	$g·kg^{-1}$ TS	$L·kg^{-1}$ TS
Sorghum silage I (n = 4)	28 ± 3	95 ± 1	20 ± 3	71 ± 7	436 ± 24	378 ± 8 (n = 3)
Sorghum silage II (n = 5)	1 ± 2	92 ± 1	27 ± 4	104 ± 10	412 ± 26	350 ± 7 (n = 3)
Cattle manure (n = 10)	9 ± 1	76 ± 1	37 ± 8	160 ± 18	259 ± 20	241 ± 9 (n = 3)

2.4. Pipe Viscometer

A pipe viscometer (see Figure 2) was developed over the course of the project and used to assess the rheological properties of the digester slurry. The slurry is drawn from one of the digesters and stored in a tempered reservoir tank (V = 1000 L). A progressive cavity pump with a controlled speed is used to convey the slurry into the viscometer, which consists of an inlet zone to obtain the laminar flow and a metering section equipped with two pressure sensors mounted at a distance of 8 m, followed by an outflow zone. Three different pipe diameters (65, 80 and 100 mm) are available in order to adapt the flow behavior to different viscosities. The volume rate is measured by a magneto-inductive flowmeter (PROMAG 55; Endress + Hauser; Switzerland), positioned in the inlet zone. After passing the outlet zone, the medium is transported back to the reservoir tank. A number of ball valves allow the purging of the entire pipeline with fresh material to remove any residues from previous measurements.

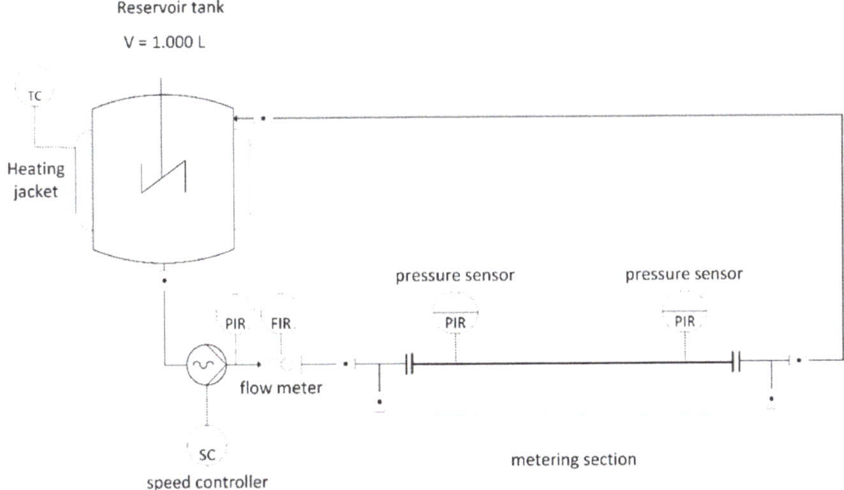

Figure 2. Scheme of the pipe viscometer.

For the validation and calibration of the viscometer, a commercial capillary viscometer type Rheotest 2 (Rheotest Medingen GmbH, Germany) is used as a reference. Non-Newtonian flow behavior is assumed for the digester slurry, and therefore validation runs are conducted with xanthan solution, which shows a similar shear-thinning and thixotropic flow characteristics [26]. The xanthan solution was used at three different concentrations (200, 250 and 287 g·L^{-1}), representing three different levels of apparent viscosity.

Measurements with the pipe viscometer followed a fixed procedure. After filling the reservoir tank with fresh slurry from the digester, samples for TS and VS analyses were taken. The temperature was kept constant (39 ± 2 °C) over the entire measuring procedure by an external heating jacket, and the homogeneity was maintained by an internal mixer. Before measurement, the pipeline system was flushed with fresh digester slurry and the losses were replenished. The measurement procedure included 5 min pre-shearing at the lowest shear rate, followed by a stepwise increase in the shear rates. Each shear rate was maintained for 5 min, and the pump drive speed, volume rate, pressure and pressure drop over the metering section were recorded every second. The procedure was repeated in triplicate.

For each shear rate, the apparent viscosity was calculated by Equation (1):

$$\eta = \frac{\Pi \cdot \Delta_P \cdot D^4}{128 \cdot Q_v \cdot L},\tag{1}$$

where:

η apparent viscosity in Pas;
Δp pressure drop over L in Pa;
D diameter of metering section in m;
Q_v volume rate in m$^3 \cdot$s^{-1};
L distance between pressure sensors in m.

With the apparent viscosity, the Reynolds number was calculated according to Equation (2) to ensure that the flow regime was laminar.

$$Re = \frac{uD\rho}{\eta},$$

(2)

where:

Re Reynolds number;
η apparent viscosity in Pas;
ρ density in kg·m^{-3};
D diameter of metering section in m;
u flow velocity in m·s^{-1}.

The apparent viscosity was displayed over the shear rate in a flow diagram. The Ostwald–de Baer law as shown in Equation (3) is suitable to describe the shear-thinning behavior [9] and was used to determine the rheological parameters.

$$\eta = K \cdot \dot{\gamma}^{(n-1)},$$

(3)

where:

η apparent viscosity in Pa·s;
$\dot{\gamma}$ shear rate in s^{-1};
K consistency factor in Pa·s^n;
n flow index.

The parameters K and n were calculated by plotting the logarithmized values for the apparent viscosity $\log \eta$ against the logarithmized corresponding shear rate $\log \dot{\gamma}$ and performing linear regression. The procedure for measuring and data processing was the same for the control measurement with xanthan as well as for the measurements with digester slurry during the enzyme application.

Unfortunately, the pipe viscometer was not available during the pre-trial phase, and therefore no data for this period are available.

2.5. Substrate Degradation Efficiency

The substrate degradation efficiency (SDE) was used as an assessment parameter to evaluate the effects of enzyme application on the biomethane yield. The biomethane potential (BMP) of a specific substrate defines the maximum amount of methane that can potentially be produced during AD. The BMP can be approximated based on the chemical substrate composition, stoichiometric calculations or discontinuous digestion (anaerobic batch or BMP) tests. Due to diverse metabolic pathways during biochemical conversion, a certain amount of substrate is also utilized for microbial growth or maintenance, which consequently lowers the BMP. Furthermore, substrate pre-treatment or disintegration can change the BMP of the investigated substrate [27]. The methane yield describes the utilized fraction of the methane potential under practical conditions during full-scale (or laboratory) continuous operated anaerobic digestion. Thus, the yield depends on numerous impact factors, such as retention time, organic loading rate, inhibitory effects or nutrient deficiency. The SDE can be calculated as the quotient of the methane yield over the BMP:

$$SDE = \frac{methane\ yield}{methane\ potential}.$$

(4)

By definition, the specific methane yield has to be lower or equal to the BMP. Thus, the degradation efficiency SDE should be in a range between 0 and 1. The specific BMP was calculated in accordance to a method published by Weißbach [28]. This method allows to us calculate the BMP of silages based on the crude fibre content. The method is described in [24]. The methane yield was measured during the trials at the DBFZ research plant described above.

3. Results and Discussion

3.1. Specific Biogas Yield and Substrate Degradation Efficiency

The experiment started with a preparation period of 35 days in which both digesters were operated identically in order to establish uniform process conditions. Steady state was defined by a uniform and constant specific gas production over at least 1 HRT and was proven by Neumann trend test statistics for both of the digesters separately [29]. The application of C1 enzymes started at day 120 in digester 3.1, whilst digester 3.2 was used as reference—i.e., with the same process conditions (same substrate and organic loading Rate (OLR)) but without enzyme application. The enzyme application was maintained over 97 days (approx. 4 hydraulic retention times (HRT)). Subsequently, a phase-out period of 20 d (0.8 HRT) followed. The process parameters as described above were monitored over the entire trial period and are shown in Table 2.

Table 2. Process parameters in digester 3.1 (enzyme addition) and 3.2 (reference) during different experimental phases.

Digester	pH	VFA g·L^{-1}	NH$_4$-N mg·L^{-1}	TS %	VS % TS	Specific CH$_4$-Yield m^3·t^{-1}
Digester 3.2 (reference) (during pre-phase)	7.6 ± 0.1 (n = 12)	2.0 ± 0.1 (n = 13)	2.0 ± 0.1 (n = 7)	10.4 ± 0.3 (n = 8)	79.5 ± 0.9 (n = 8)	226 ± 13 (n = 44)
Digester 3.2 (reference) (during enzyme addition in digester 3.1)	7.6 ± 0.1 (n = 18)	1.7 ± 0.1 (n = 19)	1.7 ± 0.1 (n = 14)	8.6 ± 0.4 (n = 12)	77.2 ± 1.1 (n = 12)	317 ± 24 (n = 68)
Digester 3.2 (reference) (during post phase)	7.6 ± 0.2 (n = 7)	1.7 ± 0.0 (n = 7)	1.7 ± 0.1 (n = 4)	8.7 ± 0.1 (n = 4)	75.7 ± 4 (n = 4)	337 ± 17 (n = 20)
Digester 3.1 (pre-phase)	7.6 ± 0.1 (n = 13)	2.0 ± 0.1 (n = 13)	2.0 ± 0.1 (n = 7)	10.2 ± 0.3 (n = 8)	79.5 ± 0.7 (n = 8)	224 ± 15 (n = 44)
Digester 3.1 (during enzyme addition)	7.7 ± 0.1 (n = 19)	1.7 ± 0.1 (n = 19)	1.6 ± 0.1 (n = 14)	8.5 ± 0.4 (n = 12)	76.6 ± 1.2 (n = 11)	312 ± 33 (n = 68)
Digester 3.1 (post-phase)	7.6 ± 0.1 (n = 7)	1.7 ± 0.1 (n = 7)	1.7 ± 0.1 (n = 4)	8.5 ± 0.1 (n = 4)	76.4 ± 0.6 (n = 4)	318 ± 16 (n = 20

Unfortunately, the sorghum silage quality changed between the used batches. Table 3 shows the differences between the two batches of sorghum silage used during the experiment. The TS and VS contents as well as the specific methane potential differ significantly. As a result, the specific biomethane yield of the mix (sorghum silage + cow manure) varied during the experiment. The biomethane potential and biomethane yields were calculated based on the input data. For a comparison of the viscosity, time periods were selected with the same feeding quality—i.e., for the same batch of sorghum silage.

Table 3. Specific methane yields and substrate degradation efficiency.

Digester	Specific CH$_4$-Yield of Mix (Sorghum Silage + Cow Manure) m^3·t^{-1} VS	SDE (Equation (4)) %
Digester 3.2 (reference) (during pre-phase)	226	67
Digester 3.2 (reference) (during enzyme addition in digester 3.1)	322	94
Digester 3.2 (reference) (during post phase)	331	96
Digester 3.1 (pre-phase)	231	68
Digester 3.1 (during enzyme addition)	323	94
Digester 3.1 (post-phase)	317	93

The increase in specific methane yield during the phase of enzyme addition occurred in both digesters to the same extent. It can therefore clearly be attributed to the change in feedstock quality rather than to enzyme application. To enable a comparison of both digesters independent of the specific methane yield of the input, the relative deviation—i.e., the difference between both digesters in relation to the reference, digester 3.2—was calculated. The result is shown in Figure 3. As can be seen, the difference between digester 3.1 with enzyme application and digester 3.2 (reference) is within the range of ±10% over the entire experiment. The relative deviation remains in the same range during the phases with and without enzyme application in digester 3.1. There is no statistically significant trend, as confirmed by a Neumann trend test at a level of significance of 95%.

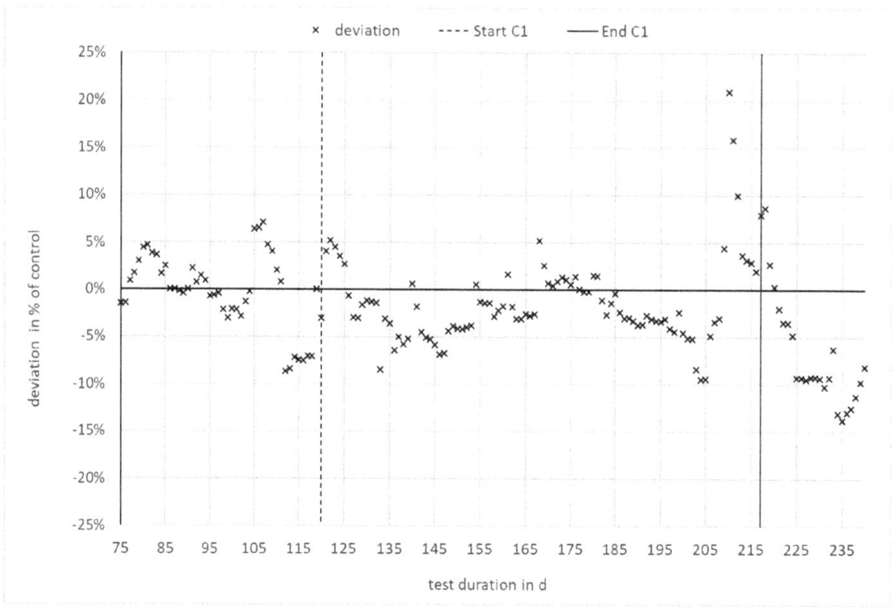

Figure 3. Deviation in specific methane yield in % of the reference digester 3.2.

Both of the digesters showed no indications of instabilities, all process parameters remained in a normal range and no differences between the digester with enzyme application and the reference digester were observed throughout the entire experiment (see also Table 2).

3.2. Calibration of Pipe Viscometer

Xanthan was used in three concentrations—200, 250 and 287 g·L^{-1}—to validate and calibrate the pipe viscometer. The flow curves of all three concentrations were measured by the reference capillary viscometer (CV) in duplicate and the pipe viscometer in triplicate. The results for the consistency factor K and flow index n are shown in Figures 4 and 5.

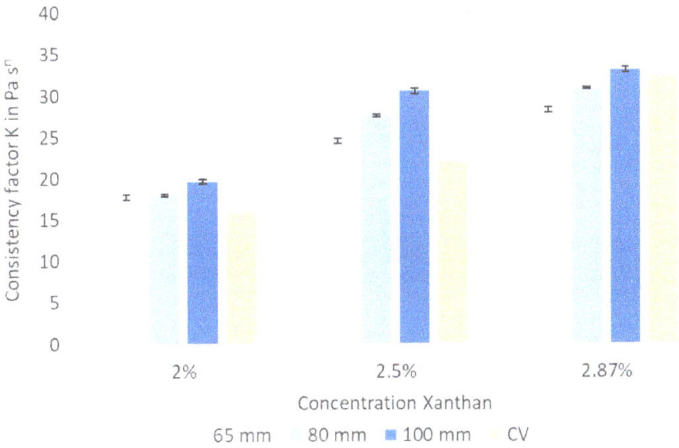

Figure 4. Xanthan solution: consistency factor K, measured with a pipe viscometer with different diameters and with a capillary viscometer (CV).

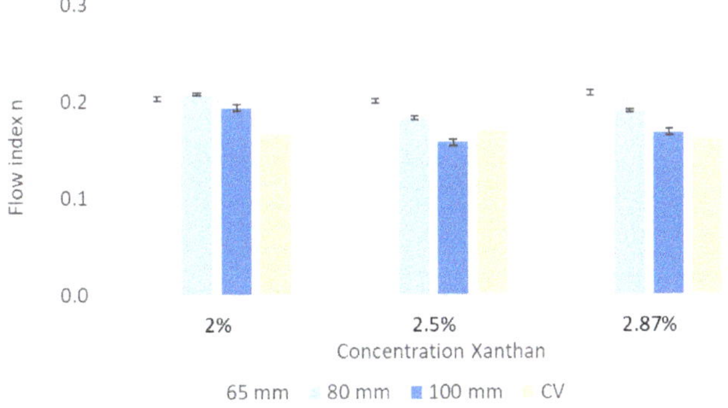

Figure 5. Xanthan solution: flow index n, measured with a pipe viscometer with different diameters and with a capillary viscometer (CV).

As expected, the consistency factor K depends on the Xanthan concentration. Furthermore, the calculated value of K increases with an increased pipe diameter. This effect can be explained by the different flow profiles. The Reynolds number was in the range of 1600 (2.87% Xanthan) to 2800

(2% Xanthan) for the 65 mm pipe diameter. For the 100 mm pipe diameter, the Reynolds number ranged between 400 (2.87% Xanthan) and 680 (2% Xanthan). Hence, a non-laminar flow was assumed for the 65 mm pipe diameter, and the 100 mm pipe was used for further measurements. A comparison of the results gained by both viscometers is shown in Table 4. The difference between the capillary viscometer, used as a reference, and the pipe viscometer with a diameter of 100 mm is less at higher absolute values of K—i.e., at a higher apparent viscosity.

Table 4. Validation runs for the pipe viscometer (results for diameter 100 mm in triplicate).

Xanthan Concentration	Pipe Viscometer D = 100 mm		Capillary Viscometer	
	Consistency Factor K	Flow Index n	Consistency Factor K	Flow Index n
2%	20 ± 0,2	0.19 ± 0,003	16	0.17
2.5%	30 ± 0,3	0.16 ± 0,003	22	0.17
2.87%	33 ± 0,2	0.17 ± 0,002	32	0.16

Measurements with the pipe viscometer showed a high reproducibility. The viscometer is therefore suitable to detect changes in the apparent viscosity of digester slurries with or without enzyme application. As the apparent viscosity of the digester slurry is expected to be in the higher range (K > 30), all the following measurements were carried out with the 100 mm pipe viscometer.

3.3. Rheological Characteristics of Digester Slurry

Flow curves of the digester slurry from both digesters (3.1/enzyme application, 3.2/reference) were measured three times after the start of the enzyme application. A typical flow diagram showing the apparent viscosity vs. the shear rate is shown in Figure 6. As can be seen from the regression parameters, the approach shown in Equation (3) delivers a good accordance with the data.

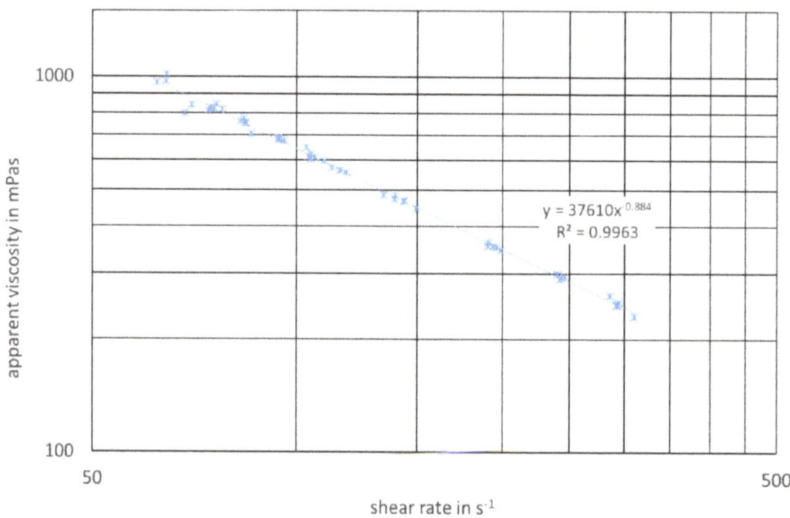

Figure 6. Flow diagram, digester 3.1/with enzyme application and regression parameters.

The results of measurements are shown in Table 5. The consistency factor K is far lower for the slurry from digester 3.1. This means that the general flow properties have been improved and the apparent viscosity of the digester slurry is far lower with enzyme application than without.

The difference in the flow index n is low but reproducible throughout all the measurements. It is slightly higher for the slurry from the digester with enzyme application. This result implies that the non-Newtonian behavior—i.e., the increased shear rate depending shear thinning—is more apparent for the digester with the enzyme additive.

Table 5. Consistency factors and flow indices of the digestate from the treated and untreated digester at different times.

Time	Digester 3.1/with Enzyme Application		Digester 3.2/Reference	
from Starting Enzyme Application	Consistency Factor K	Flow Index n	Consistency Factor K	Flow Index n
86 d	38	0.12	52	0.06
90 d	38	0.12	51	0.07
93 d	36	0.12	53	0.07
from termination of enzyme application				
25 d	37	0.12		

Additional measurements were conducted at day 122, which is 25 days (approx. 1 HRT) after the enzyme application had stopped. The parameters K and n (shown in last row of Table 5) had not increased after the termination of the enzyme application and remained at the level they were before. It is expected that this effect will subside over time. The apparent viscosity versus the shear rate is shown for the digester slurry with and without enzyme addition in Figure 7. The effect of the enzyme addition on the apparent viscosity is higher at lower shear rates. This effect almost disappears at higher shear rates.

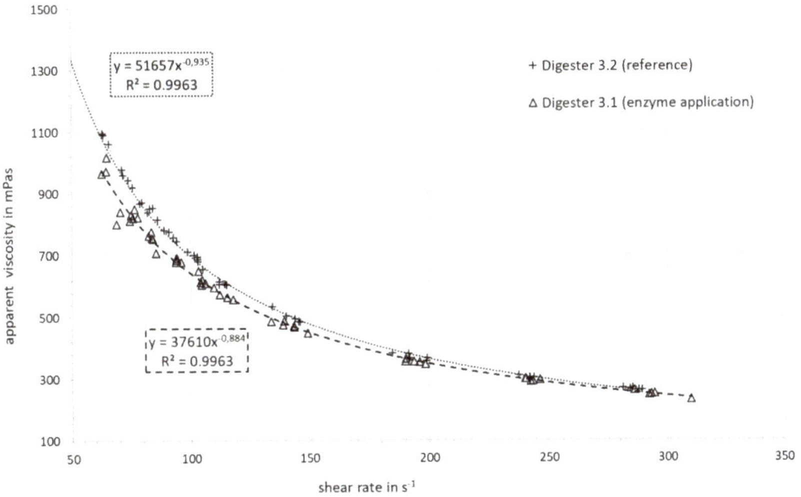

Figure 7. Apparent viscosity vs. the shear rate: comparison of digester slurry with and without enzyme addition.

The economic effects of reduced viscosity are mainly due to a lower energy demand for stirring processes. Studies [13] have shown that there is a strong correlation between the viscosity and energy consumption in biogas fermenters. A deep discussion of economic aspects would go beyond the scope of this manuscript. Energy requirement for mixing makes up a considerable part of the total energy requirement of biogas plants [30]. Hence, a positive economic effect of the enzyme application can be assumed.

4. Conclusions

The validation tests showed that a pipe viscometer is suitable to measure the rheological parameters of digester slurry under practical conditions. Validation and calibration with a non-Newtonian model fluid showed good accordance with the reference method (capillary viscometer) for a laminar flow regime at a 100 mm pipe diameter. Non-Newtonian flow behavior can be approximated by the Ostwald–de Baer equation with good accordance to the data. Measurements with digester slurry from a full-scale biogas research plant have confirmed the reproducibility of the results.

The experiment revealed the clear effects of enzyme application in full-scale operation. The enzyme product influenced the rheological parameters of the digester slurry. Slurry from the digester with enzyme application showed a significantly decreased consistency factor K compared to the reference digester. This indicates a generally decreased apparent viscosity. The flow index n is slightly higher for the slurry from the digester with enzyme application. This means that the dependence of the shear thinning on the shear rate was also influenced by the enzyme product. Thep possible economic effects are a reduced energy demand for mixers and pumps.

There was no observable effect of enzyme application on the substrate degradation efficiency or specific methane yield. Both the digesters with and without enzyme application showed nearly the same specific methane yields over the entire experimental period. Changes in the substrate degradation efficiency can be attributed to changes in the feedstock quality during the trial. The observed changes in the rheological parameters are apparently not associated with an improved methane yield.

Author Contributions: Conceptualization: L.M., N.E. and S.H.; methodology L.M. and U.M.; investigation L.M., K.R., U.M., C.K. and S.H.; writing—original draft preparation N.E.; writing—review and editing, N.E., L.M. and S.H. All authors have read and agreed to the published version of the manuscript.

Funding: The evaluated experiments were part of the EU project DEMETER. The project has received funding from the European Union's Horizon 2020 Research and Innovation program under Grant Agreement N. 720714.

Acknowledgments: Many thanks for supporting the performed trials go to Andreas Müller and Jonathan Rummel as well as Torsten Schüßler and Heiko Müller from DBFZ.

References

1. Daniel-Gromke, J.; Rensberg, N.; Denysenko, V.; Trommler, M.; Reinholz, T.; Völler, K.; Beil, M.; Beyrich, W. *DBFZ Report Nr. 30. Anlagenbestand Biogas und Biomethan- Biogaserzeugung und -nutzung in Deutschland*; Deutsches Biomasseforschungszentrum (DBFZ): Leipzig, Germany, 2017; ISBN 978-3-946629-24-5.

2. EBA- European Biogas Association. *EBA Statistical Report*. Available online: https://www.europeanbiogas.eu/eba-annual-report-2019/ (accessed on 7 April 2020).

3. *Biogas in der Landwirtschaft: Stand und Perspektiven FNR / KTBL Fachkongress vom 26. bis 27*; Kuratorium für Technik und Bau in der Landwirtschaft KTBL: Bayreuth, Germany, 2017.

4. Stafford, D.A. The effects of mixing and volatile fatty acid concentrations on anaerobic digester performance. *Biomass* **1982**, *2*, 43–55. [CrossRef]

5. Dachs, G.; Rehm, W. Der Eigenstromverbrauch von Biogasanlagen und Potenziale zu dessen Reduzierung. 2006. Available online: https://www.sev-bayern.de/wp-content/uploads/2019/01/SeV_BASE-Studie_Biogas-Eigenverbrauch_.pdf (accessed on 3 January 2020).

6. Tian, L.; Shen, F.; Yuan, H.; Zou, D.; Liu, Y.; Zhu, B.; Li, X. Reducing agitation energy-consumption by improving rheological properties of corn stover substrate in anaerobic digestion. *Bioresour. Technol.* **2014**, *168*, 86–91. [CrossRef]

7. Singh, B.; Szamosi, Z.; Siménfalvi, Z. State of the art on mixing in an anaerobic digester: A review. *Renew. Energy* **2019**, *141*, 922–936. [CrossRef]

8. The rheological characterisation of sludges. *Water Sc. Technol.* **1997**, *36*, 9–18. [CrossRef]

9. Barnes, H.A.; Hutton, J.F.; Walters, K. *An Introduction to Rheology, 3. Impression*; Elsevier: Amsterdam, The Netherlands, 1993; ISBN 0444871403.

10. El-Mashad, H.M.; Van Loon, W.K.P.; Zeeman, G.; Bot, G.P.A. Rheological properties of dairy cattle manure. *Bioresour. Technol.* **2005**, *96*, 531–535. [CrossRef]

11. Baudez, J.C.; Markis, F.; Eshtiaghi, N.; Slatter, P. The rheological behaviour of anaerobic digested sludge. *Water Res.* **2011**, *45*, 5675–5680. [CrossRef]

12. Baudez, J.-C.; Ayol, A.; Coussot, P. Practical determination of the rheological behavior of pasty biosolids. *J. Environ. Manag.* **2004**, *72*, 181–188. [CrossRef]

13. Šafarič, L.; Yekta, S.S.; Ejlertsson, J.; Safari, M.; Najafabadi, H.N.; Karlsson, A.; Ometto, F.; Svensson, B.H.; Björn, A. A comparative study of biogas reactor fluid rheology-implications for mixing profile and power demand. *Processes* **2019**, *7*, 700. [CrossRef]

14. Senge, B.; Blochwitz, R. Bentlin Simone. *Rheologische Stoffkennwerte richtig bestimmen. Dtsch. Milchwirtsch.* **2004**, *7*, 256–260.

15. Gienau, T.; Kraume, M.; Rosenberger, S. Rheological Characterization of Anaerobic Sludge from Agricultural and Bio-Waste Biogas Plants. *Chem. Ing. Tech.* **2018**, *90*, 988–997. [CrossRef]

16. Baudez, J.-C. About peak and loop in sludge rheograms. *J. Environ. Manag.* **2006**, *78*, 232–239. [CrossRef]

17. Wang, Y.; McAllister, T.A. *Rumen Microbes, Enzymes and Feed Digestion—A Review*; Agriculture and Agri-Food Canada Research Cenre: Lethbridge, AB, Canada, 2002.

18. Binner, R.; Menath, V.; Huber, H.; Thomm, M.; Bischof, F.; Schmack, D.; Reuter, M. Comparative study of stability and half-life of enzymes and enzyme aggregates implemented in anaerobic biogas processes. *Biomass Conv. Bioref.* **2011**, *1*, 1–8. [CrossRef]

19. Romano, R.T.; Zhang, R.; Teter, S.; McGarvey, J.A. The effect of enzyme addition on anaerobic digestion of JoseTall Wheat Grass. *Bioresour. Technol.* **2009**, *100*, 4564–4571. [CrossRef]

20. Quiñones, T.S.; Plöchl, M.; Budde, J.; Heiermann, M. Enhanced Methane Formation through Application of Enzymes: Results from Continuous Digestion Tests. *Energy Fuels* **2011**, *25*, 5378–5386. [CrossRef]

21. Brulé, M.; Vogtherr, J.; Lemmer, A.; Oechsner, H.; Jungbluth, T. Effect of enzyme addition on the methane yields of efluents from a full-scale biogas plant. *Landtechnik* **2011**, *66*, 50–52.

22. Schimpf, U.; Hanreich, A.; Mähnert, P.; Unmack, T.; Junne, S.; Renpenning, J.; Lopez-Ulibarri, R. Improving the Efficiency of Large-Scale Biogas Processes: Pectinolytic Enzymes Accelerate the Lignocellulose Degradation. *J. Sustain. Energy Environ.* **2013**, *4*, 53–60.

23. DEMETEEU-Project. DEMETER-Demonstrating more efficient enzyme production to increase biogas yields. Available online: www.demeter-eu-project.eu/ (accessed on 6 February 2020).

24. Liebetrau, J.; Pfeiffer, D.; Thrän, D. (Eds.) *Collection of Measurement Methods for Biogas. Methods to Determine Parameters for Analysis Purposes and Parameters that Describe Processes in the Biogas Sector*; Deutsches Biomasseforschungszentrum (DBFZ): Leipzig, Germany, 2016.

25. Holliger, C.; Alves, M.; Andrade, D.; Angelidaki, I.; Astals, S.; Baier, U.; Bougrier, C.; BuffiÃ¨re, P.; Carballa, M.; De Wilde, V.; et al. Towards a standardization of biomethane potential tests. *Water Sci. Technol.* **2016**, *74*, 2515–2522. [CrossRef]

26. Zhong, L.; Oostrom, M.; Truex, M.J.; Vermeul, V.R.; Szecsody, J.E. Rheological behavior of xanthan gum solution related to shear thinning fluid delivery for subsurface remediation. *J. Hazard. Mater.* **2013**, *244–245*, 160–170. [CrossRef]

27. Ariunbaatar, J.; Panico, A.; Esposito, G.; Pirozzi, F.; Lens, P.N.L. Pretreatment methods to enhance anaerobic digestion of organic solid waste. *Appl. Energy* **2014**, *123*, 143–156. [CrossRef]

28. Weißbach, F. Ausnutzungsgrad von Nawaros bei der Biogasgewinnung. *Landtechnik* **2009**, *64*, 18–21.

29. Yuri, A.W. *Statistics for Chemical and Process Engineers. A modern approach*; Springer: Cham, Switzerland; Heidelberg, Germany; New York, NY, USA; Dordrecht, The Netherlands; London, UK, 2015; ISBN 978-3-319-21508-2.
30. Naegele, H.; Lemmer, A.; Oechsner, H.; Jungbluth, T. Electric energy consumption of the full scale research biogas plant "unterer lindenhof": Results of longterm and full detail measurements. *Energies* **2012**, *5*, 5198–5214. [CrossRef]

Article

The Influence of Pressure-Swing Conditioning Pre-Treatment of Cattle Manure on Methane Production

Britt Schumacher [1,*]**, Timo Zerback** [1]**, Harald Wedwitschka** [1]**, Sören Weinrich** [1]**, Josephine Hofmann** [1] **and Michael Nelles** [1,2]

[1] Biochemical Conversion, DBFZ Deutsches Biomasseforschungszentrum Gemeinnützige GmbH, Torgauer Straße 116, D-04347 Leipzig, Germany; Timo.Rolf.Zerback@dbfz.de (T.Z.); Harald.Wedwitschka@dbfz.de (H.W.); Soeren.Weinrich@dbfz.de (S.W.); Josephine.Hofmann@dbfz.de (J.H.); michael.nelles@uni-rostock.de (M.N.)

[2] Department Waste and Resource Management, University of Rostock, Justus-von-Liebig Weg 6, D-18057 Rostock, Germany

* Correspondence: Britt.Schumacher@dbfz.de; Tel.: +49-341-2434-540

Received: 30 October 2019; Accepted: 20 December 2019; Published: 30 December 2019

Abstract: Cattle manure is an agricultural residue, which could be used as source to produce methane in order to substitute fossil fuels. Nevertheless, in practice the handling of this slowly degradable substrate during anaerobic digestion is challenging. In this study, the influence of the pre-treatment of cattle manure with pressure-swing conditioning (PSC) on the methane production was investigated. Six variants of PSC (combinations of duration 5 min, 30 min, 60 min and temperature 160 °C, 190 °C) were examined with regards to methane yield in batch tests. PSC of cattle manure showed a significant increase up to 109% in the methane yield compared to the untreated sample. Kinetic calculations proved also an enhancement of the degradation speed. One PSC-variant (190 °C/30 min) and untreated cattle manure were chosen for comparative fermentation tests in continuously stirred tank reactors (CSTR) in lab-scale with duplicates. In the continuous test a biogas production of 428 mL/g volatile solids (VS) (54.2% methane) for untreated manure was observed and of 456 mL/g VS (53.7% methane) for PSC-cattle-manure (190 °C/30 min). Significant tests were conducted for methane yields of all fermentation tests. Furthermore, other parameters such as furfural were investigated and discussed.

Keywords: anaerobic digestion; cattle manure; steam explosion; pre-treatment

---

Highlights:

- 109% methane yield increase through PSC-treatment of cattle manure in BMP-test;
- 160 °C/60 min and 190 °C/30 min are the best case of PSC-treatment in BMP-test;
- accelerated methane formation via PSC-treatment in BMP-test;
- temporarily enhanced methane production in semi-continuous test after treatment.

1. Introduction

Germany is the country leading by far in the primary production of biogas in European Union with 7852.4 ktoe in 2015, compared to United Kingdom with 2252.4 ktoe and Italy with 1871.5 ktoe biogasbarometer [1]. In 2016, approximately 8535 biogas plants fed with excrement and energy crops, biowaste or organic waste produced biogas in Germany (Daniel-Gromke, 2018) [2]. In contrast to other countries, the biogas is not only produced in landfills and sewage treatment plants in Germany [1], but mainly in the agricultural sector with energy crops and excrement as feedstock (Daniel-Gromke, 2018) [2]. Nevertheless, Germany has a high biogas production potential from manure with 90 PJ/year, which is not fully tapped (Scheftelowitz and Thrän, 2016) [3]. One of several reasons is the logistical challenge because the potentials are spread over a large number of farms (Scheftelowitz and Thrän, 2016) [3]. Another technical challenge is the handling of solid manure in practice and its slow degradation speed in anaerobic digestion. However, greenhouse gas (GHG) emissions could be saved by manure management via biogas generation as well as by substituting fossil fuels by biogas (Scheftelowitz and Thrän, 2016) [3].

Generally, bedding material in manure like straw consists to a large extent of ligno-cellulose. Hendriks and Zeeman (2009) [4] identified inter alia the crystallinity of cellulose, available surface area, and lignin content as limiting factors for the decomposition of ligno-cellulosic biomass. Hence, a pre-treatment of solid manure is advisable to improve the decomposition and to overcome technical challenges like floating layers in plug flow digesters or continuously stirred tank reactors (CSTR). In practice, physical pre-treatment methods are dominant, in 81% of German biogas plants (Schumacher, 2014) [5], but thermal pressure hydrolysis is a comparatively recent and seldom implemented process in the agricultural sector.

Lab-scale batch tests with cattle manure pretreated with thermal pressure hydrolysis revealed contradictory results. Risberg et al. [6] found no improved gas production, while Budde et al. [7] measured increased methane yields by up to 58% with a treatment temperature of 180 °C. Besides the positive effect of surface enlargement of a thermal pressure hydrolysis, negative effects may occur due to the formation of inhibiting substances. These substances include phenolic compounds as well as furalaldehydes, e.g., furfural and 5-hydroxymethylfurfural, which are the main degradation products derived from the dehydratation of hexoses or pentoses, respectively (Barakat et al., 2012) [8]. The negative effects of such by-products have been widely investigated for biotechnological ethanol or hydrogen production, whereas studies focusing on methane production all still very limited. One possible reason could be due to the fact that the microbial consortium involved in biogas production tends to have a higher tolerance to such inhibitory by-products (Monlau et al., 2014) [9]. For example, the treshold value for furfurals in biotechnological hydrogen production with single cultures is often given with 0.62 g/L (Siqueira and Reginatto, 2015) [10], whereas similar concentrations in biogas production generally lead to no negative effects (Pekařová et al., 2017) [11]. Further reasons could be due to lower substrate/inoculum and inhibitor/inoculum ratios as well as different incubation times (Monlau et al., 2014) [9]. Nevertheless, inhibitory impacts on biogas production could be observed especially at higher furfural concentrations too (Pekařová et al., 2017) [11]. For this reason concentrations of toxic by-products should always be taken into account during hydrothermal substrate pretreatment.

However, discontinuous biochemical methane potential (BMP) tests are limited to a basic assessment of methane yield, anaerobic biological degradation kinetics and qualitative assessment of inhibitory effects of the substrate (VDI 4630, p. 44) [12], Schumacher et al., 2019 [13]. For the assessment of process stability behavior, synergistic effects, mono-fermentation and/or limits of organic loading rate of substrates as cattle manure, continuous running anaerobic digestion (AD) tests have to be conducted (VDI 4630, p. 44) [12], Schumacher 2019 [13]. Nevertheless, continuous running AD tests have the disadvantage that they are time-consuming and labor-intensive. These are the reasons why studies about the effects of pre-treatment of cattle manure with thermal pressure hydrolysis in CSTRs are rare.

The aim of this study was to examine the influence of cattle manure's pre-treatment with pressure-swing conditioning (PSC) under varying treatment conditions (treatment temperature, treatment duration) on the methane yield in batch tests. Additionally, a semi-continuous test in CSTRs in duplicates was conducted at lab-scale. The test were operated for 2.5 hydraulic retention times, in order to evaluate process stability during the feeding of a selected PSC-sample in comparison to the reference. These labor-intensive continuous tests are beyond the investigations of most other authors. Furthermore, PSC is a specific thermal pressure hydrolysis module in pilot-scale under development of the company VENTURY GmbH Energieanlagen (Dresden, Germany).

2. Materials and Methods

Cattle manure from a German farm was collected several times and pre-treated by the company VENTURY GmbH Energieanlagen (Dresden, Germany) for BMP tests as well as for semi-continuous tests in CSTRs. The cattle manure consisted of excrements from calves and cereal straw as bedding material.

2.1. Pre-Treatment and Biomethane Potential (BMP) Testing

The cattle manure (excrements and straw together) was treated with different combinations of temperatures and treatment times with PSC. The operation of the PSC pilot plant was described also by Schumacher et al. (2019) [13]. Approximately 5 kg of manure were fed manually into the pressure tank and heated with steam to the chosen temperature. At the end of the treatment period, the solid-liquid phase was released explosively into an expansion tank. The expansion tank was equipped with a gas valve were surplus steam is released. Detailed information about the PSC-technology are available on the internet (https://biomethane-map.eu/fileadmin/Biomethane_Map_-_TDs/UPD_TD_pressure_ swing_conditioning_ventury_-_signature_website.pdf, 27 September 2019). Six variants of PSC-treated cattle manure (combinations of 5 min, 30 min, 60 min; 160 °C, 190 °C) and one untreated reference were investigated regarding the methane yields in BMP tests in laboratory scale in triplicates each. AMPTS devices (Bioprocesscontrol, Lund, Sweden, temperature set on 39 ± 1 °C) were used. BMP tests were conducted in accordance with the VDI guideline 4630 (2006). The methane yields were standardized (dry gas, 273.15 K, 1013.25 hPa).

Approximately 20 g of PSC-treated cattle manure and 400 g inoculum were weighted in for BMP-test. As inoculum served digestate gained from DBFZ's research biogas plant (fed with cattle manure and corn silage) diluted with tap water 50% w/w.

2.2. Semi-Continuous Anaerobic Digestion Tests

Based on the results of the BMP tests, one PSC treatment variant of cattle manure (190 °C/30 min) and the reference were selected for a semi-continuous AD test in CSTR in laboratory scale with duplicates per variant of treatment. The 'untreated' cattle manure had to be chopped with a cutting mill (Fritsch Pulverisette 19/mesh size 6 mm) (Fritsch GmbH, Idar-Oberstein, Germany) in order to make the supply of the fermenters and the removal of digestate feasible (Supplementary Materials).

The semi-continuous AD tests with cattle manure were carried out in four CSTRs each with a net volume of 10 L (Bräutigam Kunststofftechnik GmbH, Mohlsdorf-Teichwolframsdorf, Germany). The temperature was set at 39 °C with a thermostat (JULABO GmbH, Seelbach, Germany). Stirrer 'RZR 2102 control' (Heidolph Instruments GmbH & Co.KG, Schwabach, Germany) were used at 50 rpm to mix the digesters. The biogas volume was measured with drum-type gas meter TG05/5 (Dr.-Ing. RITTER Apparatebau GmbH & Co. KG, Bochum, Germany). The biogas quality was determined with AwiFLEX No 1185_11 (Awite Bioenergie GmbH, Langenbach, Germany). CSTR tests were conducted in accordance with the VDI guideline 4630 (2006) [12] as well. Methane and biogas yields were standardized respectively (dry gas, 273.15 K, 1013.25 hPa).

The general feeding regime for semi-continuous AD tests at DBFZ is given in Liebetrau et al. (2016, p. 162) [14]. Sieved digestate (mesh size 4 mm, pH-value = 7.46, total solids (TS) = 7.69% FM, volatile solids (VS) = 77.35% TS, total ammonia nitrogen (TAN) = 1.49 g/L) of a full-scale biogas plant (DBFZ research biogas plant, Germany, was used as inoculum for the four lab-scale fermenters. The biogas plant was supplied with cattle manure and corn silage. The set temperature in the lab reactors was 39 ± 1 °C. After 10 days without feeding, the supply of all reactors with cellulose and pellets of DDGS (Distillers' Dried Grains with Solubles) started due to a delay in the supply with cattle manure. At day 18 the feeding of all reactors with chopped cattle manure started and the organic loading rate (OLR) was raised to 2.5 g VS/(L·d) step by step. The hydraulic retention time (HRT) of 30 days was kept constant over the entire test. At the beginning (18th day) of the test 100 g fresh manure and 225 g tap water were fed daily to the reactors to adjust the ORL and HRT.

At day 57 the feeding of two reactors with PSC treated manure began, while the other two reactors remained as an untreated reference. In contrast to the reference manure, the PSC-manure was not chopped. Due to the fact that the dry matter content of the first PSC-material was a little bit lower than expected, the OLR of all four reactors had to be decreased from 2.5 to 2.4 g VS/(L·d). Hence, at day 57 two reactors were supplied with 333 g FM of PSC manure and the other two reactors with 128 g of reference manure plus 205 g tap water. In the course of the three HRT three material changes were necessary.

No additives like trace elements were used. The untreated manure and the PSC manure were stored in plastic barrels in a cooling chamber at 5 °C during laboratory tests.

2.3. Analytical Methods

Total solids (TS) and volatile solids (VS) were measured in accordance with DIN EN 12880 (2001) [15] and DIN EN 12879 (2001) [16]. The pH-value was measured with a pH device 3310 (WTW Wissenschaftlich-Technische Werkstätten GmbH, Weilheim, Germany). The analysis of furfural were conducted as described in Schumacher et al. (2019) [13]. Volatile organic acids and total inorganic carbon (VOA/TIC) and total ammonia nitrogen (TAN) were determined as described in Liebetrau et al. (2016, pp. 32, 34) [14].

2.4. Kinetic Modelling

For kinetic modelling single first-order kinetics according to Angelidakit et al., 2009 [17] have been applied to evaluate methane production kinetics of BMP tests. Based on Equation (1) the specific methane potential S_{BMP} (mL/g VS) and single first-order reaction constant k (1/d) of the investigated substrates need to be adjusted to depict progression of the cumulative methane production S (mL/g VS) over time.

$$S(t) = BMP \cdot \left(1 - e^{(-k \cdot t)}\right) \tag{1}$$

Model implementations as well as numeric parameter identification were realized in the software environment Matlab (The MathWorks, Inc., Torrance, CA, USA) as previously describe in Schumacher et al. (2019) [13]. The coefficient of determination R^2 (−) was calculated for each BMP test to evaluate model efficiency.

2.5. Statistical Analysis

2.5.1. Batch

The BMP test results (final values of methane yield after 29 days) were analyzed with the help of the software SPSS statistics Version 20 (IBM, Chicago, IL, USA). In order to determine whether there was a significant difference between the methane yield of untreated (reference) and pretreated cattle manure, final data were evaluated using a Welch's analysis of variance (ANOVA) taking a confidence level of 95% into account. If differences existed, a post hoc test according to Games Howell ($\alpha = 0.05$) was chosen, to identify where they occurred.

2.5.2. Semi-Continuous Tests

Semi-continuous tests (CSTR) were divided into pre-phase and three hydraulic retention times. All phases were analyzed separately with ANOVA and post hoc test by means of SPSS statistics Version 20 (IBM, Chicago, IL, USA). Further details about the selection of statistical tests are described by Hofmann et al. (2016) [18]. In order to meet the requirements of variance analysis with regard to their applicability, the experimental data were checked for normal distribution using the preliminary significance tests (Kolmogrov–Smirnov and Shapiro–Wilk). Furthermore, the assumption of variance homogeneity was checked by means of a preliminary Levene test. If the criterion of variance homogeneity could not be met, the significance test was performed with the help of Welch's ANOVA.

3. Results and Discussion

3.1. Biochemical Methane Potential (BMP) Testing and Kinetic Modeling

Table 1 displays TS, VS, the final methane yields, and variation coefficients of BMP test after 29 days. The PSC pretreatment with 160 °C/60 min and 190 °C/30 min showed the highest methane yields with 239 mL/g VS and 229 mL/g VS, respectively. Budde et al. (2014) [7] published methane yields for untreated manure of at least 162 mL/g OM in BMP tests (30 days) and up to 255 mL/g OM (160 °C/5 min) for cattle manure pretreated with thermal pressure hydrolysis. Furthermore, Budde et al. (2014) [7] more often stated results between 216 and 232 mLCH$_4$/g OM for treated manure. In this study the results for PSC treated manure are comparable to the data of Budde et al. (2014) [7], but the untreated manure has lower methane yields. In contrast Risberg et al. (2013) [6] found no improved gas production after pretreatment. Our study presents an increase by up to 109%, Table 1.

Table 1. Final methane yields of untreated and pressure-swing conditioning (PSC)-treated cattle manure (biochemical methane potential (BMP) test) including total solids (TS), volatile solids (VS), variation coefficient, increase of methane yield, first-order kinetic constant and coefficient of determination.

	Total Solids	Volatile Solids	Methane Yield	Variation Coefficient	Increase of CH$_4$ Yield	First-Order Kinetic Model [a]	
						k	R^2
	%FM	%TS	[mL/g VS]	[−]	[%]	[1/d]	[−]
untreated	20.87	75.52	114	0.01	–	0.0594	0.92
PSC 160 °C 5 min	13.24	81.43	193	0.07	68.43	0.1339	0.96
PSC 160 °C 30 min	13.18	84.72	224	0.01	96.11	0.2822	0.95
PSC 160 °C 60 min	16.09	84.76	239	0.22	109.20	0.2591	0.92
PSC 190 °C 5 min	13.16	73.39	128	0.07	12.16	0.0723	0.81
PSC 190 °C 30 min	10.36	79.50	229	0.04	100.53	0.1916	0.85
PSC 190 °C 60 min	8.00	79.55	227	0.02	98.94	0.1714	0.76

[a] Estimated for fixed methane potentials S$_{BMP}$ determined during BMP tests (final experimental methane yield) after 29 days.

Possible causes for divergent results in various studies are: (a) substrate-specific: divergent physical and chemical composition; (b) pretreatment technology-specific: temperature set, pretreatment duration, the heating system (electrical jacket heating or with steam inside of the pressure tank), the chosen relaxation phase (gaseous or liquid), the heating speed and relaxation speed (cooling down slowly or explosive release) (c) BMP test-specific: microbiological consortium of the utilized inoculum, the temperature and the implementation of mixing, see also Schumacher et al. (2019) [13].

The tremendous enhancement of degradation speed through PSC is visualized in Figure 1. Due to time limitations within the project the test termination criterion (daily biogas rate is equivalent to only 1% of the total volume of biogas produced up to that time, VDI 4630 2006 [12] for the untreated manure was not reached, see Figure 1.

To investigate degradation kinetics in more detail a first-order model was applied to depict experimental data. To ensure realistic simulation results the unknown first-order kinetic constants were estimated for fixed biochemical methane potentials. Thus, the final experimental value of the cumulative methane production after 29 days equals the biochemical methane potential S$_{BMP}$ in the applied model structure (Equation (1)). As shown in Table 1 the resulting kinetic constant during pretreatment increases in comparison to the untreated sample. Furthermore, increasing intensity (length) of pretreatment from 5 to 30 or 60 min clearly shows an additional benefit in faster degradation kinetics during batch operation.

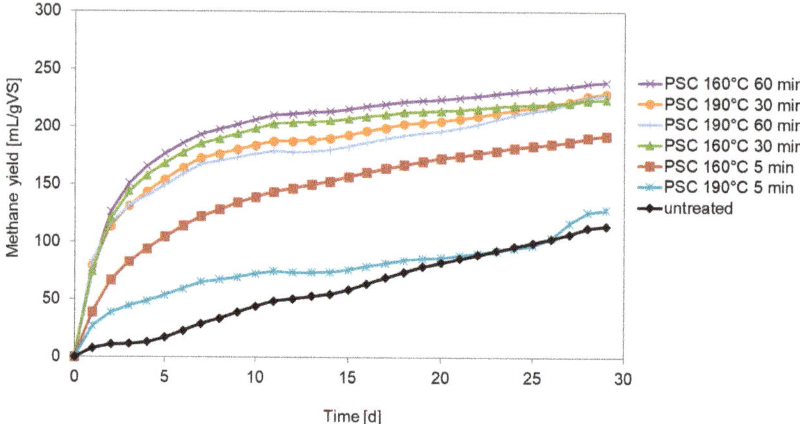

Figure 1. Methane yields in BMP tests.

The statistical evaluation of the BMP results, using Welch's ANOVA, delivered a significant result ($p < 0.000$). This means, the null hypothesis 'There is no significant difference between the methane yield of untreated and PSC treated cattle manure' can be rejected at a confidence level of 95%. In addition, the results from the Games–Howell post hoc test given under Table 2, show significant effects ($p < 0.05$) between the methane yield of control and the parameter combinations b, c, e and f.

Table 2. Significance test for BMP.

Control	Parameter Combination	Mean Difference [mL/g VS]	Significance
	a: PSC 160 °C 5 min	−78.29	0.195
	b: PSC 160 °C 30 min	−109.87 *	0.002
	c: PSC 160 °C 60 min	−124.60 *	0.001
untreated	d: PSC 190 °C 5 min	−13.93	0.836
	e: PSC 190 °C 30 min	−115.12 *	0.049
	f: PSC 190 °C 60 Min	−113.13 *	0.013

* Significant difference between the methane yield reference and pretreated cattle manure (confidence level of 95%).

However, the 68% or 12% increases in methane yield, associated with the PSC variants 160 °C/5 min and 190 °C/5 min, could not be confirmed statistically significant ($p > 0.05$).

3.2. Furfural Concentrations (BMP)

The following Table 3 shows the furfural concentrations of the liquid phase of pretreated cattle manure and the corresponding severity factor R_0.

Table 3. Furfural concentrations of pretreated cattle manure and severity factor.

	Furfural Concentration [mg/L]	Severity Factor, log R_0 [−]
PSC 160 °C 5 min	2.00	2.47
PSC 160 °C 30 min	3.71	3.24
PSC 160 °C 60 min	-	3.54
PSC 190 °C 5 min	2.08	3.35
PSC 190 °C 30 min	5.83	4.13
PSC 190 °C 60 min	8.15	4.43

In all cases, the determined furfural concentrations were showing a relatively low level. The lowest value amounted to 2 mg/L whereas the highest concentration of 8.15 mg/L was corresponding with a severity factor R_0 4.43. Due to an insufficient amount of hydrolysate, it was not possible to determine the furfural concentration of the PSC variant 160 °C/60 min.

It is interesting to note that the release of furfural seems to correlate with the severity factor. Thus, a continuous increase of furfural concentration can be seen both over the pretreatment period and temperature. The latter seems to affect the severity (R_0) and thus the formation of the inhibitor in a particular degree. For comparison: At the end of the PSC pretreatment at 160 °C, an increase of 85.5% of furfural was observed after 30 min compared to the five minutes, whereas the furfural concentration at 190 °C increased by about 180% from 5 min to 30 min sample.

In comparison to other publications, dealing with inhibitory effects of toxic compounds, an inhibition of the biogas process can probably be ruled out. As an example: Pekařová et al. (2017) [11] reported an inhibitory effect only at a furfural concentration between 1–2 g/L by using sodium acetate as a carbon substrate. By contrast, concentrations below 1 g/L seemed to have a stimulating effect on methane production. The inhibition was manifested through an increase in lag phase but after a certain time methane production was nearly restored with a comparable production rate to the control group.

Similar results were provided by Barakat et al., who also found no inhibition of the fermenter biology using furfural and 5-hydroxymethylfurfural (5-HMF) as single substrates (c = 2 g/L). The fermentation with 5-HMF showed a methane yield of 450 mL/g VS, whereas the digestion of furfural was 430 mL/g VS. The final level of methane production was reached after 10 days for furfural and 14 days for 5-HMF (Barakat et al., 2012) [8]. Both substances were initially associated with delays in methane production, which was explained by the prior conversion of the inhibitors to furfural alcohol (Barakat et al., 2012) [8] and (Rivard and Grohmann, 1991) [19].

3.3. Semi-Continuous Anaerobic Digestion Tests

3.3.1. Methane Production

On the base of the BMP test, the PSC-variant 190 °C/30 min (variation coefficient lower than variant 160 °C/60 min) was chosen for the semi-continuous fermentation test. A positive effect of PSC treatment on the methane production was observed at the beginning of the first HRT until the first change of PSC treated substrate. During the second HRT the methane production of PSC manure decreased to the reference level, while the variance of the methane production of PSC manure is visibly higher than the variance of the reference. After the third material change and at the beginning of the third HRT the methane production of PSC manure decreased below the level of the reference. During the test period new substrate had to be collected and treated due to limited cooling capacity and the lack of stability of the substrate. This seems to cause a varying composition and/or structure of the treated substrate at different times, which could be a reason for varying methane production. The average methane production of PSC-treated manure was 239 mL/g VS from the beginning of feeding with treated substrate until the end of the semi-continuous fermentation test and 235 mL/g VS of the reference, respectively.

Statistical Tests

For the statistical evaluation of the semi-continuous test period, the distribution of the measured methane production is illustrated with the help of boxplots. This allows presumptions on the result of every single phase and on the fulfillment of the assumptions for statistical tests according to Hofmann et al. (2016) [18]. Figure 2 shows the boxplots of the biogas yields for the pre-phase and three following hydraulic retention times. The boxplots displays four reactors where two are assigned to the reference (U = untreated) and two to the PSC treatment. The median of the measurement data is represented by the band in the middle of the box and is limited by the first and third quartiles. Furthermore the whiskers, are illustrated with 1.5 times the interquartile distance (IQR), which can be used to approximate the distribution of the variances. The variances can be considered homogeneous, if whiskers have an equal length. In addition, outliers can be displayed. A distinction is made between mild and extreme outliers. Mild outliers have a distance to the first or third quartile of 1.5× IQR to 3.0× IQR. In a SPSS box plot, these values are marked with individual dots.

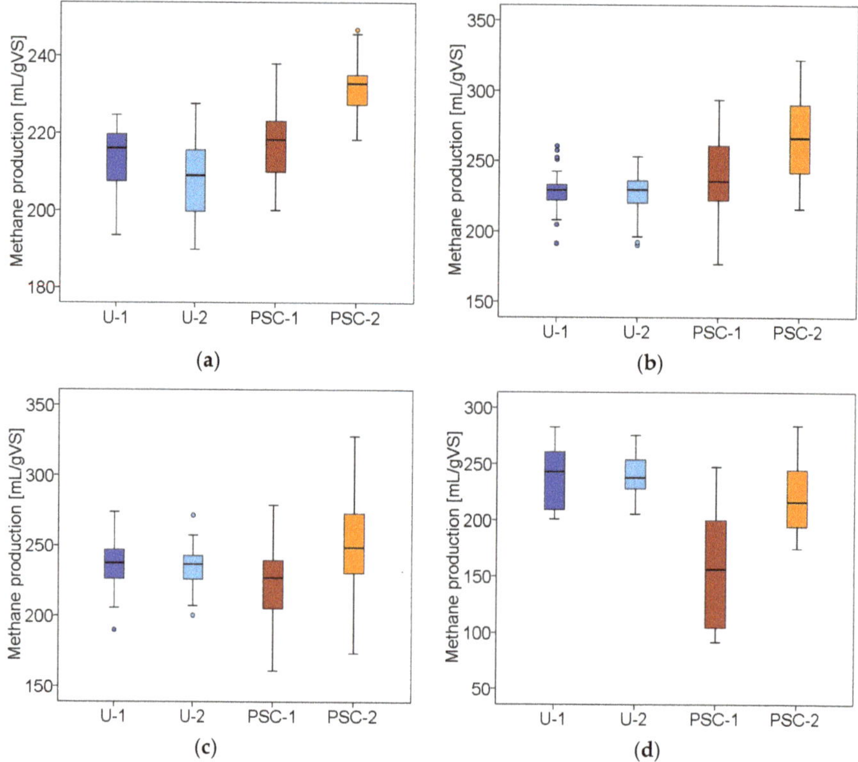

Figure 2. Boxplots of methane production (**a**) pre-phase; (**b**) hydraulic retention time (HRT) 1; (**c**) HRT 2; (**d**) HRT 3.

Pre-Phase

The boxplot in Figure 2a illustrates the experimental data of the pre-phase where no PSC-treatment of cattle manure occurred. It is expected, that neither the methane production between the reactors nor between the approaches will differ in this phase. However, this expectation seems to be refuted by the first visual impression. Instead, boxplot shows a slight difference between the methane yields between the two approaches. With the help of ANOVA, the first impression was confirmed ($p = 0.000$) meaning, the null hypothesis 'There is no significant difference in methane yield between the approaches' had to be rejected. To identify where the differences occurred, a Bonferroni post-hoc test was used. The test procedure confirmed that reactor PSC-2 significantly differed from U-1 ($p = 0.000$), U-2 ($p = 0.000$) and PSC-1 ($p = 0.000$). Furthermore, a significant difference between the methane yields of the second reactor of the control group U-2 ($p = 0.015$) compared to the methane yield of the first reactor of the PSC approach, PSC-1 could be revealed. A possible reason for these deviations could be that most digesters had not reached stable conditions by the end of the pre-phase. A regression analysis showed trends of the methane yield over time in three out of four cases which questions the assumption of steady-state conditions.

Hydraulic Retention Time (HRT) 1

As already mentioned, the positive effect of PSC-treatment on methane production could only be observed at the beginning of the first hydraulic retention time until the first change of pretreated material. Afterwards, the methane value of both PSC-reactors clearly decreased. This result is reflected in Figure 2b. Regarding the boxplots, only small effects on the methane production by the PSC-treatment can be observed. In order to assess differences between the methane productions of reference- and PSC-approaches, methane values were considered over the entire period. As the Levene-statistics indicate heterogeneity of variances a Welch's ANOVA was used for the statistical evaluation. Comparing the methane production of the pre-treatment related to the reference a significant difference ($p = 0.000$) between the two approaches could be observed. Nevertheless, the results must be called into question at this point. On the one hand a following Games–Howell test showed that the difference primarily depends on PSC-2 while no significant differences between PSC-1 and U-1 ($p = 0.211$) or U-2 ($p = 0.094$) could be determined. On the other hand, a regression analysis revealed a trend over time which also indicates that steady state conditions were not reached.

In order to make a statement about the influence of PSC-approach, the effect of the substrate disintegration adjusted for the effect of time, was determined by means of linear regression taking a confidence level of 95% into account (Table 4).

Table 4. Linear regression for substrate disintegration adjusted for the effect of time.

	Non-Standard Coefficients		Standard		
	Regression Coeficient	Standard Error	Coefficient Beta	T	Significance
(Constant)	211.371	8.076		26.172	0.000
Measuring Time	−0.621	0.257	−0.193	−2.417	0.017
PSC-Approach	25.733	4.445	0.463	5.790	0.000

Here, the regression coefficient 'Measuring time' represents the influence of the time factor on the methane production of the PSC approach. If the influence of the disintegration approach is adjusted for this value, a statement about the pure effect of the disintegration can be made. Thus, the analysis showed that methane production increased by an average of 25.733 mL/g VS compared to the reference approach.

Nonetheless, the influence of the time factor cannot be ignored in the final assessment. Moreover, possible interaction effects between disintegration approach and time should not be disregarded. Therefore, in a second regression analysis it was examined whether there is an interaction effect between the time of measurement and the approach and to what extent this influences the methane production of the PSC approach (Table 5).

Table 5. Linear regression to determine the interaction effect between PSC-approach and time.

	Non-Standard Coefficients		Standard		
	Regression Coeficient	Standard Error	Coefficient Beta	T	Significance
(constant)	138.600	11.959		11.589	0.000
Measuring time	4.074	0.674	1.269	6.048	0.000
PSC-approach	74.247	7.564	1.336	9.816	0.000
Interaction (time/PSC-approach)	−3.310	0.426	−1.771	−7.346	0.000

The analysis showed significant interaction effect between measuring time and the methane production of the PSC-approach at $p = 0.000$. A regression coefficient of −3.110 indicates a decreasing effect over time. In other words, the increase in methane yield from 25.733 mL/g VS could not be maintained until the end of the first retention time.

HRT 2

Nearly no trend over time and, therefore, steady-state conditions were confirmed for the methane production of all digesters given in Figure 2c. The box plots do not show any differences between the methane productions of the approaches. As the Levene-statistics indicate heterogeneity of variances a Welch's ANOVA verified the first impression by retaining the null hypothesis of 'There is no significant difference between the methane production of different approaches' at $p = 0.907$. Only a slight deviation is indicated within the PSC-approach. A further Welch's ANOVA delivered a significant difference between PSC-1 and PSC-2 at $p = 0.040$. However, the positive effect of disintegration on the methane production which could still be observed at the beginning of pre-treatment is no longer discernible towards the end of the first hydraulic retention time.

HRT 3

The final results the third hydraulic retention time are shown in Figure 2d. Taking the regression analysis into account, stationary conditions in methane production could only be determined for the reactors of the reference approach, whereas the fermenters, which were fed with pre-treated cattle manure, showed a clear decrease in the methane production indicating a trend over time. Since the Levene test indicated a heterogeneous distribution of the experimental data, the methane production of both the approaches and the reactors were statistically examined with the help of Welch's ANOVA. In both cases, the analysis revealed a significant result at $p = 0.000$. Considering the course of methane production shown in Figure 3a this result was expected. The following Games–Howell post-hoc test showed that the difference between the approaches was due to PSC-1, which differed significantly from PSC-2 ($p = 0.006$), U-1 ($p = 0.000$) and U-2 ($p = 0.000$), respectively. The positive influence of the pre-treatment, which had already disappeared towards the end of the first hydraulic retention time, could no longer be demonstrated at the end of the test.

3.3.2. pH-Value, Volatile Organic Acids and Total Inorganic Carbon (VOA/TIC) and Total Ammonia Nitrogen (TAN)

Figure 3a shows methane production of the duplicates for the PSC variant and the reference over the pre-phase and the 3 HRT. At the 127th day the pH-value dropped from 7.17 to 6.94 and from 7.28 to 6.70 in the PSC digestate, Figure 3b. The reference's digestate decreased less sharply from 7.22 to 7.08 and 7.21 to 7.11 between the days 120 to 127, Figure 3b. At the same time, the VOA (6.59 g/L) and VOA/TIC (1.16 gVOA/gCaCO$_3$) strongly increased particularly in reactor PSC-2, Figure 3c. During the test the TAN constantly decreased in the PSC reactors and in the references, Figure 3d. At 127th day 40 g of ammonium hydrogen carbonate were added to avoid acidification, but at day 134/135 the PSC reactors had to be finished due to the acidification.

3.4. Comparison of BMP and Semi-Continuous Anaerobic Digestion Tests

The methane yields of PSC manure in BMP and the average in semi-continuous AD were comparable, but the reference showed strong differences in methane yields. A reason for this was the chopping of the manure with a cutting mill (mesh size 6 mm) for the semi-continuous test, to facilitate feeding and sampling of the digesters in contrast to the uncrushed manure in BMP.

3.5. Outlook

The statistical analyses revealed that it would be advisable to check the steady-state conditions and the variances of methane production between the reactors in the pre-phase before changing to pre-treated substrate in semi-continuous tests. The storage conditions of manure and the treatment frequency will always influence the results. An optimized combination of temperature/treatment duration for PSC, the observation of effects of PSC on microbiological community, improved supply with nitrogen and trace elements might be a starting point for further investigations with the objective of stable methane production in (semi-)continuous AD-tests at lab-scale or full-scale.

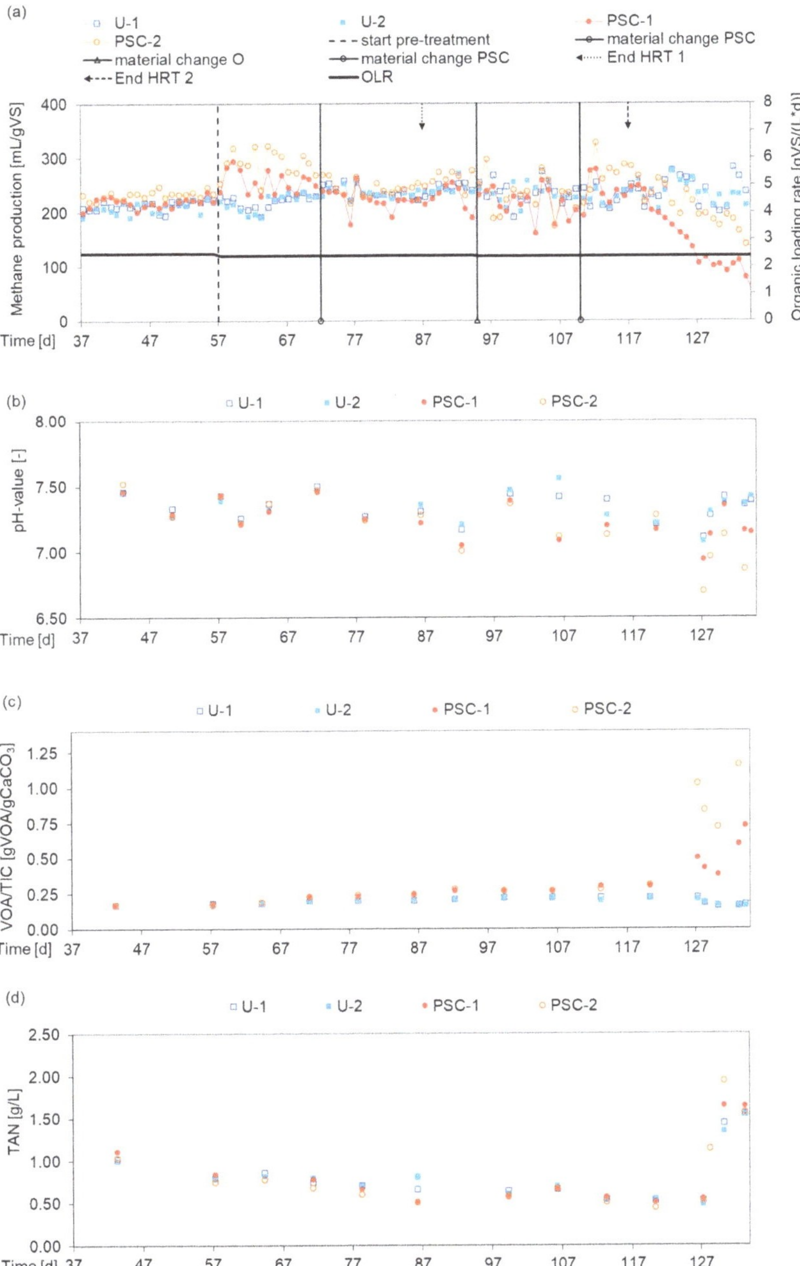

Figure 3. (**a**) Methane production of cattle manure in semi-continuously stirred tank reactor (CSTR) (U = untreated, PSC = pressure swing conditioning); (**b**) pH-value; (**c**) volatile organic acids and total inorganic carbon (VOA/TIC); (**d**) total ammonia nitrogen (TAN).

4. Conclusions

During the BMP test of PSC treatment of cattle manure, the methane yield increased by a minimum 12% and maximum 109% compared to the untreated reference; 160 °C/60 min and 190 °C/30 min showed the highest methane yields. The degradation was accelerated by PSC treatment and the increase of the methane yield after 29 days statistically significant, if the treatment duration was 30 or 60 min for both chosen temperatures (160 °C, 190 °C).

The methane yields of PSC manure in BMP and the average in semi-continuous AD were at the same level. The reference showed strong differences in methane yields because of extra chopping of the manure with a cutting mill for the semi-continuous AD, to facilitate feeding and sampling of the digesters. In semi-continuous AD tests in lab-scale CSTR the methane production of PSC treated cattle manure increased only temporarily during the first HRT, but a positive effect of PSC in steady state could not be proven.

However, for stable continuous processes substrate-specific parameter sets or ranges for PSC have to be determined and applied. Also, the framework conditions like nitrogen and trace element supply have to be optimal. It is presumable that the enhancement of the methane yields through PSC under full-scale conditions is higher than under lab-scale conditions because of the extra mechanical treatment of the reference. However, PSC has the potential to facilitate the handling, enhance degradation speed and, at best, to increase the methane yield of ligno-cellulosic substrates in full-scale biogas plants.

Supplementary Materials: The following are available online at http://www.mdpi.com/2306-5354/7/1/6/s1: Figure S1: Untreated for batch test; PCS-treated for batch /continuous tests, 'Untreated' chopped for continuous test, f.l.t.r. (Source: DBFZ).

Author Contributions: Conceptualization and methodology, B.S., H.W., S.W. and T.Z.; validation, H.W. and J.H.; formal analysis, B.S., T.Z., H.W. and S.W.; investigation, B.S. and H.W.; writing—original draft preparation, B.S., T.Z. and S.W.; writing—review and editing, S.W., J.H. and M.N.; visualization, B.S., H.W. and T.Z.; supervision, M.N.; project administration, B.S.; funding acquisition, B.S. All authors have read and agreed to the published version of the manuscript.

Funding: This research project SE.Biomethane (funding code 22028412) was funded by the Federal Ministry of Food and Agriculture based on a decision of the Parliament of the Federal Republic of Germany in the frame of 6th ERA-NET Bioenergy call.

Acknowledgments: The authors would also like to thank Henri Wernecke, Felix Friedrich and Robert Kühn of the company VENTURY GmbH (Dresden, Germany) for the technical support in conducting the pressure swing conditioning. The authors express their appreciation to colleagues from DBFZ-laboratory as well as to the student Michael Schmidt for his valuable contribution during the execution of semi-continuous AD-tests in lab-scale CSTR.

Conflicts of Interest: The authors declare no conflict of interest. The funders had no role in the design of the study; in the collection, analyses, or interpretation of data; in the writing of the manuscript; or in the decision to publish the results.

References

1. EurObserv'Er. Biogas Barometer 2017. November 2017. Available online: https://www.eurobserv-er.org/biogas-barometer-2017/ (accessed on 13 September 2019).
2. Daniel-Gromke, J.; Rensberg, N.; Denysenko, V.; Stinner, W.; Schmalfuß, T.; Scheftelowitz, M.; Nelles, M.; Liebetrau, J. Current Developments in Production and Utilization of Biogas and Biomethane in Germany. *Chem. Ing. Tech.* **2018**, *90*, 17–35. [CrossRef]
3. Scheftelowitz, M.; Thrän, D. Unlocking the Energy Potential of Manure—An Assessment of the Biogas Production Potential at the Farm Level in Germany. *Agriculture* **2016**, *6*, 20. [CrossRef]
4. Hendriks, A.T.W.M.; Zeeman, G. Pretreatments to enhance the digestibility of lignocellulosic biomass. *Bioresour. Technol.* **2009**, *100*, 10–18. [CrossRef] [PubMed]
5. Schumacher, B.; Wedwitschka, H.; Hofmann, J.; Denysenko, V.; Lorenz, H.; Liebetrau, J. Disintegration in the biogas sector—Technologies and effects. *Bioresour. Technol.* **2014**, *168*, 2–6. [CrossRef] [PubMed]
6. Risberg, K.; Sun, L.; Levén, L.; Horn, S.J.; Schnürer, A. Biogas production from wheat straw and manure—Impact of pretreatment and process operating parameters. *Bioresour. Technol.* **2013**, *149*, 232–237. [CrossRef] [PubMed]

7. Budde, J.; Heiermann, M.; Suárez Quiñones, T.; Plöchl, M. Effects of thermobarical pretreatment of cattle waste as feedstock for anaerobic digestion. *Waste Manag.* **2014**, *34*, 522–529. [CrossRef]

8. Barakat, A.; Monlau, F.; Steyer, J.-P.; Carrere, H. Effect of lignin-derived and furan compounds found in lignocellulosic hydrolysates on biomethane production. *Bioresour. Technol.* **2012**, *104*, 90–99. [CrossRef]

9. Monlau, F.; Sambusiti, C.; Barakat, A.; Quemeneur, M.; Trably, E.; Steyer, J.-P.; Carrere, H. Do furanic and phenolic compounds of lignocellulosic and algae biomass hydrolyzate inhibit anaerobic mixed cultures? A comprehensive review. *Biotechnol. Adv.* **2014**, *32*, 934–951. [CrossRef] [PubMed]

10. Siqueira, M.R.; Reginatto, V. Inhibition of fermentative H_2 production by hydrolysis byproducts of lignocellulosic substrates. *Renew. Energy* **2015**, *80*, 109–116. [CrossRef]

11. Pekařová, S.; Dvořáčková, M.; Stloukal, P.; Ingr, M.; Šerá, J.; Koutny, M. Quantitation of the Inhibition Effect of Model Compounds Representing Plant Biomass Degradation Products on Methane Production. *Bioresources* **2017**, *12*, 2421–2432. [CrossRef]

12. *VDI 4630: 2006. Fermentation of Organic Materials—Characterisation of the Substrate, Sampling, Collection of Material Data, Fermentation Tests*; Beuth Verlag GmbH: Berlin, Germany, 2006.

13. Schumacher, B.; Wedwitschka, H.; Weinrich, S.; Mühlenberg, J.; Gallegos, D.; Oehmichen, K.; Liebetrau, J. The influence of pressure swing conditioning pre-treatment of chicken manure on nitrogen content and methane yield. *Renew. Energy* **2019**, *143*, 1554–1565. [CrossRef]

14. Liebetrau, J.; Pfeiffer, D.; Thrän, D. (Eds.) 2016 Collection of Measurement Methods for Biogas—Methods to determine parameters for analysis purposes and parameters that describe processes in the biogas sector. In *Series of the Funding Programme "Biomass Energy Use"*; Volume 07, Leipzig—ISSN online—2364-897X; Available online: https://www.energetische-biomassenutzung.de/fileadmin/user_upload/Downloads/Ver%C3%B6ffentlichungen/07_MMS_Biogas_en_web.pdf (accessed on 30 October 2019).

15. DIN. *EN 12880. Characterization of Sludges—Determination of Dry Residue and Water Content*; DIN Deutsches Institut für Normung e. V.: Berlin, Germany, 2001.

16. DIN. *EN 12879. Characterization of Sludges—Determination of the Loss on Ignition of Dry Mass*; DIN Deutsches Institut für Normung e. V.: Berlin, Germany, 2001.

17. Angelidaki, I.; Alves, M.; Bolzonella, D.; Borzacconi, L.; Campos, J.L.; Guwy, A.J.; Kalyuzhnyi, S.; Jenicek, P.; van Lier, J.B. Defining the biomethane potential (BMP) of solid organic wastes and energy crops: A proposed protocol for batch assays. *Water Sci. Technol.* **2009**, *59*, 927–934. [CrossRef] [PubMed]

18. Hofmann, J.; Peltri, G.; Sträuber, H.; Müller, L.; Schumacher, B.; Müller, U.; Liebetrau, J. Statistical Interpretation of Semi-Continuous Anaerobic Digestion Experiments on the Laboratory Scale. *Chem. Eng. Technol.* **2016**, *39*, 643–651. [CrossRef]

19. Rivard, C.J.; Grohmann, K. Degradation of Fufural (2-Furaldehyde) to Methane and Carbon Dioxide by an Anaerobe Consortium. *Appl. Biochem. Biotech.* **1991**, *28–29*, 285–295. [CrossRef] [PubMed]

 *bioengineering*

Article

Comparison of Dry Versus Wet Milling to Improve Bioethanol or Methane Recovery from Solid Anaerobic Digestate

Florian Monlau [1,*], Cecilia Sambusiti [2] and Abdellatif Barakat [2]

1 APESA, Pôle Valorisation, Cap Ecologia, Avenue Fréderic Joliot Curie, 64230 Lescar, France
2 UMR, IATE, CIRAD, Montpellier SupAgro, INRA, Université de Montpellier, 34060 Montpellier, France; cecilia.sambusiti@gmail.com (C.S.); abdellatif.barakat@inra.fr (A.B.)
* Correspondence: florian.monlau@apesa.fr; Tel.: +33688491845

Received: 26 July 2019; Accepted: 15 August 2019; Published: 6 September 2019

Abstract: Biogas plants for waste treatment valorization are presently experiencing rapid development, especially in the agricultural sector, where large amounts of digestate are being generated. In this study, we investigated the effect of vibro-ball milling (VBM) for 5 and 30 min at a frequency of $20\ s^{-1}$ on the physicochemical composition and enzymatic hydrolysis ($30\ U\ g^{-1}$ total solids (TS) of cellulase and endo-1,4-xylanase from *Trichoderma longibrachiatum*) of dry and wet solid separated digestates from an agricultural biogas plant. We found that VBM of dry solid digestate improved the physical parameters as both the particle size and the crystallinity index (from 27% to 75%) were reduced. By contrast, VBM of wet solid digestate had a minimal effect on the physicochemical parameters. The best results in terms of cellulose and hemicelluloses hydrolysis were noted for 30 min of VBM of dry solid digestate, with hydrolysis yields of 64% and 85% for hemicelluloses and cellulose, respectively. At the condition of 30 min of VBM, bioethanol and methane production on the dry solid separated digestate was investigated. Bioethanol fermentation by simultaneous saccharification and fermentation resulted in an ethanol yield of $98\ g_{eth}\ kg^{-1}$ TS (corresponding to 90% of the theoretical value) versus $19\ g_{eth}\ kg^{-1}$ TS for raw solid digestate. Finally, in terms of methane potential, VBM for 30 min lead to an increase of the methane potential of 31% compared to untreated solid digestate.

Keywords: anaerobic digestion; solid digestate; milling process; sugars recovery; energy balances; bioethanol production

---

1. Introduction

Agricultural wastes represent a vast biomass resource that can readily be converted into sustainable biofuels, chemicals, and other economic by-products by use of a biorefinery concept. As anaerobic digestion has become an efficient technology for waste treatment and renewable energy production, it has resulted in a substantial increase in the use of agricultural anaerobic digesters throughout the EU [1,2]. Anaerobic digestion (AD) is a well-established biological process that has been in use for a long time to anaerobically degrade organic materials into biogas, which is a mixture of CH_4 (50–70%) and CO_2 (30–50%), and digestate [3]. After a cleaning process, biogas can be further converted into heat and electricity through cogeneration heat and power (CHP) systems [4]. In parallel, biogas can also be upgraded into biomethane through several technologies (e.g., chemical absorption, membrane separation, water scrubbing, pressure swing adsorption, etc.) as a substitute of natural gas or its use as a transport fuel [3]. Digestate is mainly comprised of water (over 90%), residual undegraded fibers (e.g., cellulose, hemicelluloses, and lignin), and inorganic compounds (e.g., ash), and it is generally mechanically separated into liquid and solid fractions that are stored and handled separately [5,6]. Nevertheless, due to the low hydraulic retention times (HRTs) that are generally applied in industrial biogas plants, after which biogas production starts to decrease, part of the organic matter remains in the solid phase of the digestate [7,8].

Consequently, research on alternative valorization routes for solid separated digestate to reduce the environmental impact of disposal and to improve the economic profitability of AD plants is gaining a great interest from the scientific community [5,6]. To date, the valorization strategies that have been investigated are thermochemical processes such as torrefaction and hydrothermal carbonization [9], pyrolysis [4,10], and gasification [11]. Few studies to date have investigated how to use the remaining organic matter present into the solid digestate for sugar platform production and to generate further biofuels such as bioethanol or methane [12,13]. As reported by Santi et al., (2015), digestate produced by commercial agricultural biogas plants still contains a considerable quantity of cell wall polymers [8]. These could potentially be back to AD after a pretreatment step [7] or used in biorefinery processes, for sugar platform production (from C_5 and C_6) by enzymatic hydrolysis and further generation of biofuels such as bioethanol [14,15].

Nonetheless, due to the recalcitrant properties of the organic matter remaining in solid digestate, a pretreatment step is necessary to improve its biodegradability and further enzymatic hydrolysis and/or biofuel production such as bioethanol or methane by recirculation into the AD process [16–18]. Several pretreatment technologies have been tested with agricultural wastes such as physical, thermal, thermochemical, and biological pretreatment, on their own or in combination with each other [17]. Thermal and thermochemical pretreatments have been extensively investigated in terms of their ability to enhance enzymatic hydrolysis and further bioethanol production, although they can also result in potential inhibitors (e.g., furans and polyphenols) generation for microorganisms and strains [19,20]. Another promising option is the use of mechanical pretreatment, which avoids the production of potential inhibitors in addition to improving the physical properties (e.g., the crystallinity, accessible surface area, and particle size) for further enzymatic hydrolysis and/or biofuel production [21,22]. Over the past decades, a number of mechanical size-reduction processes have been developed and investigated including ball mills, vibratory mills, hammer mills, knife mills, colloids mills, two-roll mills, and extruders [21,22].

Although mechanical processes are commonly used in the biorefinery process [21], few studies have investigated the use of mechanical fractionation on solid separated digestate of agricultural biogas plants [12]. Among these fractionation methods, vibro-ball milling (VBM) has the advantages of already being used at the industrial level and of having the capacity to process both dry and wet biomasses. Indeed, when the biogas is valorized through a CHP system, excess heat can be available for drying of the solid digestate [4]. VBM has previously been shown to be a promising technology for alteration of the physical properties (e.g., the particle size and the crystallinity index) and improvement of enzymatic hydrolysis [23–25]. Therefore, the aims of this study were the following ones: (i) to evaluate the impact of vibro-ball milling on the physicochemical properties of both dry and wet solid digestates; (ii) to evaluate the effect of vibro-ball milling on the enzymatic hydrolysis performances of dry and wet solid anaerobic digestates; (iii) to derive the energetic balances of the various scenarios investigated; and (iv) to assess the best conditions for bioethanol and methane production compared to raw solid digestate.

2. Materials and Methods

2.1. Samples Preparation and Mechanical Pretreatment

Solid separated digestate (SS-DIG) was collected from a mesophilic full-scale AD plant in the Lombardy region of northern Italy. The plant was fed with 95 tons fresh matter per day, composed of 42 wt% of maize silage, 5 wt% of cow manure, and 53 wt% of cow sewage. Table 1 lists the main operational characteristics of the anaerobic digestion plant. SS-DIG was recovered from the solid-to-liquid separator (helical screw press). Once collected, a quantity of SS-DIG sample (referred to as "dry-SS-DIG") was dried at 37 °C for 48 h. Another sample (referred to as "wet-SS-DIG") was stored in gas-tight containers at 3 °C before being used. The initial solid digestate was composed of 22.4 (± 1.1) g TS/100 g of raw material and of 89.2 (± 2.3) g VS/100 g TS.

Table 1. The main operational characteristics and performances of the agricultural biogas plant unit.

Anaerobic Digester Parameters	
Number of reactors	2 digesters, 1 post-fermenter, 1 storage tank
Reactors Volume (m^3)	Digesters: 2 × 2400 Post-fermenter: 2700 Storage tank: 2700
Feeding (t FM day^{-1}) [a]	95
HRT (d) [a]	80
pH [a]	7.9
Temperature (°C) [a]	40
Biogas	
Biogas (Nm^3 day^{-1})	12,000
Methane (%)	53
Total Energy (kWh day^{-1}) [b]	63,600

[a] Referred to digesters and post-fermenter; [b] 10 kWh Nm^{-3} methane.

Two grams (in equivalent of dry matter) of both dry SS-DIG and wet SS-DIG samples were milled using a vibro-ball mill "VBM" (MM400, Retsch, Düsseldorf, Germany) at a frequency of 20 s^{-1}, at ambient temperature, for 5 or 30 min. For each condition, the experiment was repeated ten times to overcome heterogeneity of the SS-DIG samples. This apparatus consists of two 20-mL milling cups containing two stainless steel balls (2 cm diameter) each.

2.2. Enzymatic Hydrolysis

Enzymatic hydrolysis of untreated and milled dry and wet-SS-DIG samples was performed using a mixture of enzymes (cellulase and endo-1,4-xylanase from *Trichoderma longibrachiatum*, Sigma-Aldrich®, Saint Louis, Missouri, United States) at 30 U/g TS, each. Enzymatic hydrolysis was carried out at a solids concentration of 50 g TS/L in 50 mM sodium acetate buffer (pH 5.0), 0.5 g TS L^{-1} chloramphenicol (Sigma–Aldrich®, Saint Louis, Missouri, United States) at 37 °C for 72 h, with stirring [12]. Each test was performed in triplicate. Liquid samples were withdrawn at 72 h and, after centrifugation, the supernatants were analyzed to determine the concentrations of monomeric sugars (i.e., glucose, xylose, and arabinose) using an HPLC (Waters Corporation, Milford, Connecticut, USA) equipped with a BioRad HPX-87H column at 40 °C, a refractive index detector at 40 °C, and 0.005 M H$_2$SO$_4$ as the solvent at 0.3 mL/min [12]. These analyses were performed in duplicate.

The cellulose and hemicelluloses hydrolysis yields were defined as follows:

$$\text{Cellulose yield\%} = \frac{[\text{Glucose}]}{1.111f} \times 100\%, \tag{1}$$

$$\text{Hemicelluloses yield\%} = \frac{[\text{Xylose} + \text{Arabinose}]}{1.136f} \times 100\%, \tag{2}$$

where [Glucose] is the glucose concentration (g/100 g TS); [Xylose + Arabinose] is the xylose and arabinose concentration (g/100 g TS); and f is the cellulose or hemicellulose fraction in the dry biomass (g/100 g TS).

2.3. Bioethanol Fermentation

The ethanol yields of raw and pretreated SS-DIG were evaluated and compared by simultaneous saccharification and fermentation experiments (SSF). The SSF tests were performed in duplicate on unsterilized samples, using 7 mL flasks (a working volume of 5 mL) sealed with rubber septa and equipped with an air vent system, comprising a sterilized needle and filter, in order to evacuate the CO_2 produced during fermentation. A lyophilized yeast *Saccharomyces cerevisiae* strain was used as the inoculum. For this, lyophilized cells were washed and suspended in sterilized distilled water at a final concentration of 30 g TS L^{-1} [12]. Each flask contained: 0.25 g TS of sample (at 50 g TS L^{-1}); cellulose and endo-1,4-xylanase enzymes (Sigma–Aldrich®, Saint Louis, Missouri, United States) at 20 U g^{-1} TS each; 0.25 mL of yeast (30 g TS L^{-1}); and 0.5 mL of nutrient solution containing 50 g TS L^{-1} yeast extract (Yeast Extract, Technical, BactoTM, Becton, Dickinson and Company, Rutherford, NJ, USA), 4 g TS L^{-1} urea (Sigma–Aldrich®, Saint Louis, Missouri, United States), 0.5 g TS L^{-1} chloramphenicol (Sigma–Aldrich® , Saint Louis, Missouri, United States), and 50 mM acetate buffer (pH 5.0). The flasks were incubated at 37 °C for 72 h with stirring at 500 rpm [12].

Samples (200 µL each) were withdrawn at 0, 2, 6, 24, 48, and 72 h and the cell-free supernatants were analyzed to determine the ethanol, cellobiose, glucose, xylose, and arabinose concentrations using an HPLC system (Waters Corporation, Milford, Connecticut, USA) equipped with a BioRad HPX-87H column at 40 °C, a refractive index detector at 40 °C, and 0.005 M H_2SO_4 as the solvent at 0.3 mL min^{-1}. These tests were performed in duplicate.

Ethanol yields (g ethanol kg^{-1} TS) were calculated according to:

$$\text{Ethanol yield (t)} = [\text{Ct ethanol} / \text{C solid}] \times 1000, \tag{3}$$

where Ct ethanol (g ethanol L^{-1}) is the concentration of ethanol produced during SSF, at time t; C solid (g TS L^{-1}) is the total solids concentration in the flask.

2.4. Biochemical Methane Potential

Treated and untreated SS-DIG were digested in batch mesophilic anaerobic flasks. The volume of each flask was 1 L, with a working volume of 600 mL, the remaining 400 mL volume serving as headspace. SS-DIG and SS-DIG pretreated were introduced into the flasks with the inoculum ata substrate to inoculum ratio of 0.5 g VS/g VS. The inoculum was an inoculum acclimated in our laboratory with a feed composed of wastewater sludge and grasses. The inoculum contained 34 g/L total solids (TS) and 23 g/L volatile solids (VS). Once the flasks were prepared, degasification with nitrogen was carried out to obtain anaerobic conditions and the bottles were closed. Triplicate bottles were incubated at 35 °C. Assays with only inoculum were performed also in triplicate as blank control in the same condition. Biogas volume was monitored by pressure measurement using a manometer (Digitron 2023P, Digital Instrumentation Ltd., Corby, UK). Biogas composition was determined using a gas chromatograph (Varian GC-CP4900, Agilent, Ratingen, Germany) equipped with two columns: the first (Molsieve 5A PLOT, Agilent, Ratingen, Germany) was used at 110 °C to separate O_2, N_2 and CH_4, the second (HayeSep A, Agilent, Ratingen, Germany) was used at 70 °C to separate CO_2 from other gases. The injector temperature was 110 °C and the detector was 55 °C. The detection of gaseous compounds was done using a thermal conductivity detector. The calibration was carried out with two standard gasses composed of 35% CO_2, 5% O_2, 20% N_2 and 40% CH_4 for the first one and 9.5% CO_2, 0.5% O_2, 80% N_2 and 10% CH_4 for the second one (special gas Air Liquide®, Paris, France). All the results were expressed as biogas volume produced at normal conditions (in NL at 0 °C, 1013 hPa).

2.5. Analytical Determinations

Total Solids (TS), Volatile Solids (VS), and the ash content were analyzed according to a protocol used in previous studies [24]. The N content was determined with an elemental analyzer (VarioMacroCube, Elementar, Frankfurt, Germany). Protein levels were estimated by multiplying the

N content by 6.25. Structural-carbohydrates from cellulose and hemicelluloses were measured using a strong acid hydrolysis method as described elsewhere [4]. All of the monosaccharides (i.e., glucose, xylose, and arabinose) were analyzed using an HPLC system (12620 Infinity II, Agilent, Ratingen, Germany) equipped with a column at 40 °C (Hi-Plex H, Agilent, Ratingen, Germany), a refractive index detector at 40 °C, and 0.005 M H_2SO_4 as the solvent at 0.3 mL/min. The system was calibrated with glucose, xylose, and arabinose standards (Sigma-Aldrich®, Saint Louis, Missouri, United States). The particle size distribution of untreated SS-DIG was determined using a vibratory sieving apparatus (Analytical Sieve Shaker AS 200, Retsch®, Düsseldorf, Germany) equipped with seven sieves of different sizes (2.5; 1; 0.71; 0.63; 0.45; 0.32 and 0.13 mm). The particle sizes of milled SS-DIG samples were analyzed using a laser granulometry (Mastersizer 2000, Malvern Instrument, Worcestershire, UK). Cellulose crystallinity (CrI) was determined using X-ray diffractometer (D8 Advance, Bruker, Aubrey, Texas, USA) as described elsewhere [26]. All of the analytical determinations were performed in duplicate.

2.6. Energy Balances of the Scenarios

Energy balances were computed by considering the additional electric and heat energy requirements for both scenarios: vibro-ball milling of dry SS-DIG and vibro-ball milling of wet SS-DIG.

The thermal energy requirement (kWh_{th} kg^{-1} TS) for drying SS-DIG samples (E_{th_drying}) before the vibro-ball milling process was calculated using Equations (4)–(8):

$$E_{th\_drying} = E_{Heat} + E_{Evaporation}, \tag{4}$$

where E_{Heat} (kWh t^{-1} TS) is the energy requirement to increase the temperature of water and digestate from 25 °C to 100 °C, $E_{Evaporation}$ (kWh_{th} day^{-1}) is the energy of evaporation of water at 105 °C. E_{Heat} and $E_{Evaporation}$ were calculated according to the equations below:

$$E_{Heat} = m \times Cp \times [T_{Final} - T_{Initial}]/3600, \tag{5}$$

where m is the mass of water and digestate (equivalent to 1 kg of TS), Cp is the water specific heat (4.18 kJ kg^{-1} $°C^{-1}$); $T_{Initial}$ (°C) is the initial temperature of the substrate suspension, assumed to be 25 °C; T_{Final} (°C) is the final temperature of 105 °C.

$$E_{Evaporation} = [m_{water} \times Lv]/3600, \tag{6}$$

where Lv = 2380 kJ kg^{-1} and m is the mass of water in the digestate sample (equivalent to 1 kg of TS).

The electric energy requirement ($E_{el\_milling}$, kWh kg^{-1} TS) to mill SS-DIG was evaluated by using a watt-meter. The active power (watts), active electric energy (watts), frequency (hertz), and time (min) were logged into a PC card at 1-s intervals. The energy requirement was calculated according to Equation (7) below:

$$Eel\_milling = \int_0^t (Pt)dt \, / \, m, \tag{7}$$

where Pt is the power consumed in watts at time t and m is the mass of biomass to be ground in kg. Both Pt were measured in triplicate for each sample.

The energy efficiency (η, kg C6 $_{monomeric\ sugar}$/kWh) was calculated using Equation (8):

$$\eta = C6\ Monomeric\ sugar\ yield\ /\ E, \tag{8}$$

where the monomeric sugar yield (kg/kg TS) is the amount of glucose produced during enzymatic hydrolysis of feedstock or digestate and E (kWh kg^{-1} TS) corresponding to the electrical and thermal energy requirements ($Eel\_milling$ and/or E_{th_drying}). The two energy efficiency coefficients investigated were "η_{el}", which is the ratio of soluble sugars (C6) and the electrical requirement, and "η_{tot}", which is the ratio of soluble sugars (C6) and the overall energy requirement including both thermal and electrical requirements.

3. Results and Discussion

3.1. Physicochemical Characteristics of Untreated and Pretreated Digestate

The chemical compositions of untreated and VBM solid digestates, after mechanical screw separation, are listed in Table 2. There was a high volatile solids (VS) content of 89.2 g/100 g TS, mainly due to residual proteins and fibers (i.e., cellulose hemicelluloses, and lignin) that were not degraded during the AD process. The ash content was in accordance with the values obtained in our previous study [4], but lower than those of Ref. [27] who reported ash contents that varied from 23 to 38 g/100 g TS for four digestates derived from agricultural mesophilic AD plants that mainly treated manures, slurries, and silages [27]. The protein content (approximately 10.5 g/100 g TS), derived mainly by co-digestion of maize silage with cow manure and sewage, was in accordance with the values obtained by [4]. Similarly, Ref. [27] reported protein contents that varied from 10.5–12 g/100 g TS for various digestates from agricultural biogas plants [27]. In parallel, there was a large amount of residual carbohydrate polymers (cellulose, hemicelluloses), because during AD a large part of the biodegradable matter is not degraded, mainly due to physicochemical barriers that limit processing by bacteria [8,28]. In terms of the carbohydrate composition, there was a slightly higher cellulose content than hemicelluloses, as has also been observed by other authors who demonstrated that the anaerobic digestion process consumes hemicelluloses at a faster rate than cellulose [12]. Finally, there was a large amount of Klason lignin after the AD process (27 g/100 g TS). Such polymers are not degraded during the AD process and consequently accumulate in solid digestate [8,29].

Table 2. Physicochemical characteristics of untreated and pretreated digestate by vibro-ball milling of dry and wet solid separated digestate (SS-DIG).

Parameters	SS-DIG	VBM Dry—5 min	VBM Dry—30 min	VBM Wet—5 min	VBM Wet—30 min
Volatile Solids (g/100 g TS)	89.2 ± 2.3	91.3 ± 0.4	91.7 ± 0.5	91.0 ± 0.2	90.0 ± 0.2
Proteins (g/100 g TS)	nd	10.5 ± (0.1)	9.9 ± (0.3)	9.7 ± (0.6)	10.8 ± (0.0)
Cellulose (g/100 g TS)	nd	22.9 ± (3.1)	20.2± (0.4)	22.2 ± (1.5)	21.1 ± (1.5)
Hemicelluloses (g/100 g TS)	nd	16.6 ± (1.9)	16.5 ± (0.2)	16.9 ± (1.5)	17.9 ± (2.2)
Lignin (g/100 g TS)	nd	26.9 ± (0.3)	28.2± (0.8)	26.3 ± (0.3)	29 ± (1.3)
Ash (g/100 g TS)	10.1 ± 1.9	8.2 ± (0.4)	8.1 ± (0.9)	10.1 ± (1.9)	9.9 ± (0.2)
Particle size (D50, μm)	-	35 (± 7)	48 (± 2)	202 (± 28)	162 (± 34)
Cri (%)	44	32	11	46	41

nd = not determined.

In terms of the chemical composition (cellulose, hemicelluloses, lignin, and protein), no significant changes were observed before and after VBM of both dry and wet solid separated digestate (SS-DIG), as shown in Table 2. By contrast, VBM had a significant effect on the physical properties (i.e., particle size, CrI). First of all, VBM led to a significant decrease in the particle size of both dry and wet SS-DIG compared to untreated samples (Table 2 and Figure 1). However, there was a much more pronounced decrease in particle size with dry milling than with wet milling, thus confirming that the moisture content affects the decrease in particle size, as noted previously by [21]. Similar results were also noted by comparing dry and wet planetary ball milling as the pretreatment of *Pennisetum hybrid* [30]. Indeed, Ref. [30] found that dry milling had a greater impact on dry than wet biomass [D_{50} (μm)] of 131 μm and 243 μm were observed for dry and wet biomass, respectively, after 6 hours of milling) [30]. The results reveal that VBM pretreatment led to a reduction in the CrI index for dry SS-DIG, with a decrease in the CrI that varied from 27% to 75% after 5 min and 30 min of vibro-ball milling, respectively. Ref. [31] also observed a significant decrease in the CrI index after ball milling (30–120 min) of both sugarcane bagasse and straw [31]. Similarly, Ref. [23]reported significant decreases in the CrI of 55% and 96% compared to untreated sugarcane bagasse after 1 h and 3 h of VBM, respectively [23]. During the ball milling process, the shearing and compressive forces lead to a decrease in crystallinity [22,32,33]. By contrast, VBM had little or no impact on the CrI of wet SS-DIG. The lower degree of crystallinity of dry-milled biomass compared to wet may be due to the higher mechanical force from the dry milling process causing considerable change in the crystalline structure [32,34].

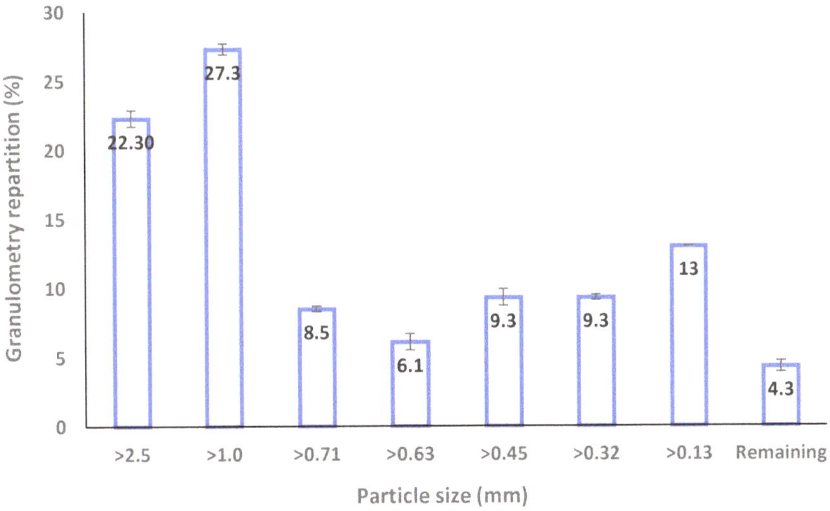

Figure 1. Granulometry repartition (in % of dry biomass) of the raw solid digestate.

3.2. Enzymatic Hydrolysis

The results show that VBM enhanced the enzymatic hydrolysis of the cellulose and the hemicelluloses fractions in both the dry and the wet mode (Figure 2). Raw digestate led to poor hydrolysis yields of cell carbohydrate polymers, with yields of 17% and 13% for cellulose and hemicelluloses, respectively. Similarly, a previous study reported that AD fibers were poor substrates for saccharification, with a sugar yield of only 12% of the original value on a dry basis [35]. In the case of VBM of dry SS-DIG, an increase in the duration of the pretreatment led to a higher level of enzymatic hydrolysis for both cellulose and hemicelluloses (Figure 2). Indeed, the yield of cellulose hydrolysis increased from 40% to 85% for 5 and 30 min of VBM, respectively. Similarly, the yield of hemicelluloses hydrolysis increased from 36% to 63% for 5 and 30 min of VBM, respectively. This improvement can be correlated to the decrease in both the crystallinity and the particle size when the duration of the ball milling of dry SS-DIG increased. This observation is supported by a number of authors who have demonstrated that during enzymatic hydrolysis of cellulose the readily accessible regions (i.e., the amorphous regions) are hydrolyzed more efficiently than the crystalline regions [36,37]. Furthermore, the surface area and the number of reactive sites increased due to reduction of the size of the substrate, thereby facilitating the adsorption of enzyme and hence the initial rate of hydrolysis [36]. This increase in enzymatic hydrolysis after mechanical processing has been reported previously in several investigations [23,31,38]. For instance, Ref. [23] reported a cellulose hydrolysis yield of 95% after 3 hours of VBM of sugarcane bagasse, whereas Ref. [39] reported a cellulose hydrolysis yield of 56% after 30 min of VBM of corn stover. The effect of VBM on wet digestate was less pronounced with a similar duration. Indeed, at 30 min of VBM, cellulose hydrolysis yields of 37% and 85% and hemicelluloses hydrolysis yields of 23% and 63% were obtained for wet and dry solid digestates, respectively. Such results can be explained by the lower impact of VBM on the particle size and the cellulose crystallinity of wet digestate, as indicated previously in Table 2.

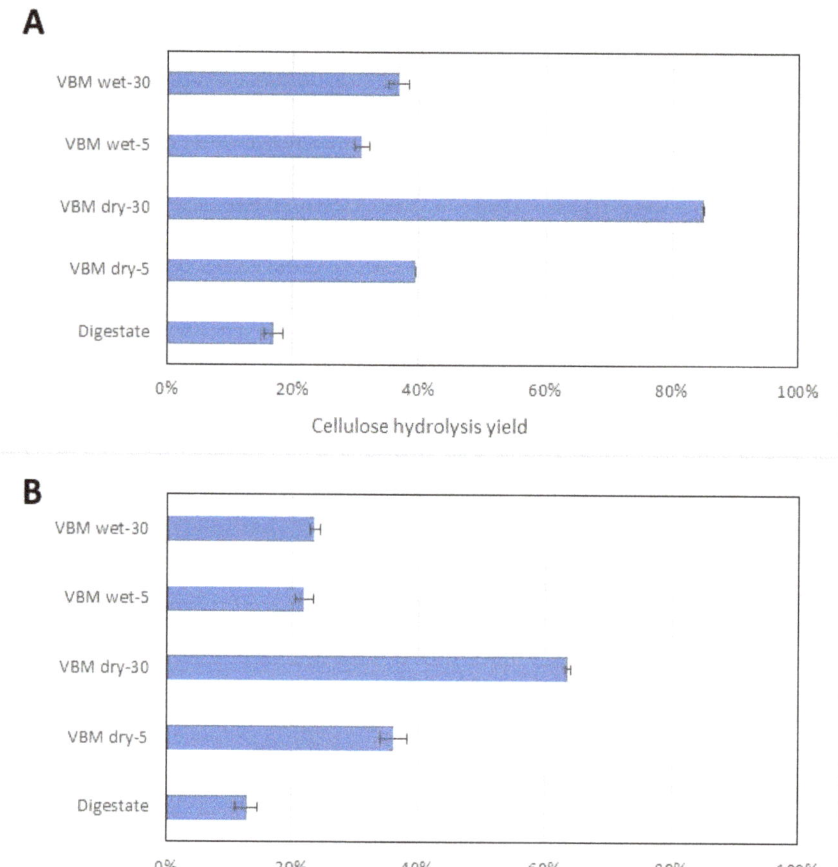

Figure 2. Enzymatic hydrolysis performances in terms of cellulose hydrolysis yield (**A**) and the hemicelluloses hydrolysis yields (**B**) for raw SS-DIG and vibro-ball milling (VBM)-pretreated samples in wet and dry modes.

3.3. Energy Balance and Energy Efficiency Considerations

The total energy requirement for the drying and milling process of SS-DIG (Eel_milling and/or E_{th_drying}) is presented in Table 3. A thermal energy requirement "E_{th_drying}" of 3 kWh$_{th}$ kg^{-1} TS was necessary to dry the solid digestate for further mechanical fractionation in dry mode. As demonstrated previously, if the biogas is converted by a CHP system, the excess heat produced during AD can meet the drying needs for the solid digestate [4,12]. In parallel, the electrical energy requirements (Eel_milling) for the VBM process were 9.9 and 54.4 kWh$_{el}$ kg^{-1} TS for a duration of 5 min and 30 min, respectively. Similar values have been previously reported on various mechanical pretreatments even if the technologies used (e.g., ball milling, vibro-ball milling, hammer milling, and knife milling) and the substrates' origin can influence the results [21,22]. For instance, Ref. [25] have reported energy requirements for dry ball milling of rice straw and consumption of 2.5 kWh$_{el}$ kg^{-1} TS and 30 kWh$_{el}$ kg^{-1} TS were observed for respectively 5 min and 30 min of pretreatment [25]. Some other mechanical systems (hammer mill, knife mill) have reported lower energy consumption on lignocellulosic biomasses but the final size particle of the substrates was higher than reported in our study [22,40]. In addition to enzymatic hydrolysis, in order to compare the various VBM modalities, the energy efficiency coefficients were calculated, as shown in Table 3 [41]. In general, the highest η corresponds to the most effective pretreatment [39,41]. Both total sugar recovery and pretreatment energy efficiency should be used in evaluating and comparing the performance of pretreatment processes [41,42]. Interestingly, the best η_{el} for the dry modality was obtained at VBM for 5 min, with a value of 0.102 kg sugars kWh_{el}^{-1}. Considering the thermal energy requirement as imput needed for SS-DIG drying, the energy efficiency η_{tot} was reduced from 0.102 to 0.078 kg sugars kWh^{-1}. Aside from a higher level of enzymatic hydrolysis of dry solid digestate at VBM for 30 min, the η_{el} was worse due to the higher energy requirement for VBM. In regard to wet SS-DD, η_{el} of 0.076 kg sugars kWh_{el}^{-1} and 0.015 kg sugars kWh_{el}^{-1} were noted for 5 min and 30 min of VBM, respectively. Such values are in agreement with previous one that reported η_{el} from 0.011 to 0.078 kg sugars kWh_{el}^{-1} after ball milling of rice straw for 60 and 5 min respectively [21,25].

Table 3. Electrical and thermal energy requirements and energy efficiencies of the various milling modes applied to dry and wet SS-DIG.

	Electrical Consumption (kWh$_{el}$ kg^{-1} TS)	Thermal Energy (kWh$_{th}$ kg^{-1} TS)	C6 Sugars (g kg TS^{-1})	Efficiency η_{el} (kg Sugars kWh$_{el}^{-1}$)	Efficiency η_{tot} (kg Sugars kWh^{-1})
SS-DIG	-	-	44		
VBM dry—5 min	9.9	3.0	101	0.102	0.078
VBM dry—30 min	54.4	3.0	191	0.035	0.033
VBM wet—5 min	9.9	-	76	0.076	0.076
VBM wet—30 min	54.4	-	86	0.015	0.015

In terms of industrial scale-up, the choice of the best scenario will be directly dependent on the on-site valorization of the biogas of the agricultural biogas plant. If the biogas is valorized through a CHP system, dry modalities of VBM can be pertinent for digestate valorization, as part of the heat excess from the cogeneration unit can be used to dry the digestate. In the case where biogas is upgraded into biomethane for gas injection, VBM to improve bioethanol or methane recovery from solid digestate has less merit. When optimizing VBM conditions, it is important to find a good compromise between the enzymatic hydrolysis performances and energy aspects represented by the coefficient of efficiency. As there was little difference between the η_{tot} of dry SS-DIG at 5 min and 30 min, the 30 min conditions were hence selected for bioethanol production.

3.4. Bioethanol Fermentation and Methane Potential

Bioethanol production was tested on the dry SS-DIG at the 30 min VBM conditions that led to the best level of cellulose hydrolysis (Figure 2 and Table 3). Bioethanol on raw digestate led to a low ethanol yield of 19 g_{eth} kg^{-1} TS corresponding to only 17% of the theoretical value, as shown in Figure 3. VBM of dry SS-DIG for 30 min led to an enhancement of the ethanol yield, with 98 g_{eth} kg^{-1} TS, corresponding to 90% of the theoretical value. Lower yields were reported by Ref. [18], with an ethanol yield of 37 g_{eth} kg^{-1} TS with dry SS-DIG that was mechanically pretreated (centrifugal milling). Other studies have demonstrated the benefit of pretreated SS-DIG to improve bioethanol production [14,15]. For instance, thermo-alkaline pretreatment has led to bioethanol recoveries from solid digestate of 75% to 80% compared to the theoretical value [13,14]. In future studies, it would be interesting to test other strains or bacteria capable of producing bioethanol from C_5 sugar sources to improve bioethanol recovery from SS-DIG hydrolysate, as C_5 sugars were not consumed in the present assay, as shown in Figure 3.

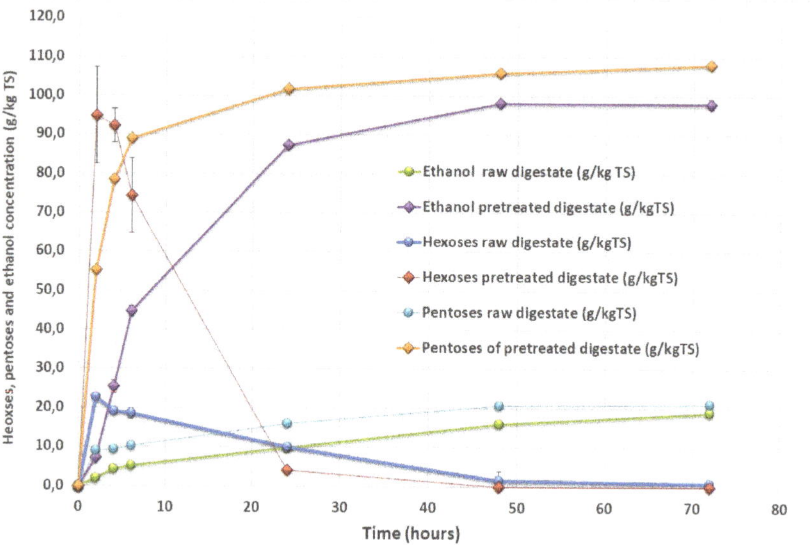

Figure 3. Hexose consumption and ethanol production for raw solid separated digestate and pretreated solid separated digestate with VBM for 30 min.

In parallel, the impact of VBM pretreatment (for 30 min) have also been investigated on the methane potential of the SS-DIG in batch mesophilic conditions tests. Results are presented in Figure 4. SS-DIG exhibited a methane potential of 101.5 (± 4.3) NL CH_4 kg^{-1} VS which is in agreement with previous studies that investigated the methane potential of solid separated digestate [7,18]. For instance, Ref. [18] have reported a methane potential of 90 (± 1.2) NL CH_4 kg^{-1} VS for solid separated digestate from an agricultural biogas plant located in Italy. Similarly, Ref. [27] have reported methane potentials varying from 71.4 (± 5.3) NL CH_4 kg^{-1} VS to 156.9 (± 7.4) NL CH_4 kg^{-1} VS for three solid separated digestates from agricultural biogas plant. Results were also found in accordance with those of Ref. [7] who investigated the residual methane potential of whole digestate from 21 full-scale digesters and they reported methane yields varying from 24 to 126 NL CH_4 kg^{-1} VS. The application of VBM pretreatment on the dry solid separated digestate has permitted significant improvement of the methane potential. Indeed, a methane potential of 133.6 (± 3.3) NL CH_4 kg^{-1} VS was observed corresponding to an improvement of the methane potential of 31% compared to untreated SS-DIG. Such results are in agreement with previous ones that have highlighted the benefits of applying a post-treatment to improve methane recovery from the SS-DIG [18,27]. For instance, Ref. [27] have reported that thermal pretreatment (120 °C, 30 min) lead to an improvement of the methane potential from 71.4 (± 5.3) NL CH_4 kg^{-1} VS to 153.7 (± 20.6) NL CH_4 kg^{-1} VS from the solid separated digestate of an agricultural biogas plant fed with swine slurry, grass silage and maize silage. Similarly, Ref. [43] reported an improvement of the methane potential of a solid separated digestate from a full-scale biogas plant from 21 (± 0.0) NL CH_4 kg^{-1} VS to 57.7 (± 4.5) NL CH_4 kg^{-1} VS after 10 min of ball milling pretreatment.

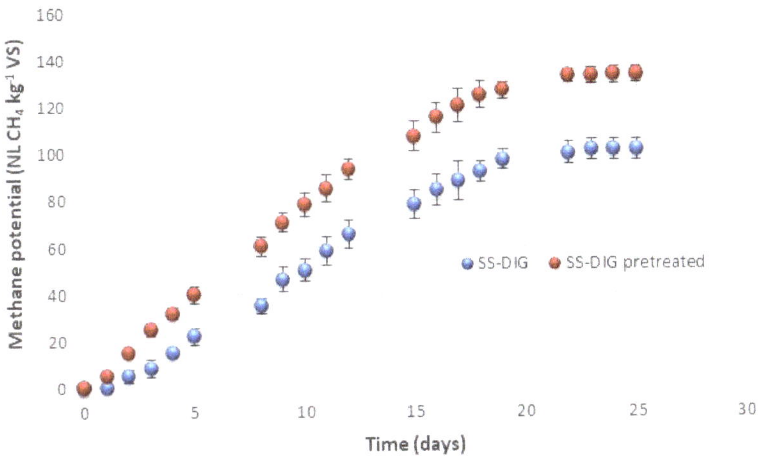

Figure 4. Methane potential (NL CH$_4$ kg^{-1} VS) of raw SS-DIG and pretreated SS-DIG by VBM at 30 min.

According the results obtained, it is obvious that the vibro-ball milling is a promising technology for improving the biodegradability of the SS-DIG and improving the recovery of bioethanol and methane. In future work, such results have to be confirmed at pilot-scale as the energy requirement of the milling step is often overestimated at laboratory-scale and can be used only as rough estimation for the optimization of the operational parameters. Currently, the pretreatment of the SS-DIG for further recirculation into the AD process seems to be simpler and more interesting from an energetic point of view (1.22 kWh kg^{-1} TS for AD process compared to 0.72 kWh kg^{-1} TS for bioethanol production). Nonetheless, AD/bioethanol answers better to the prerequisites of a biorefinery, by diversifying biofuels production (biogas, bioethanol).

4. Conclusions

VBM appears to be a promising technology to improve sugar recovery after enzymatic hydrolysis from dry solid anaerobic digestate. VBM of dry anaerobic digestate led to a significant decrease in the particle size and crystallinity. The best results in terms of cellulose and hemicelluloses hydrolysis were noted with VBM for 30 min, with hydrolysis yields of 64% and 85% for hemicelluloses and cellulose, respectively. Bioethanol fermentation (SSF) under this condition led to an ethanol yield of 98 g$_{eth}$ kg^{-1} TS (corresponding to 90% of the theoretical value) compared to 19 g$_{eth}$ kg^{-1} TS for raw solid digestate. In parallel, VBM for 30 min led to an improvement of the methane potential of 31% compared to untreated SS-DIG.

Author Contributions: F.M. has participated to the conceptualization, methodology, investigation, supervision, writing—review and editing. C.S. has participated to the conceptualization, methodology investigation, writing—review and editing. A.B. has participated to the conceptualization, writing—review and editing and supervision.

Funding: The authors gratefully acknowledge the European Commission Directorate-General for Research & Innovation for funding the project "NoAW: No Agricultural wastes". This project received funding from the European Union's Horizon 2020 research and innovation programme under grant agreement No 688338.

Conflicts of Interest: The authors declare no conflict of interest.

Abbreviations

AD	Anaerobic digestion
CHP	Combined Heat and Power
Cp	Specific heat of water
E_{DY}	Energy requirement for drying
E_{Heat}	Energy requirement for heating
$E_{Evaporation}$	Energy requirement for evaporation
Lv	Latent heat of vaporization
HRT	Hydraulic Retention Time
SS-DIG	Solid Separated Digestate
TS	Total Solids
VS	Volatile Solids
wt.	Weight

References

1. Angelidaki, I.; Alves, M.; Bolzonella, D.; Borzacconi, L.; Campos, J.L.; Guwy, A.J.; Kalyuzhnyi, S.; Jenicek, P.; van Lier, J.B. Defining the biomethane potential (BMP) of solid organic wastes and energy crops: A proposed protocol for batch assays. *Water Sci. Technol.* **2009**, *59*, 927–934. [CrossRef] [PubMed]
2. Dinuccio, E.; Balsari, P.; Gioelli, F.; Menardo, S. Evaluation of the biogas productivity potential of some Italian agro-industrial biomasses. *Bioresour. Technol.* **2010**, *101*, 3780–3783. [CrossRef]
3. Angelidaki, I.; Treu, L.; Tsapekos, P.; Luo, G.; Campanaro, S.; Wenzel, H.; Kougias, P.G. Biogas upgrading and utilization: Current status and perspectives. *Biotechnol. Adv.* **2018**, *36*, 452–466. [CrossRef]
4. Monlau, F.; Sambusiti, C.; Antoniou, N.; Barakat, A.; Zabaniotou, A. A new concept for enhancing energy recovery from agricultural residues by coupling anaerobic digestion and pyrolysis process. *Appl. Energy* **2015**, *148*, 32–38. [CrossRef]
5. Monlau, F.; Sambusiti, C.; Ficara, E.; Aboulkas, A.; Barakat, A.; Carrère, H. New opportunities for agricultural digestate valorization: Current situation and perspectives. *Energy Env. Sci.* **2015**, *8*, 2600–2621. [CrossRef]
6. Sheets, J.P.; Yang, L.; Ge, X.; Wang, Z.; Li, Y. Beyond land application: Emerging technologies for the treatment and reuse of anaerobically digested agricultural and food waste. *Waste Manag.* **2015**, *44*, 94–115. [CrossRef] [PubMed]
7. Ruile, S.; Schmitz, S.; Mönch-Tegeder, M.; Oechsner, H. Degradation efficiency of agricultural biogas plants – A full-scale study. *Bioresour. Technol.* **2015**, *178*, 341–349. [CrossRef] [PubMed]
8. Santi, G.; Proietti, S.; Moscatello, S.; Stefanoni, W.; Battistelli, A. Anaerobic digestion of corn silage on a commercial scale: Differential utilization of its chemical constituents and characterization of the solid digestate. *Biomass Bioenergy* **2015**, *83*, 17–22. [CrossRef]
9. Sawatdeenarunat, C.; Nam, H.; Adhikari, S.; Sung, S.; Khanal, S.K. Decentralized biorefinery for lignocellulosic biomass: Integrating anaerobic digestion with thermochemical conversion. *Bioresour. Technol.* **2018**, *250*, 140–147. [CrossRef]
10. Fabbri, D.; Torri, C. Linking pyrolysis and anaerobic digestion (Py-AD) for the conversion of lignocellulosic biomass. *Curr. Opin. Biotechnol.* **2016**, *38*, 167–173. [CrossRef]
11. Chen, G.; Guo, X.; Cheng, Z.; Yan, B.; Dan, Z.; Ma, W. Air gasification of biogas-derived digestate in a downdraft fixed bed gasifier. *Waste Manag.* **2017**, *69*, 162–169. [CrossRef] [PubMed]
12. Sambusiti, C.; Monlau, F.; Barakat, A. Bioethanol fermentation as alternative valorization route of agricultural digestate according to a biorefinery approach. *Bioresour. Technol.* **2016**, *212*, 289–295. [CrossRef] [PubMed]
13. Yue, Z.; Teater, C.; MacLellan, J.; Liu, Y.; Liao, W. Development of a new bioethanol feedstock—Anaerobically digested fiber from confined dairy operations using different digestion configurations. *Biomass Bioenergy* **2011**, *35*, 1946–1953. [CrossRef]
14. Teater, C.; Yue, Z.; MacLellan, J.; Liu, Y.; Liao, W. Assessing solid digestate from anaerobic digestion as feedstock for ethanol production. *Bioresour. Technol.* **2011**, *102*, 1856–1862. [CrossRef] [PubMed]
15. MacLellan, J.; Chen, R.; Kraemer, R.; Zhong, Y.; Liu, Y.; Liao, W. Anaerobic treatment of lignocellulosic material to co-produce methane and digested fiber for ethanol biorefining. *Bioresour. Technol.* **2013**, *130*, 418–423. [CrossRef] [PubMed]

16. Carrere, H.; Antonopoulou, G.; Affes, R.; Passos, F.; Battimelli, A.; Lyberatos, G.; Ferrer, I. Review of feedstock pretreatment strategies for improved anaerobic digestion: From lab-scale research to full-scale application. *Bioresour. Technol.* **2016**, *199*, 386–397. [CrossRef] [PubMed]

17. Monlau, F.; Barakat, A.; Trably, E.; Dumas, C.; Steyer, J.-P.; Carrère, H. Lignocellulosic materials into biohydrogen and biomethane: Impact of structural features and pretreatment. *Crit. Rev. Env. Sci. Technol.* **2013**, *43*, 260–322. [CrossRef]

18. Sambusiti, C.; Monlau, F.; Ficara, E.; Musatti, A.; Rollini, M.; Barakat, A.; Malpei, F. Comparison of various post-treatments for recovering methane from agricultural digestate. *Fuel Process. Technol.* **2015**, *137*, 359–365. [CrossRef]

19. Taherzadeh, M.J.; Karimi, K. Pretreatment of Lignocellulosic Wastes to Improve Ethanol and Biogas Production: A Review. *Int. J. Mol. Sci.* **2008**, *9*, 1621–1651. [CrossRef]

20. Hendriks, A.T.W.M.; Zeeman, G. Pretreatments to enhance the digestibility of lignocellulosic biomass. *Bioresour. Technol.* **2009**, *100*, 10–18. [CrossRef]

21. Barakat, A.; de Vries, H.; Rouau, X. Dry fractionation process as an important step in current and future lignocellulose biorefineries: A review. *Bioresour. Technol.* **2013**, *134*, 362–373. [CrossRef] [PubMed]

22. Kratky, L.; Jirout, T. Biomass Size Reduction Machines for Enhancing Biogas Production. *Chem. Eng. Technol.* **2011**, *34*, 391–399. [CrossRef]

23. Licari, A.; Monlau, F.; Solhy, A.; Buche, P.; Barakat, A. Comparison of various milling modes combined to the enzymatic hydrolysis of lignocellulosic biomass for bioenergy production: Glucose yield and energy efficiency. *Energy* **2016**, *102*, 335–342. [CrossRef]

24. Sambusiti, C.; Licari, A.; Solhy, A.; Aboulkas, A.; Cacciaguerra, T.; Barakat, A. One-pot dry chemo-mechanical deconstruction for bioethanol production from sugarcane bagasse. *Bioresour. Technol.* **2015**, *181*, 200–206. [CrossRef] [PubMed]

25. Hideno, A.; Inoue, H.; Tsukahara, K.; Fujimoto, S.; Minowa, T.; Inoue, S.; Endo, T.; Sawayama, S. Wet disk milling pretreatment without sulfuric acid for enzymatic hydrolysis of rice straw. *Bioresour. Technol.* **2009**, *100*, 2706–2711. [CrossRef] [PubMed]

26. Basset, C.; Kedidi, S.; Barakat, A. Chemical- and Solvent-Free Mechanophysical Fractionation of Biomass Induced by Tribo-Electrostatic Charging: Separation of Proteins and Lignin. *Acs Sustain. Chem. Eng.* **2016**, *4*, 4166–4173. [CrossRef]

27. Menardo, S.; Balsari, P.; Dinuccio, E.; Gioelli, F. Thermal pre-treatment of solid fraction from mechanically-separated raw and digested slurry to increase methane yield. *Bioresour. Technol.* **2011**, *102*, 2026–2032. [CrossRef]

28. Monlau, F.; Sambusiti, C.; Barakat, A.; Guo, X.M.; Latrille, E.; Trably, E.; Steyer, J.-P.; Carrere, H. Predictive models of biohydrogen and biomethane production based on the compositional and structural features of lignocellulosic materials. *Env. Sci. Technol.* **2012**, *46*, 12217–12225. [CrossRef]

29. Barakat, A.; Monlau, F.; Steyer, J.-P.; Carrere, H. Effect of lignin-derived and furan compounds found in lignocellulosic hydrolysates on biomethane production. *Bioresour. Technol.* **2012**, *104*, 90–99. [CrossRef]

30. Pengyu, D.; Lianhua, L.; Feng, Z.; Xiaoying, K.; Yongming, S.; Yi, Z. Comparison of dry and wet milling pre-treatment methods for improving the anaerobic digestion performance of the Pennisetum hybrid. *Rsc Adv.* **2017**, *7*, 12610–12619. [CrossRef]

31. da Silva, A.S.A.; Inoue, H.; Endo, T.; Yano, S.; Bon, E.P.S. Milling pretreatment of sugarcane bagasse and straw for enzymatic hydrolysis and ethanol fermentation. *Bioresour. Technol.* **2010**, *101*, 7402–7409. [CrossRef]

32. Yuan, X.; Liu, S.; Feng, G.; Liu, Y.; Li, Y.; Lu, H.; Liang, B. Effects of ball milling on structural changes and hydrolysis of lignocellulosic biomass in liquid hot-water compressed carbon dioxide. *Korean J. Chem. Eng.* **2016**, *33*, 2134–2141. [CrossRef]

33. Zhang, Q.; Jérôme, F. Mechanocatalytic Deconstruction of Cellulose: An Emerging Entry into Biorefinery. *ChemSusChem* **2013**, *6*, 2042–2044. [CrossRef] [PubMed]

34. Leewatchararongjaroen, J.; Anuntagool, J. Effects of Dry-Milling and Wet-Milling on Chemical, Physical and Gelatinization Properties of Rice Flour. *Rice Sci.* **2016**, *23*, 274–281. [CrossRef]

35. Vancov, T.; Schneider, R.C.S.; Palmer, J.; McIntosh, S.; Stuetz, R. Potential use of feedlot cattle manure for bioethanol production. *Bioresour. Technol.* **2015**, *183*, 120–128. [CrossRef] [PubMed]

36. Gupta, R.; Lee, Y.Y. Mechanism of cellulase reaction on pure cellulosic substrates. *Biotechnol. Bioeng.* **2009**, *102*, 1570–1581. [CrossRef]

37. Jeoh, T.; Ishizawa, C.I.; Davis, M.F.; Himmel, M.E.; Adney, W.S.; Johnson, D.K. Cellulase digestibility of pretreated biomass is limited by cellulose accessibility. *Biotechnol. Bioeng.* **2007**, *98*, 112–122. [CrossRef]
38. Mais, U.; Esteghlalian, A.R.; Saddler, J.N.; Mansfield, S.D. Enhancing the Enzymatic Hydrolysis of Cellulosic Materials Using Simultaneous Ball Milling. *Appl. Biochem. Biotechnol.* **2002**, *98*, 815–832. [CrossRef]
39. Loustau-Cazalet, C.; Sambusiti, C.; Buche, P.; Solhy, A.; Bilal, E.; Larzek, M.; Barakat, A. Innovative deconstruction of biomass induced by dry chemo-mechanical activation: Impact on enzymatic hydrolysis and energy efficiency. *Acs Sustain. Chem. Eng.* **2016**, *4*, 2689–2697. [CrossRef]
40. Mani, S.; Tabil, L.G.; Sokhansanj, S. Effects of compressive force, particle size and moisture content on mechanical properties of biomass pellets from grasses. *Biomass Bioenergy* **2006**, *30*, 648–654. [CrossRef]
41. Zhu, J.Y.; Pan, X.; Zalesny, R.S. Pretreatment of woody biomass for biofuel production: Energy efficiency, technologies, and recalcitrance. *Appl. Microbiol. Biotechnol.* **2010**, *87*, 847–857. [CrossRef] [PubMed]
42. Zhu, J.Y.; Zhu, W.; Obryan, P.; Dien, B.S.; Tian, S.; Gleisner, R.; Pan, X.J. Ethanol production from sporl-pretreated lodgepole pine: Preliminary evaluation of mass balance and process energy efficiency. *Appl. Microbiol. Biotechnol.* **2010**, *86*, 1355–1365. [CrossRef] [PubMed]
43. Lindner, J.; Zielonka, S.; Oechsner, H.; Lemmer, A. Effects of mechanical treatment of digestate after anaerobic digestion on the degree of degradation. *Bioresour. Technol.* **2015**, *178*, 194–200. [CrossRef] [PubMed]

 *bioengineering*

Article

Inoculum Source Determines Acetate and Lactate Production during Anaerobic Digestion of Sewage Sludge and Food Waste

Jan Moestedt [1,2], **Maria Westerholm** [3], **Simon Isaksson** [3] **and Anna Schnürer** [1,3,*]

[1] Department of Thematic Studies–Environmental Change, Linköping University, SE 581 83 Linköping, Sweden; jan.moestedt@tekniskaverken.se
[2] Department R&D, Tekniska verken i Linköping AB, SE 581 15 Linköping, Sweden
[3] Department of Molecular Sciences, Swedish University of Agricultural Sciences, BioCenter, SE 750 07 Uppsala, Sweden; maria.westerholm@slu.se (M.W.); simon.isaksson@slu.se (S.I.)
* Correspondence: anna.schnurer@slu.se

Received: 15 November 2019; Accepted: 18 December 2019; Published: 23 December 2019

Abstract: Acetate production from food waste or sewage sludge was evaluated in four semi-continuous anaerobic digestion processes. To examine the importance of inoculum and substrate for acid production, two different inoculum sources (a wastewater treatment plant (WWTP) and a co-digestion plant treating food and industry waste) and two common substrates (sewage sludge and food waste) were used in process operations. The processes were evaluated with regard to the efficiency of hydrolysis, acidogenesis, acetogenesis, and methanogenesis and the microbial community structure was determined. Feeding sewage sludge led to mixed acid fermentation and low total acid yield, whereas feeding food waste resulted in the production of high acetate and lactate yields. Inoculum from WWTP with sewage sludge substrate resulted in maintained methane production, despite a low hydraulic retention time. For food waste, the process using inoculum from WWTP produced high levels of lactate (30 g/L) and acetate (10 g/L), while the process initiated with inoculum from the co-digestion plant had higher acetate (25 g/L) and lower lactate (15 g/L) levels. The microbial communities developed during acid production consisted of the major genera *Lactobacillus* (92–100%) with food waste substrate, and *Roseburia* (44–45%) and *Fastidiosipila* (16–36%) with sewage sludge substrate. Use of the outgoing material (hydrolysates) in a biogas production system resulted in a non-significant increase in bio-methane production (+5–20%) compared with direct biogas production from food waste and sewage sludge.

Keywords: acetate; lactate; inoculum; food waste; sewage sludge; lactic acid bacteria

---

1. Introduction

Anaerobic digestion (AD) can be applied for industrial purposes to produce bio-methane (CH_4), which is used as renewable energy in transportation or for heat and power production. AD can play a key role in reducing fossil fuel use in transportation and industry, while at the same time handling organic waste and producing renewable fertilizer [1]. AD is a complex process that requires several different microbial steps and entails the formation of various intermediary components. The process typically begins with hydrolysis, followed by acidogenesis of complex organic macromolecules into, e.g., volatile fatty acids (VFA), CO_2, and H_2 [2]. VFA can be further converted into acetate, CO_2, and H_2 in a reaction step called acetogenesis. As a final step in AD, these components are converted into bio-methane by methanogens. The hydrolysis, acidogenesis, and acetogenesis steps are performed by fast-growing bacteria that, in many cases, also thrive at low pH, while methanogenesis is performed by slower-growing bacteria that thrive best at neutral pH. These differences in reaction speed and pH optimum allow AD to be divided into two separate production steps, where H_2 and VFA are generated in a primary acid reactor and bio-methane in a secondary reactor. This two-stage approach has been shown to optimize the overall AD process and increase methane yield [3–8]. It also allows additional applications, such as the production of bio-hydrogen or of a range of VFA that can be extracted and used as "green" chemical feedstock for further conversion [8,9]. The typical VFA composition in an acidic AD reactor is a mixture dominated by acetate, propionate, butyrate, and valerate [10–13], but it can also be dominated by acetate, butyrate, and H_2 [14], or mainly consist of lactate and/or acetate [15,16]. The amount and type of acid produced depend on both chemical, technical, and microbiological parameters, often interlinked. Parameters shown to be of importance include, for example, reactor configuration, substrate composition and pre-treatment, hydraulic retention time (HRT), temperature, pH, as well as the choice of inoculum and final microbial composition [10,11,17–22].

Among the different acids that can be produced in a two-stage set-up, acetate is a highly interesting compound. In addition to being an energy carrier between different stages within AD, it can be used for the production of value-added products such as biopolymers, commodity chemicals, or fuels, or as a carbon source for denitrification steps at wastewater treatment plants (WWTP) and growing single-cell cultures [23,24]. Today, acetate is predominantly produced (12.9 million metric tons/year) using petrochemical feedstocks through various chemical processes [24]. Production of acetate via microbiological processes during AD can proceed via different metabolic routes, involving different types of microorganism. Various carbohydrates and proteins can be used by fermentative bacteria, resulting in acetate formation but also other longer acids and alcohols, CO_2, and H_2 [25]. Monomeric sugars can also be converted by lactic acid bacteria (LAB) to acetate and lactate, in different proportions [15,26]. Another interesting group for acetate production is bacteria performing acetogenesis. These bacteria can use a variety of different organic substrates and inorganic gases (CO, CO_2, H_2) and have in common that they use the Wood–Ljungdahl pathway while producing acetate as the main end-product [27]. Depending on growing conditions, some acetogens can also redirect their metabolism toward more reduced end-products such as organic acids and different alcohols [27]. Acetogens can also oxidize VFA (C > 2) to acetate and hydrogen, a process requiring low H_2 partial pressure to become thermodynamically favorable [28]. In a single-stage process, a low level of hydrogen is ensured by the activity of hydrogenotrophic methanogens. In a hydrolysis reactor with no or low methanogenic activity, hydrogen could potentially be both produced and consumed by different groups of acetogens.

Knowledge on acid production per se is quite good as it has been studied in many different processes for both single compounds and complex materials and with both single and mixed cultures of microorganisms [8,17,25]. However, less is known about how to direct the process towards the production of a specific VFA from a complex substrate matrix, such as food waste or sewage sludge, owing to the complexity of AD and the wide number of different possible pathways. Moreover, most previous studies have been performed during batch cultivation and only a few studies have evaluated VFA production in continuous mode [20,29]. There is, therefore, a need to identify management strategies for achieving high levels of a specific acid, such as acetate.

This study investigated the importance of inoculum source (different microbial composition) and substrate composition for the production of VFA, with the focus on acetate. Four semi-continuously fed, laboratory-scale processes were inoculated with sludge from a co-digestion biogas plant or an AD process treating primary and secondary sludge (mixed sludge). Two different materials (food waste and mixed sludge) were used as substrates in a set-up resulting in a distinct combination of inoculum source and substrate. The effects of inoculum origin (expected to have significantly different initial community structure) and substrate on acetate production were then evaluated. VFA production and microbial community structure were followed over time throughout an operating period of 65 days, with the most promising process continued to be operated for an additional 160 days. Moreover, bio-methane potential (BMP) of the resulting hydrolysate was determined and compared with that achieved when using food waste and mixed sludge directly as a substrate. The results obtained were used to estimate the potential for VFA and/or methane production in full-scale processes.

2. Materials and Methods

2.1. Source of Inoculum

Of the four laboratory continuously stirred tank reactors (CSTR) used in the study, two (CO-F and CO-S) were inoculated with material from a co-digestion plant, and the other two (WW-F and WW-S) were inoculated with material from an AD process fed mixed sludge. The laboratory reactors had an active volume of 6 L. Inoculum for the CO reactors was collected from a co-digestion plant in Linköping, Sweden, that operates a CSTR at 42 °C, an average organic loading rate (OLR) of about 4 kg $VS/m^3/d$, and HRT of about 35 days. The co-digestion plant receives food waste from households (50% of incoming wet weight), organic industrial residues (25%), and slaughterhouse waste (25%). Inoculum for the WW reactors was collected from a WWTP anaerobically degrading a mixture of primary and secondary sludge produced during the treatment of wastewater from the city of Linköping (around 150,000 person-equivalents). Its CSTR is operated at an average OLR of 2 kg $VS/m^3/d$ and HRT of about 20 days at 38 °C.

2.2. Experimental Set-Up

All laboratory-scale reactors were fed semi-continuously (once a day, seven days per week). The CO-F and WW-F reactors were fed food waste (total solids (TS) 14.7%, volatile solids (VS) 92%, dissolved organic carbon (DOC) 24,900 mg/L, and total organic carbon (TOC) 39,000 mg/L), which was retrieved at the co-digestion plant. The food waste (annual treatment of approx. 55,000 tons) is macerated into a slurry with particle size <12 mm, diluted with water or waste liquor from food industries, such as washing liquor from a large diary industry, and treated with a Bio-Sep®technique to remove impurities such as metals and plastics. The food waste slurry was collected from a sampling point of the circulation circuit at a heated storage tank (65 °C, HRT 3 days). For the first 20 days after start-up, the reactors were operated at an OLR of 27 kg $VS/m^3/d$ and HRT of 5 days (week average) in order to rapidly obtain an acidic process (Table 1). These parameters were changed to HRT 10 days (week average) and OLR 13.5 kg $VS/m^3/d$, based on results from earlier experiments [6]. Reactors CO-S and WW-S received mixed sludge (TS 8.8%, VS 81%, DOC 5500 mg/L and 21,000 mg/L). The sewage sludge was retrieved from a dewatering band (dewatering mixed sludge, e.g., a mixture of primary- and secondary sludge with TS approx. 2% to 8%). The reactors were operated at OLR 14.2 kg $VS/m^3/d$ and HRT 5 days (week average) during the initial 50 days. HRT was then decreased to 3 days (week average) and OLR was increased to 23.7 kg $VS/m^3/d$, in order to maintain acidic conditions (Table 1). In each set of experiments, one single batch of substrate (kept frozen until use) was used during the entire experiment except in the first three HRTs, when fresh material from the same batch was fed to the reactors. This was done in order to allow for initial inoculation with potential active microorganisms present in the substrate. All reactors were operated for 65 days except for CO-F, which was operated for an additional 160 days, giving in total 225 days of operation. Between days 200 and 225, pure glucose was added to the substrate mixture to observe the effect of adding easily available feed to VFA production. Addition of glucose to CO-F increased the OLR to 16.1 kg $VS/m^3/d$, due to an additional ingoing DOC content of 10,700 mg/kg.

Table 1. Origin of inoculum, substrate, organic loading rate (OLR), and hydraulic retention time (HRT) for the different reactors (week averages). Values in brackets refer to after day 20 for reactors CO-F and WW-F, and after day 50 for reactors CO-S and WW-S.

	WW-S	CO-S	WW-F	CO-F
Inoculum source	WWTP	Co-digestion	WWTP	Co-digestion
Substrate	Sewage sludge	Sewage sludge	Food waste	Food waste
OLR (kg VS/m^3/d)	14.2 (23.7)	14.2 (23.7)	27.0 (13.5)	27.0 (13.5)
HRT (days)	5 (3)	5 (3)	5 (10)	5 (10)

2.3. Analytical Methods

Reactor volume adjustment and sampling were performed five days per week, prior to daily feeding. Volumetric gas production was measured online with a Ritter milligas counter (MGC-10, Ritter, Waldenbuch, Germany), and methane concentration was determined with a gas sensor (BlueSens, Herten, Germany). Gas was normalized for standard temperature and pressure (1.01325 bar and temperature 273.2 K). Gas composition (CH_4, CO_2, H_2S, O_2) was further analyzed using a Biogas 5000 device (Geotech Instruments, Coventry, UK). The content of H_2 in the gas was analyzed with a Micro IV sensor (GfG GmbH, Berlin, Germany). VFA content was analyzed with a Clarus 550 gas chromatograph (Perkin Elmer, Waltham, MA, USA) with a packed Elite-FFAP column (Perkin Elmer, USA) for acidic compounds [30]. Lactic acid was analyzed with HPLC, using a method described elsewhere [31]. Total ammonium nitrogen (NH_4^+-N) was analyzed as the sum of NH_4^+-N (aq) + ammonia-N (NH_3-N) (aq) by distillation (Kjeltec 8200, FOSS in Scandinavia, Sweden) in an acidic solution (H_3BO_3) and NH_4^+-N was then determined by titration with HCl (Titro 809, Metrohm, Herisau, Switzerland) according to the Tecator method for Kjeltec ISO 5664. Kjeldahl-nitrogen was determined using the same procedure and equipment as NH_4^+-N, with the exception that the samples were pre-treated with H_2SO_4 and subsequently heated to 410 °C for 1 h. The pH was measured with a potentiometric pH meter at 25 °C using a Hamilton electrode (WTW Inolab, Houston, TX, USA). TOC was analyzed according to method SS-EN 1484. The samples were homogenized and acidified with 2 M HCL and 10% H_3PO_4 to drive off inorganic carbon, and TOC was determined with Analytik Jena N/C 3100 during combustion of organic carbon and detected as CO_2 with an NDIR detector. DOC was determined by filtration (0.45 µm) prior to the TOC method.

Methane potential was determined using an automatic methane potential test system, AMPTS II (Bioprocess Control, Lund, Sweden). The inoculum was collected from the WWTP at Uppsala, Sweden, and degassed for 4 days prior to the test. Each bottle was loaded with inoculum and substrate in a 3:1 ratio (VS basis). To reach an OLR of 3 g VS/L and a working volume of 400 mL (total volume 600 mL), distilled water was added to the mixture. All substrates were operated in parallel triplicates together with three positive controls (crystalline cellulose) and three negative controls (no substrate added). The reactors were then incubated at mesophilic temperature (37 °C) for 15 days while being stirred in cycles of 1 minute, followed by 1 minute of rest.

All confidence intervals presented are calculated with Student's *t*-test ($\alpha = 0.05$).

2.4. Calculation of Efficiency of Hydrolysis, Acidogenesis, Acetogenesis, and Methanogenesis

The degree of hydrolysis (Equation (1)) and the degree of acidogenesis (Equation (2)) were calculated according to [3], but also taking gas production into consideration according to [32], with the addition of also including CO_2 production, and based on DOC analyses for liquid phase:

$$Hydrolysis = \frac{DOC_{res} + TC_{Gases} - DOC_{in}}{TOC_{tot} - DOC_{in}} \times 100 \ (\%) \tag{1}$$

$$Acidogenesis = \frac{TOC_{VFA} + TC_{Gases}}{DOC_{res} + TC_{Gases}} \times 100 \ (\%) \tag{2}$$

where

DOC_{res} = residual dissolved organic carbon,

DOC_{in} = ingoing dissolved organic carbon,

TC_{Gases} = total carbon in produced gases,

TOC_{VFA} = total organic carbon in volatile fatty acids,

TOC_{tot} = total ingoing organic carbon.

The degree of acetogenesis was calculated as the fraction of $TOC_{acetate}$ in total TOC_{VFA}. Raw protein was calculated as (Kjel-N − NH_4^+-N) × 6.25 and protein hydrolysis as a fraction of raw protein in hydrolysate as compared to substrates [33].

2.5. Additional Experiments to Determine Inhibitory Effects of VFA

Possible inhibitory effects of high acetate levels were evaluated by diluting reactor material from CO-F taken at day 200 with tapwater, in a tapwater:hydrolysate ratio of 0:100, 10:90, 25:75, 50:50, and 90:10. A 500 mL subsample of each final dilution of hydrolysate was inoculated in an anaerobic environment using reactors from AMPTSII (Bioprocess Control) with stirring at 38 °C for 48 hours in batch mode. Thereafter, the VFA concentration (specifically acetate) was analyzed. The undissociated (toxic) concentration of acetate was calculated according to Oswald´s law of dilution [34].

2.6. Sample Collection, Molecular Analyses, and Sequence Data Processing

Reactor sludge samples for molecular analyses were collected from each reactor on four occasions (days 2, 19, 33, and 57). Four additional samples were taken from reactor CO-F during the prolonged experimental period (i.e., days 135, 156, 177, 226). All samples were stored at −20 °C until further use. DNA extraction, construction of 16S amplicon libraries, and Illumina MiSeq sequencing were carried out on triplicate samples from each sampling point and reactor, as described previously [35]. Sequence data processing was performed as described by Westerholm [36]. In short, contaminating sequences were removed by Cutadapt [37] version 1.13, and sequences were further processed with the software package Divisive Amplicon Denoising Algorithm 2 (DADA2) [38], version 1.4, running in an HPC environment in R, version 3.4.0. Sequences were processed according to the DADA2 pipeline tutorial v. 1.4 with modification according to Appendix B. The forward and reverse reads were truncated at positions 250 and 200 bp, respectively. A maximum expected error of 2 was used to remove low-quality reads, and trimming and filtering were performed jointly on paired reads. Assignment of taxonomy was performed with the DADA2 taxonomy classification, using the Silva training set v128 to classify ribosomal sequence variants (RSVs). The phyloseq package [39] was used to organize the data into a single data object and for the production of graphics in R Studio version software (http://www.r-project.org, TeamR RStudio, 2016) as described previously [36]. In addition, a plot of richness estimates was created using the *plot_richness* function and a heatmap was created using the plot_heatmap function in the phyloseq package. Principal coordinate analysis (PCoA) plots of microbial community profiles were generated using Bray–Curtis weighted UniFrac distance measures. Constrained ordination was performed to evaluate associations between reactor parameters and changes in community composition. Permutational ANOVA (PERMANOVA) was performed to evaluate the effect of operating parameters on microbial community structure using the adonis functions in the vegan package [40] and significant ($p < 0.05$) parameters were included in canonical correspondence analysis (CCA) plotting. The ordination axes were constrained to linear combinations of reactor variables and plotted as arrows onto the ordination. The statistical significance of differences between reactors and over time was determined using ANOVA. Phylogenetic assignment at the genus level was evaluated using the Basic Local Alignment Search Tool (BLAST) algorithm [41] provided by the National Center for Biotechnology Information (NCBI; http://www.ncbi.nlm.nih.gov).

Raw sequences were submitted to the NCBI Sequence Read Archive (SRA) under the study accession number PRJNA575652.

3. Results and Discussion

The food waste-degrading reactors (CO-F and WW-F) had very low methane content (<1%) during the initial nine days when operating at 27 kg VS/m^3/d and HRT 5 days. pH decreased quickly in both reactors with some differences depending on the initial inoculum, from 7.8 to 5.2 in CO-F and 4.6 in WW-F. In this period, H$_2$ was produced with a concentration peaking at 20% and 15% of gas produced in CO-F and WW-F, respectively. After day 20, the H$_2$ level stabilized and represented around 1% of the gas produced. The high H$_2$ level coincided with a VFA profile dominated by acetate and butyrate (Figure 1), which is typical for dark fermentation [26]. Propionate and heptonate were also formed in high concentrations (Figure 1). However, from around day 20, in line with the drop in H$_2$ level, acetate and lactate became the dominant acids in both reactors, but with differences in the total amounts depending on the initial inoculum. At this time, the pH reached 3.8 ± 0.1 in WW-F and 4.0 ± 0.2 in CO-F and total gas production was very low (<1 g C$_{in\ gas}$/L$_{substrate}$ d in WW-F; <2 g C$_{in\ gas}$/L$_{substrate}$ d in CO-F), in line with previous studies on hydrolysis of food waste [6,10,20,42,43]. In contrast, in the sludge reactors (CO-S and WW-S), methane content and pH did not drop to similarly low levels when operating at initial parameters of OLR 14.2 kg VS/m^3/d and HRT 5 days. The methane content of the gas produced declined after 2 HRT (10 days), from 60% to stabilize at 15–20% in both reactors, while pH remained at 6.1 ± 0.2 throughout the main part of the experiment and the gas produced contained <2% H$_2$. However, from day 50 until 65 pH increased in WW-S (6.7 ± 0.7) while it remained stable in CO-S. The increase in pH occurred simultaneously as VFA-concentration decreased, and methane production started to increase. The VFA profile for both sludge reactors was different from that of the food waste reactors and included acetate, propionate, butyrate, and valerate, but no lactate (Figure 2). In contrast to the food waste processes, this profile also remained stable, although the absolute concentration decreased at the end of the experiment in WW-S (Figure 2).

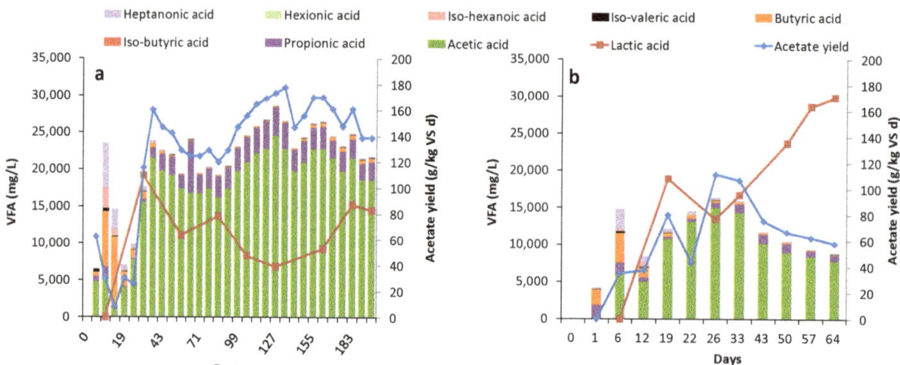

Figure 1. Volatile fatty acid (VFA) composition and acetate yield in reactors (**a**) CO-F (reactor inoculated from co-digestion plant fed with food waste) and (**b**) WW-F (reactor inoculated from wastewater treatment plant fed with food waste).

As further discussed below, one possible explanation for the differences in the observed VFA profiles between the processes operating with the different substrates was the differences in pH, suggested in several studies to be a key factor determinative for the type of organic acid produced [17,20,21].

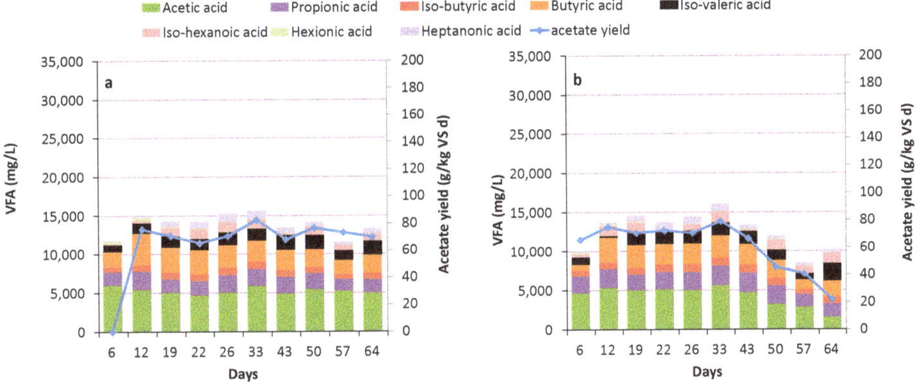

Figure 2. Volatile fatty acid (VFA) composition and acetate yield in reactors (**a**) CO-S (reactor inoculated from co-digestion plant fed with sewage sludge) and (**b**) WW-S (reactor inoculated from wastewater treatment plant fed with sewage sludge).

3.1. Hydrolysis Efficiency

The ingoing TOC for mixed sludge was 21,000 mg/L and dissolved carbon in the substrate measured as DOC was 5500 mg/L. These results revealed that about 26% of the carbon in the mixed sludge was in dissolved form when entering the reactors (Table 2). In both mixed sludge processes (WW-S and CO-S), the degree of hydrolysis was around 15%. For WW-S, a decrease in the degree of hydrolysis was seen from day 50; however this was likely due to the consumption of released DOC caused by the increasing methanogenic activity (Table 2). Specifically, hydrolysis of proteins into free ammonium-nitrogen appeared to be efficient and the level reached 53–59% for WW-S and 57–65% for CO-S. In total, the soluble fraction of TOC, including both ingoing DOC from the substrate and additionally released DOC from the acidic stage, was 41% for both reactors on days 0–50. The outgoing total soluble fraction of TOC was similar to the hydrolysis efficiency (42%) found in a previous study during semi-continuous acidification and hydrolysis of mixed sludge [32,44]. In that study, the hydrolysis efficiency increased with increasing pH and maximum acidification, and hydrolysis efficiency was reached at pH 8.9 and 9.9, respectively.

Table 2. Efficiency of the different steps in anaerobic digestion for the reactors fed mixed sludge (WW-S and CO-S) and the reactors fed food waste (WW-F and CO-F). TOC = total organic carbon, DOC = dissolved organic carbon, VFA = volatile fatty acids, nd = not determined.

	WW-S		CO-S		WW-F		CO-F		
Time period	0–50	50–65	0–50	50–65	0–20	20–65	0–20	20–200	days
HRT [a]	5	3	5	3	5	10	5	0	days
Hydrolysis	15%	10%	15%	15%	1%	2%	6%	7%	of TOC in
Acidogenesis	66%	48%	71%	61%	38%	65%	61%	56%	of DOC
Acetogenesis	38%	25%	37%	40%	37%	21%	40%	58%	of VFA
Protein hydrolysis	53%	59%	65%	57%	-	15%	-	17%	of raw protein in
Methanogenesis (1st stage)	4%	7%	2%	2%	1%	0%	1%	0%	of TOC in
Methanogenesis (2nd stage)	nd	59%	nd	60%	nd	44%	nd	46%	of TOC in

[a] The HRT for WW-S and CO-S was decreased from 5 to 3 days after 50 days, and for WW-F and CO-F, it increased from 5 to 10 days after 20 days.

For the processes fed food waste (CO-F and WW-F), the overall hydrolysis efficiency was considerably lower (2% and 7% in WW-F and CO-F respectively) than in the sludge-fed processes (Table 2). Furthermore, only 15–17% of the protein content was hydrolyzed, irrespective of the inoculum source, which was considerably lower than the fraction of proteins hydrolyzed in the sludge reactors (53–65%) (Table 2). This result was unexpected, as the protein mass in food waste ought to be more bioavailable than the material in mixed sludge, which is repeatedly pre-processed in various degradation steps in both the human gastrointestinal tract and the WWTP before entering the AD process. However, in food waste, the majority (64%) of the ingoing TOC (39,000 mg/L) was already in dissolved form (25,000 mg/L). Thus, even though the hydrolysis efficiency was low, the total outgoing DOC from the food waste processes was higher than for the sludge processes. The ingoing DOC of 64% and an additional 2–7% resulted in almost 70% total soluble fraction of TOC in the food waste reactors, indicating very high substrate availability for further treatments. The degree of hydrolysis was still much lower than observed before by Feng et al. [20] and Wu et al. [3], who reported overall hydrolysis efficiency of ca 40% in the AD of food waste. However, in these studies, the ingoing DOC represented less than 50% of total TOC and, therefore, the total outgoing soluble fraction of TOC was similar to that achieved for food waste in the acidic stage in the present study. The reason for low hydrolysis with food waste could be low activity by hydrolytic bacteria. Since almost no ammonium-nitrogen (<200 mg NH_4^+-N/L) was present in the hydrolysate, the vast majority of proteins passed through this process without being hydrolyzed into smaller molecules (e.g., NH_4^+). Similarly, Yin et al. [4] observed higher hydrolysis from carbohydrates than from proteins in the hydrolysis of food waste.

3.2. Acidogenesis and Acetogenesis

The mixed sludge reactors initially had acidogenesis efficiency corresponding to 66–70% (Table 2), with no significant difference between the processes. This indicates that the majority of DOC released from hydrolysis and DOC present in the substrate were converted into acids. However, over time the two processes deviated and for WW-S, the acidogenesis efficiency finally decreased to 48%. In this process, the VFA concentration at day 33 was 16.0 g/L, but it decreased to 10.2 g/L at day 58, of which acetate represented only 1.5 g/L. Simultaneously, methane production increased (Figures 1–3; Table 2). In CO-S, the acidogenesis efficiency was more stable and slightly higher than in WW-S (Table 2), with a total acid concentration of around 14.0 ± 0.9 g/L, acetate around 5 g/L and pH 5.9–6.1 throughout the experiment. The acetate fraction in CO-S represented 37% of total VFA and the yield corresponded to 70 g/kg VS. In WW-S, an equivalent level of acetate was reached during the first half of the experiment, but thereafter it declined to 22 g/kg VS, representing 15% of total VFA, by the end of the experiment. In the sludge reactors, the HRT was lowered after 50 days of operation in an attempt to wash out methanogens in WW-S. However, methanogenesis prevailed and reached 7% of ingoing TOC in WW-S, even at HRT of 3 days. Moreover, the pH increased from 6.1 to 7.0 by the end of the experiment, indicating that acidification of mixed sludge without pH regulation is difficult to obtain, particularly with inoculum from WWTP. High production of acetate was hence difficult to obtain with mixed sludge, and mixed-acid fermentation instead dominated. These results somewhat contradict previous findings of acetate levels up to 50–60% of total VFA in AD with mixed sludge [45–47]. However, those studies reached much lower total VFA concentrations (3–8 g/L) [45–47], especially compared with CO-S (14.0 ± 0.9 g/L), and hence the actual acetate concentration was similar or even higher in the present study. The observed difference between the processes, using the same substrate and operating conditions, suggests that the inoculum source played a major role in the efficiency of acidogenesis of the sludge. Differences in inoculum could thus also be one explanation for observed differences in process performance between studies.

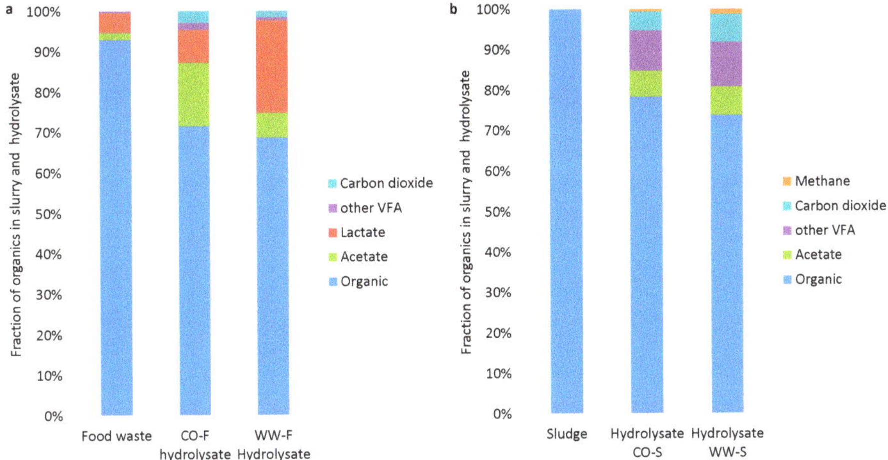

Figure 3. Composition of ingoing substrate and outgoing hydrolysate from reactors fed (**a**) food waste (CO-F, WW-F) and (**b**) sewage sludge (CO-S, WW-S). Mass of outgoing gases is included in the hydrolysate.

The reactors fed food waste had somewhat different acidogenesis efficiency in the initial phase of operation, with a lower value for the process initiated with the WWTP inoculum. However, from day 20 onwards, the processes became similar and reached 56–65% from day 20 onward (Table 2). The acid yields were higher than the sewage sludge reactors since food waste had a higher DOC concentration than the sludge (Figures 1 and 2). Several studies evaluating the effect of pH on acidogenesis and acetogenesis from food waste have found correlations between low pH and high hydrolysis of carbohydrates, while high pH increases hydrolysis of proteins, and thus release of ammonia [10,11,42]. Similarly, reactors CO-F and WW-F, with pH around 4, showed high acidogenesis of carbohydrates, while the release of NH_4^+-N was low. The importance of the inoculum source for acetate production was illustrated by a considerably higher yield in CO-F than WW-F reaching 151 and 75 g/kg VS, respectively. The acetate concentration was very high in both reactors (86–89% of total VFA, excluding lactic acid) from day 20. This supports the suggestion that low pH (in this case, uncontrolled) could be a key factor for reaching a high fraction of acetate. Differences in pH could also have caused the difference in acetogenesis efficiency between the sludge and food waste reactors, with the latter having both a lower fraction of acetate and higher pH (pH > 5.9). The high acetate levels and low pH in the food waste reactors could also have been the result of the high concentration of lactic acid in those reactors. Since lactate has low pKa, the pH will naturally decrease below 4, which could benefit acetate production. In WW-F with a pH of 3.8, the lactate concentration increased throughout the experimental period, to reach 30 g/L (Figure 3), which corresponded to 76% of DOC in acidogenesis. Lactate was lower in CO-F with pH 4.0 (10 g/L) and corresponded to 29% of DOC in acidogenesis (Figure 3).

Limiting Factors—Undissociated Acetate and Ingoing DOC

Reactor CO-F had the highest acetate yield and was kept in operation for 200 days (20 HRTs) in order to evaluate the long-term stability of the process. The process was quite stable, converting around 15% of the ingoing TOC to acetate and with an outgoing acetate concentration varying between 17 and 28 g/L. A possible limitation for increased acetate yields could be inhibitory effects of undissociated VFA. The average concentration of 20 g acetate/L is equivalent to 18.6 g undissociated acetate/L at pH 3.9, which is far above the value in other studies showing inhibitory effects [48,49]; e.g., Xiao et al. [48] observed that about 2 g/L at pH 5.5 could reduce acetate production by 60% compared with when a smaller fraction of undissociated acids was present. To obtain information about possible toxicity effects, a batch dilution experiment was performed. The results showed a linear correlation ($R^2 = 0.9997$) between acetate concentration after two days of incubation and the fraction of hydrolysate (Figure 4a). Hence dilution of acetate (i.e., reducing the undissociated, toxic concentration) did not result in higher acetate production, which theoretically would have been the case in case the acetate level was limited by a toxic effect (Figure 4a). This shows that the concentration of undissociated acids was not detrimental to reaching higher acetate concentration in the processes in this study.

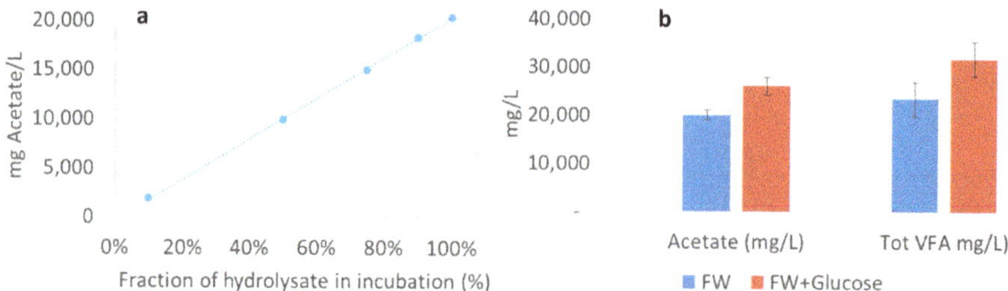

Figure 4. (**a**) Correlation between fraction of hydrolysate after two days of inoculation and absolute acetate concentration ($R^2 = 0.9997$). (**b**) Concentration of acetate and total volatile fatty acids (VFA) with only food waste (FW) (blue) compared with FW plus extra dissolved organic carbon (red).

Another limiting factor for optimizing acetate production from the food waste could be the level of available DOC in the substrate. One possible way to increase available DOC would be to pretreat the substrate, shown in a previous study to increase the DOC and VFA production from source-separated organic waste [43]. To test the effect of higher DOC on acetate production, additional DOC in the form of glucose was added to the substrate for CO-F during the last 25 days of operation. The ingoing DOC was already 25,000 mg/L, but following the glucose addition, the concentration was increased to 35,700 mg/L, which significantly increased both acetate and residual VFA concentration (Figure 4b). This confirms that it was not the concentration of undissociated acids that was the limiting factor, but rather the amount of easily convertible substrate. Still, while the absolute concentration of acetate increased with additional DOC in the substrate, the actual acetate yield from the added substrate (mg/g VS) was constant. This shows that it is possible to obtain higher acetate concentration with a substrate containing a large fraction of DOC, although in this case, the expected yield per amount of added organic material appeared to be constant (~150 g/kg VS).

3.3. Microbial Community Structure

3.3.1. Importance of Inoculum Source

The microbiology of the biogas plants from where the inocula were taken has been thoroughly analyzed in previous studies [50,51]. These investigations have shown that the high-ammonia co-digestion plant has syntrophic acetate oxidation (SAO) and hydrogenotrophic methanogenesis as the dominant reaction pathway for methane formation, whereas the low-ammonia AD process of mixed sludge is dominated by acetotrophic methanogens. The effect of inoculum source in processing food waste for acid production has also been evaluated in previous studies, although mainly in batch mode. Those studies identified inoculum source as a detrimental factor for process performance [4,18,21,42], as also found in the present study. However, based on the Illumina sequencing results, the inoculum source appeared to have minor effects on the development of the microbial community structure. The analyses demonstrated relatively similar overall microbial community structure at two days of operation, with dominance of the phylum Firmicutes and minor levels of Thermotagae (8–14%), Synergistetes (2–3%), Bacterioidetes (5%), and Atribacteria (1–5%) in the processes with co-digestion plant inoculum (Figures 5 and 6). In the processes inoculated with sludge, Firmicutes (79%) was accompanied by Actinobacteria (17%) in WW-F, whereas Firmicutes (38%), Bacteroidetes (31%), and Aegiribacteria (14%) dominated in WW-S. However, differences in the microbial community caused by the inoculum in CO and WW reactors diminished over the course of the operation and the community structure became more structured based on the substrate (Figure 5), as discussed further in the next section.

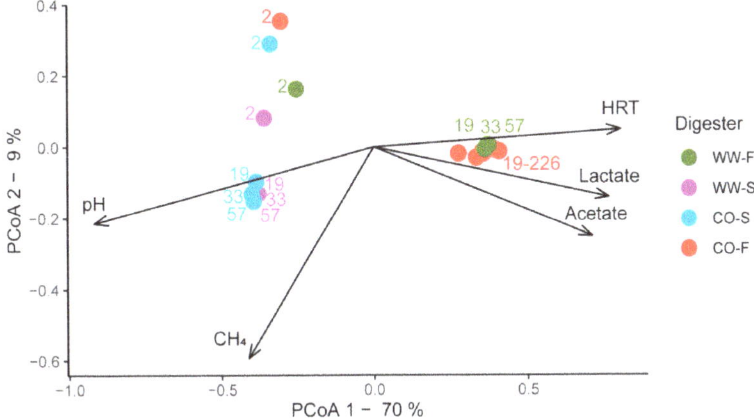

Figure 5. Principal coordinate analysis (PCoA) plot of microbial community structure using weighted UniFrac. Reactor parameters significantly associated with changes in microbial community structure are plotted as vectors, where the length and direction indicate the contribution of the variable to the principal components. Variable HRT stands for hydraulic retention time.

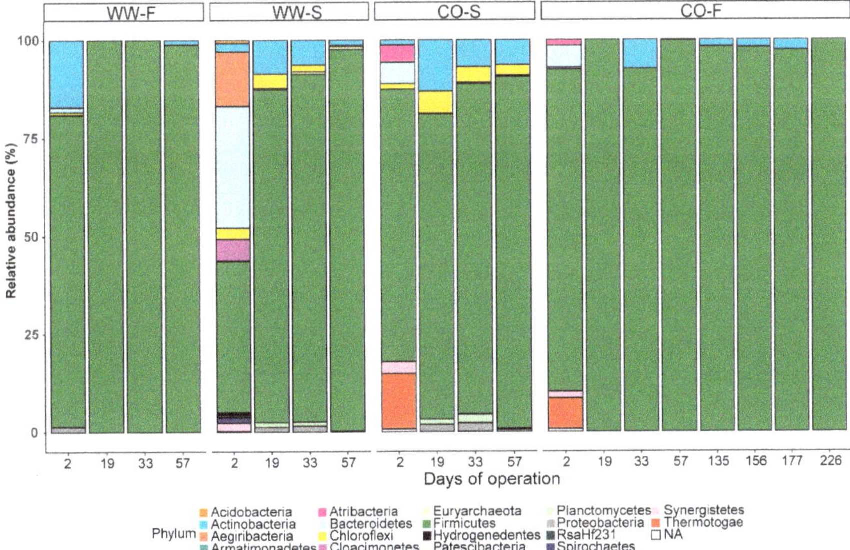

Figure 6. Relative abundance of microbial phyla (based on total bacterial and archaeal sequences) in reactors inoculated with sludge from a co-digestion plant (CO) or an anaerobic digestion process fed mixed sludge (WW) and fed food waste (F) or sewage sludge (S). Days of operation at the point of sampling are given on the *x*-axis.

Although no strong effect of inoculum source on microbial community structure was observed, there were differences in process performance, such as higher methanogenic activity in WW-S (7% of TOC) compared with CO-S (2% of TOC) (Table 2). This difference indicates that the methanogenic community originating from an inoculum dominated by SAO and hydrogenotrophic methanogenesis (i.e., CO-S) was easier to wash out at high OLR than an acetotrophic community (i.e., WW-S). SAO-driven processes are known to have slower growth rates than acetotrophic methanogenesis [52], a possible explanation for the observed results. Still, as mixed sludge is a rather recalcitrant material where hydrolysis limits acid production [53], it is clearly difficult to load the process to sufficiently high level to reach overload of the methanogenic step, irrespective of methanogenesis pathway. Consequently, even though the acetate to total VFA ratio was similar for the two processes fed sludge, the higher methanogenic activity in WW-S likely explains the lower total VFA yield. For food waste, with a higher degree of DOC in the ingoing TOC, hydrolysis was less important for reaching high VFA production. In these processes, a difference was noted for the processes started with the different inocula, with the inoculum from the co-digestion plant resulting in higher acetate yield. A possible explanation for this is the higher hydrogen consumption rate by acetogenesis, which would facilitate oxidation of longer-chain VFA to acetate [54].

3.3.2. Effect of Substrate on Microbial Community Structure

Although the microbial communities in all four reactors were similar at the phylum level, with a vast majority belonging to Firmicutes (90–100%; Figure 6), differences depending on substrate became obvious at lower taxonomic rank after 19 days of operation (Figures 5 and 7). The microbial community in the reactor inoculated with sludge and fed food waste (WW–F) were already diverged at day 2 from the other process inoculated with similar sludge (WW-S) and became highly similar to the community in the CO-F reactor (Figure 5). Similarly, from day 19 onwards, the reactors inoculated with material from the co-digestion plant diverged and became highly similar to the process fed the same substrate (Figure 5).

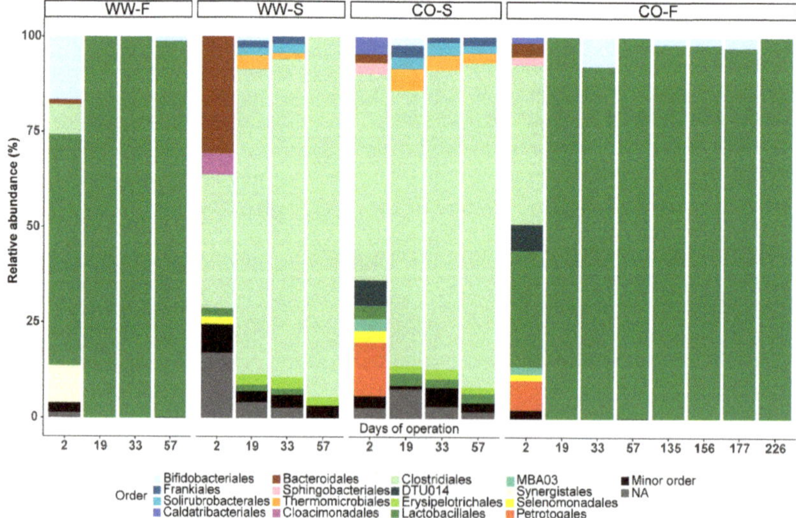

Figure 7. Relative abundance of microbial order (based on total bacterial and archaeal sequences) in reactors inoculated with sludge from the co-digestion plant (CO) or an anaerobic digestion process fed mixed sludge (WW) and fed food waste (F) or sewage sludge (S). Days of operation at the point of sampling are given on the *x*-axis.

3.3.3. Food Waste Reactors

In both WW-F and CO-F, the genus *Lactobacillus* (order Lactobacillales) clearly dominated (92–100%) and was only accompanied by minor levels of the genus *Aeriscardovia* (order Bifidobacteriales <5% from day 33 (Figure 7 and Supplementary Figure S1). The genus *Lactobacillus* was represented by sequence variants, for which Blast searches revealed a relationship (93–97% gene sequence similarity) to *Lactobacillus amylolyticus*. This species has been isolated from beer malt [55] and is known to efficiently convert carbon (including starchy materials) to lactic acid under anaerobic conditions [56]. Similarly, in an earlier study of acid production from food waste during semi-continuous operation, *Lactobacillus* dominated the processes, with a smaller fraction of Bifidobacteriaceae (*Aeriscardovia* at genus level). In that study, a combination of acetic acid and/or lactic acid was produced [15]. Likewise, Lactobacillus was also highly abundant during continuous VFA production from food waste at low pH (3.2–4.5) in the study by Feng et al. [20]. Lactobacillus has also been also shown to dominate during batch-wise VFA production from food waste, with comparably higher abundance at acidic compared or neutral and alkaline conditions [21,57]. The observed high abundance of *Lactobacillus* is likely due to the character of the food waste, with high levels of soluble organic matter. However, the substrate per se also represents a source of active microorganisms, which is an additional factor potentially influencing the microbial composition and consequently the fermentation products obtained [13,16]. In the study by Yin et al. [13], lactate was produced as a major product from non-treated food waste, while no lactate was formed after thermal treatment (and thus sterilization) of the food waste. Similarly, since the reactors in the present study were fed non-hygienized food waste, the high lactate concentration could be due to the activity of microorganisms originating from the substrate.

Lactobacillus spp. can utilize a wide range of sugars as the carbon source, while producing either mainly lactate (homofermentation) or lactate, acetate, and ethanol (heterofermentation), while one species within this genus is able to convert lactate to acetate and formate (and in rare cases form H_2) [58]. Thus, the formation of acetate from lactate has been observed while fermenting food waste at regulated pH 7, with associated H_2 production [59]. These three routes may, therefore, have occurred simultaneously to differing extents in the food waste reactors, explaining the divergent patterns of lactic and acetic acid production in the reactors (Figure 2). From day 57 onward, CO-F produced a molar ratio of acetic acid:lactic acid of about 2:1. In WW-F, this ratio was instead about 1:2.5. Neither of these patterns can be explained by the occurrence of simply homo- or heterofermentation, which would theoretically yield only lactate or a 1:1 ratio of lactate to acetate. Thus, some of the lactic acid produced was most probably degraded into acetate to different extents, yielding the observed molar ratios. The lactic acid concentration increased in WW-F after day 33, which correlated with a drop in pH from 3.9 to on average 3.6 between days 33 and 64. In CO-F, the pH remained between 3.9 and 4.2 and here, the acetate level was comparably higher. A similar pattern was seen in the study by Feng et al. [20], where the proportion of lactate to acetate during continuous fermentation of food waste increased with decreasing pH (<4, controlled). In that study, the acetate level increased at pH above 4.5, in line with an increase in the relative abundance of Bifidobacteria. This connection between acetate production and abundance of Bifidobacteria could not be confirmed in the present study, showing similar levels in both CO-F and WW-F. The decrease in pH could have been a consequence of the increased lactic acid production, or of low pH decreasing the capability for lactic acid conversion into acetate. This has been observed for another *Lactobacillus* species (*L. bifermentans*), which requires pH > 4 for activity [58,60]. A similar correlation between acetic acid or lactic acid as the major product and pH has also been observed during acid production from food waste [15].

3.3.4. Sewage Sludge Reactors

The communities in both reactors fed mixed sludge diverged from those in the reactors fed food waste. They were dominated by the order Clostridiales (representing up to 85–94% at day 57), followed by smaller fractions of Lactobacillales, Erysipelotrichales (phylum Firmicutes), Thermomicrobiales (Chloroflexi), Solirubrobacterales, and Frankiales (Actinobacteria) (<3% each) (Figure 7). Investigations at genus rank revealed an increased relative abundance of *Roseburia* (family Lachnospiraceae) and *Fastidiosipila* (family Ruminococcaceae), representing 44–45% and 16–36% of the total community at day 57, respectively (Supplementary Figures S1 and S2). *Roseburia* is able to utilize carbohydrates and acetic acid and produce butyric and lactic acids [61–63], and could thus have been the main producer of the high level of butyric acid found in the sludge reactors (Figure 1). *Fastidiosipila* may be involved in production of acetic and butyric acids [64] and has previously been identified as a major genus in mesophilic (35–37 °C) anaerobic digesters treating a variety of wastes, including the organic fraction of municipal solid waste [65], food waste [66,67], landfill leachate [68], and sewage sludge [69].

As indicated at the family level, minor levels (≤7%) of Peptostreptococcaceae, Family XIII, and Eubacteriaceae (all belonging to the order Clostridiales) were detected in reactors WW-S (days 19–33) and WW-F (days 19–57; Figure S2). In WW–S, several of the low-abundance bacteria (Peptostreptococcaceae, Family XIII, Eubacteriaceae) decreased below the detection level at the last sampling point (day 57). Members of the families Peptostreptococcaceae and Eubacteriaceae are known to ferment carbohydrates and proteins to various acids or sugars into acetate, respectively [70,71]. Peptostreptococcaceae are able to produce acetic acid from a variety of sugars and glycerol, as well as from CO_2 and H_2. Hence, at the last sampling point, when both carbohydrate- and protein-hydrolyzing families disappeared, hydrolysis, acidogenesis, and acetogenesis decreased in efficiency in the WW-S reactor (Table 2). This shift co-occurred with the increase in pH from 6.0–6.1 to 7.0. Since all Peptostreptococcaceae have optimum pH around 7, it is possible that the increasing pH was a consequence of the microbial shift (and lower acidogenic efficiency), rather than the converse.

Microbial community richness and evenness were relatively similar in all processes, and no drastic change over time was observed (Supplementary Figure S3).

3.4. Methane Potential

In order to evaluate the effect of two-stage AD on the overall methane yield, each hydrolysate and the respective untreated substrate were evaluated using BMP tests. Since CO_2 and bio-methane were produced during the acidic stage (resulting in loss of mass) and organic material had partly been converted to volatile compounds, the comparison between fresh substrate and hydrolysate was based on volumetric production (mL CH_4/g material), rather than per g VS. Compensation for mass loss by the gases produced is important to avoid misinterpretation of the BMP result. Bio-methane production from the substrates and hydrolysates was significantly higher for food waste (47–52 mL CH_4/g material) than for mixed sludge (21–25 mL CH_4/g material; Figure 8). Differences in methane yield potential between hydrolysate and substrate in the present study were non-significant, due to high variation in the results. However, these results can still assist in interpreting the effect of operating two-stage AD on bio-methane yield. The hydrolysate from all reactors gave 5–20% higher methane yield than the untreated material, indicating that two-stage AD could be very positive for the overall methane yield for both mixed sludge and food waste. Similarly, Luo et al. [7] obtained an 11% increase in overall energy output on applying two-stage operation, but with hydrogen production in the first stage.

For the processes treating mixed sludge, which showed a higher degree of hydrolysis than when food waste was used, different gas production kinetics for treated and untreated material were expected, with faster kinetics from hydrolysate due to release of DOC in the acidic stage. In fact, the untreated mixed sludge actually gave a higher gas production rate during the initial five days (Figure 8). One reason for this could be that persistent methanogenic activity during the acidic stage of operation in WW-S and CO-S consumed most readily available DOC before the BMP test. This assumption is supported by the faster methane production kinetics in CO-S compared with WW-S, with the former having lower bio-methane production in the previous semi-continuous experiment. For the food waste system, the kinetics were similar for all fractions during the first four days, but thereafter, the hydrolysates resulted in higher bio-methane production [15].

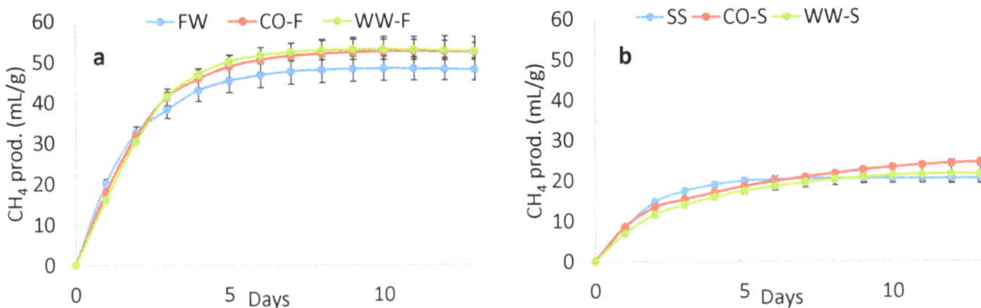

Figure 8. Volumetric bio-methane production from (**a**) food waste (FW) and hydrolysate from reactors CO-F and WW-F, and (**b**) sewage sludge (SS) and hydrolysate from reactors CO-S and WW-S. Error bars indicate the calculated standard deviation for daily values.

3.5. Potential Acid Production

Results for reactor CO-F, with the highest acetate yield per kg VS, were used to estimate acetate production potential from food waste in an industrial-scale scenario. In the co-digestion plant (Linköping) from which the substrate and inoculum were collected, about 50,000 tons of food waste are treated annually. On average, about 28% of the ingoing material (weight basis) is organic material, i.e., VS. Reactor CO-F had an average conversion efficiency of 16.2% of the ingoing VS, and hence an expected 2300 tons of acetate could be produced from the food waste treated in the Linköping co-digestion plant by applying a two-stage operation. The market value of this quantity of acetate is substantial (2.7–3.8 million USD) [25]. However, production and extraction of acetate would decrease the biogas production equivalent to 625,000 kg CH_4. This corresponds to 5,650,000 kWh of energy and a market value of about 1.3 million USD. The production of acetic acid could hence be profitable if it could be extracted in an efficient manner from the rather complex matrix of the hydrolysate, an issue that needs to be further studied.

4. Conclusions

This study clearly demonstrates that both substrate and inoculum have major effects on the acidic stage of AD. High VFA and acetate yields were difficult to obtain when treating sewage sludge, indicating that the efficiency of hydrolysis and acidogenesis was too low to achieve acidification of the process, although protein hydrolysis efficiency was higher with sewage sludge than with food waste. When food waste was used as feedstock, acidification and low pH were obtained. At low pH, protein degradation was very low, while available DOC in the form of carbohydrates and fat was apparently readily converted into VFA. The process using inoculum from WWTP produced high levels of lactate (30 g/L) and acetate (10 g/L), while the process using co-digestion plant inoculum had higher acetate (25 g/L) than lactate (15 g/L) levels. Despite the reactors having similar microbial communities, the acids produced differed between the food waste reactors, indicating that the resulting lower pH in WW-F was the cause of the higher lactate production. The microbial communities developed during acid production mainly consisted of the genera *Lactobacillus* (92–100%) when fed food waste, and *Roseburia* (44–45%) and *Fastidiosipila* (16–36%) when fed sewage sludge waste. Use of the hydrolysates in a biogas production system resulted in a non-significant increase in bio-methane production compared with directly using the substrate for biogas production (5–20% increase).

Supplementary Materials: The following are available online at http://www.mdpi.com/2306-5354/7/1/3/s1, Figure S1: Heatmap illustrating the relative abundance of genera in reactors inoculated with sludge from co-digestion plant (CO) or from an anaerobic digestion process fed mixed sludge (WW), and fed food waste (F) or sewage sludge (S). Genera discussed in the text are shown in red. Days of operation at point of sampling are given on the x-axis, Figure S2: Relative abundance of microbial families (based on total bacterial and archaeal sequences) in reactors inoculated with sludge from co-digestion plant (CO) or from an anaerobic digestion process fed mixed sludge (WW), and fed food waste (F) or sewage sludge (S). Days of operation at point of sampling are given on the x-axis, Figure S3: Microbial richness and evenness indices in the digesters inoculated with sludge from co-digestion plant (CO) or from an AD process fed mixed sludge (WW) and fed food waste (F) or sewage sludge (S). Day number in the study period is given on the x-axis.

Author Contributions: Conceptualization, A.S. and J.M.; methodology, A.S., J.M. and M.W.; formal analysis, J.M., M.W. and S.I.; data curation, S.I.; writing—original draft preparation, J.M.; writing—review and editing, A.S., J.M. and M.W.; supervision, A.S.; project administration, A.S.; funding acquisition, A.S. All authors have read and agreed to the published version of the manuscript.

Funding: J.M., A.S., and S.I. were supported by Formas grant number 2016-00311. M.W. was supported by department funding. The funders had no role in study design, data collection and analysis, decision to publish, or preparation of the manuscript.

Acknowledgments: Illumina sequencing and analysis were supported by the SNP&SEQ Technology Platform, Uppsala, as part as the Swedish National Genomic Infrastructure (NGI), Science for Live Laboratory (SciLifeLab) and the SLU Bioinformatics Infrastructure (SLUBI). Thanks to laboratory technician Yasna Calderon at Tekniska verken, Linköping AB for feeding and sampling the reactors.

Conflicts of Interest: The authors declare no conflict of interest. The funders had no role in the design of the study; in the collection, analyses, or interpretation of data; in the writing of the manuscript, or in the decision to publish the results.

References

1. Hagman, L. The role of biogas solutions in sustainable biorefineries. *J. Clean Prod.* **2018**, *172*, 3982–3989. [CrossRef]

2. Plugge, C.M. Biogas. *Microb. Biotechnol.* **2017**, *10*, 1128–1130. [CrossRef] [PubMed]

3. Wu, Q.L.; Guo, W.Q.; Zheng, H.S.; Luo, H.C.; Feng, X.; Yin, R.L.; Ren, N.Q. Enhancement of volatile fatty acid production by co-fermentation of food waste and excess sludge without pH control: The mechanism and microbial community analyses. *Bioresour. Technol.* **2016**, *216*, 653–660. [CrossRef] [PubMed]

4. Yin, J.; Yu, X.; Zhang, Y.; Shen, D.; Wang, M.; Long, Y.; Chen, T. Enhancement of acidogenic fermentation for volatile fatty acid production from food waste: Effect of redox potential and inoculum. *Bioresour. Technol.* **2016**, *216*, 996–1003. [CrossRef] [PubMed]

5. Chu, C.F.; Li, Y.Y.; Xu, K.Q.; Ebie, Y.; Inamori, Y.; Kong, H.N. A pH- and temperature-phased two-stage process for hydrogen and methane production from food waste. *Int. J. Hydrog. Energy* **2008**, *33*, 4739–4746. [CrossRef]

6. Moestedt, J.; Nordell, E.; Hallin, S.; Schnürer, A. Two stage anaerobic digestion for reduced hydrogen sulphide production. *J. Chem. Technol. Biotechnol.* **2015**, *91*, 1055–1062.

7. Luo, G.; Xie, L.; Zhou, Q.; Angelidaki, I. Enhancement of bioenergy production from organic wastes by two-stage anaerobic hydrogen and methane production process. *Bioresour. Technol.* **2011**, *102*, 8700–8706. [CrossRef]

8. Strazzera, G.; Battista, F.; Garcia, N.H.; Frison, N.; Bolzonella, D. Volatile fatty acids production from food wastes for biorefinery platforms: A review. *J. Environ. Manag.* **2018**, *226*, 278–288. [CrossRef]

9. Zhou, M.; Yan, B.; Wong, J.W.C.; Zhang, Y. Enhanced volatile fatty acids production from anaerobic fermentation of food waste: A mini-review focusing on acidogenic metabolic pathways. *Bioresour. Technol.* **2018**, *248*, 68–78. [CrossRef]

10. Feng, L.Y.; Yan, Y.Y.; Chen, Y.G. Co-fermentation of waste activated sludge with food waste for short-chain fatty acids production: Effect of pH at ambient temperature. *Front. Environ. Sci. Eng. China* **2011**, *5*, 623–632. [CrossRef]

11. Jiang, J.G.; Zhang, Y.J.; Li, K.M.; Wang, Q.; Gong, C.X.; Li, M.L. Volatile fatty acids production from food waste: Effects of pH, temperature, and organic loading rate. *Bioresour. Technol.* **2013**, *143*, 525–530. [CrossRef] [PubMed]

12. Shen, D.; Yin, J.; Yu, X.; Wang, M.; Long, Y.; Shentu, J.; Chen, T. Acidogenic fermentation characteristics of different types of protein-rich substrates in food waste to produce volatile fatty acids. *Bioresour. Technol.* **2017**, *227*, 125–132. [CrossRef] [PubMed]

13. Yin, J.; Wang, K.; Yang, Y.Q.; Shen, D.S.; Wang, M.Z.; Mo, H. Improving production of volatile fatty acids from food waste fermentation by hydrothermal pretreatment. *Bioresour. Technol.* **2014**, *171*, 323–329. [CrossRef] [PubMed]

14. Liu, D.W.; Liu, D.P.; Zeng, R.J.; Angelidaki, I. Hydrogen and methane production from household solid waste in the two-stage fermentation process. *Water Res.* **2006**, *40*, 2230–2236. [CrossRef] [PubMed]

15. Moestedt, J.; Müller, B.; Nagavara, N.Y.; Schnürer, A. Acetate and lactate production during two-stage anaerobic digestion of food waste driven by Lactobacilli and Aeriscardovia. *Bioenergy Biofuels*. (Submitted 2019, under review).

16. Tang, J.; Wang, X.; Hu, Y.; Zhang, Y.; Li, Y. Lactic acid fermentation from food waste with indigenous microbiota: Effects of pH, temperature and high OLR. *Waste Manag.* **2016**, *52*, 278–285. [CrossRef] [PubMed]

17. Atasoy, M.; Owusu-Agyeman, I.; Plaza, E.; Cetecioglu, Z. Bio-based volatile fatty acid production and recovery from waste streams: Current status and future challenges. *Bioresour. Technol.* **2018**, *268*, 773–786. [CrossRef] [PubMed]

18. Atasoy, M.; Eyice, O.; Schnurer, A.; Cetecioglu, Z. Volatile fatty acids production via mixed culture fermentation: Revealing the link between pH, inoculum type and bacterial composition. *Bioresour. Technol.* **2019**, *292*, 121889. [CrossRef]

19. Yu, L.; Zhang, W.; Liu, H.; Wang, G.; Liu, H. Evaluation of volatile fatty acids production and dewaterability of waste activated sludge with different thermo-chemical pretreatments. *Int. Biodeterior. Biodegrad.* **2018**, *129*, 170–178. [CrossRef]

20. Feng, K.; Li, H.; Zheng, C. Shifting product spectrum by pH adjustment during long-term continuous anaerobic fermentation of food waste. *Bioresour. Technol.* **2018**, *270*, 180–188. [CrossRef]

21. Tang, J.; Wang, X.C.; Hu, Y.; Zhang, Y.; Li, Y. Effect of pH on lactic acid production from acidogenic fermentation of food waste with different types of inocula. *Bioresour. Technol.* **2017**, *224*, 544–552. [CrossRef] [PubMed]

22. Bhatia, S.K.; Yang, Y.H. Microbial production of volatile fatty acids: Current status and future perspectives. *Rev. Environ. Sci. Bio/Technol.* **2017**, *16*, 327–345. [CrossRef]

23. Khan, M.A.; Ngo, H.H.; Guo, W.S.; Liu, Y.; Nghiem, L.D.; Hai, F.I.; Deng, L.J.; Wang, J.; Wu, Y. Optimization of process parameters for production of volatile fatty acid, biohydrogen and methane from anaerobic digestion. *Bioresour. Technol.* **2016**, *219*, 738–748. [CrossRef] [PubMed]

24. Lim, H.G.; Lee, J.H.; Noh, M.H.; Jung, G.Y. Rediscovering Acetate Metabolism: Its Potential Sources and Utilization for Biobased Transformation into Value-Added Chemicals. *J. Agric. Food Chem.* **2018**, *66*, 3998–4006. [CrossRef] [PubMed]

25. Vidra, A.; Németh, Á. Bio-produced Acetic Acid: A Review. *Period. Polytech. Chem. Eng.* **2018**, *62*, 245–256. [CrossRef]

26. Hawkes, F.R.; Hussy, I.; Kyazze, G.; Dinsdale, R.; Hawkes, D.L. Continuous dark fermentative hydrogen production by mesophilic microflora: Principles and progress. *Int. J. Hydrog. Energy.* **2007**, *32*, 172–184. [CrossRef]

27. Schuchmann, K.; Müller, V. Energetics and Application of Heterotrophy in Acetogenic Bacteria. *Appl. Environ. Microbiol.* **2016**, *82*, 4056–4069. [CrossRef]

28. Schink, B. Energetics of Syntrophic Cooperation in Methanogenic Degradation. *Microbiol. Mol. Biol. Rev.* **1997**, *61*, 262–280.

29. Garcia-Aguirre, J.; Esteban-Gutiérrez, M.; Irizar, I.; González-Mtnez, J.; Aymerich, E. Continuous acidogenic fermentation: Narrowing the gap between laboratory testing and industrial application. *Bioresour. Technol.* **2019**, *282*, 407–416. [CrossRef]

30. Jonsson, S.; Borén, H. Analysis of mono- and diesters of o-phthalic acid by solid-phase extractions with polystyrene-divinylbenzene-based polymers. *J. Chromatogr. A* **2002**, *963*, 393–400. [CrossRef]

31. Westerholm, M.; Hansson, M.; Schnurer, A. Improved biogas production from whole stillage by co-digestion with cattle manure. *Bioresour. Technol.* **2012**, *114*, 314–319. [CrossRef] [PubMed]

32. Chen, Y.; Jiang, X.; Xiao, K.; Shen, N.; Zeng, R.J.; Zhou, Y. Enhanced volatile fatty acids (VFAs) production in a thermophilic fermenter with stepwise pH increase–Investigation on dissolved organic matter transformation and microbial community shift. *Water Res.* **2017**, *112*, 261–268. [CrossRef] [PubMed]

33. FAO. Food Energy-Methods of Analysis and Conversion Factors. Available online: http://www.fao.org/3/Y5022E/Y5022E00.htm (accessed on 15 November 2019).

34. Stein, U.H.; Wimmer, B.; Ortner, M.; Fuchs, W.; Bochmann, G. Maximizing the production of butyric acid from food waste as a precursor for ABE-fermentation. *Sci. Total. Environ.* **2017**, *598*, 993–1000. [CrossRef] [PubMed]

35. Müller, B.; Sun, L.; Westerholm, M.; Schnürer, A. Bacterial community composition and *fhs* profiles of low- and high-ammonia biogas digesters reveal novel syntrophic acetate-oxidising bacteria. *Biotechnol. Biofuels.* **2016**, *9*, 1–18. [CrossRef]

36. Westerholm, M.; Isaksson, S.; Karlsson-Lindsjö, O.; Schnürer, A. Microbial community adaptability to altered temperature conditions determines the potential for process optimisation in biogas processes. *Appl. Energy* **2018**, *226*, 838–848. [CrossRef]

37. Martin, M. Cutadapt removes adapter sequences from high-throughput sequencing reads. *EMBnet J.* **2011**, *17*, 10–12. [CrossRef]

38. Callahan, B.J.; McMurdie, P.J.; Rosen, M.J.; Han, A.W.; Johnson, J.A.; Holmes, S.P. DADA2: High-resolution sample inference from Illumina amplicon data. *Nat. Methods* **2016**, *13*, 581–583. [CrossRef]

39. McMurdie, P.J.; Holmes, S. Phyloseq: An R package for reproducible interactive analysis and graphics of microbiome census data. *PLoS ONE* **2013**, *8*, e61217. [CrossRef]

40. Oksanen, J.; Blanchet, G.; Friendly, M.; Kindt, R.; Legendre, P.; McGlinn, D.; Minchin, P.R.; O'Hara, R.B.; Simpson, G.L.; Solymos, P.; et al. Vegan: Community Ecology Package. Available online: https://www.researchgate.net/publication/282247686_Vegan_Community_Ecology_Package_R_package_version_20-2 (accessed on 15 November 2019).

41. Altschul, S.F.; Gish, W.; Miller, W.; Myers, E.W.; Lipman, D.J. Basic local alignment search tool. *J. Mol. Biol.* **1990**, *215*, 403–410. [CrossRef]

42. Wang, K.; Yin, J.; Shen, D.; Li, N. Anaerobic digestion of food waste for volatile fatty acids (VFAs) production with different types of inoculum: Effect of pH. *Bioresour. Technol.* **2014**, *161*, 395–401. [CrossRef]

43. Kakar, F.; Koupaie, E.H.; Hafez, H.; Elbeshbishy, E. Effect of Hydrothermal Pretreatment on Volatile Fatty Acids Production from Source-Separated Organics. *Processes* **2019**, *7*, 576. [CrossRef]

44. Liu, H.; Han, P.; Liu, H.; Zhou, G.; Fu, B.; Zheng, Z. Full-scale production of VFAs from sewage sludge by anaerobic alkaline fermentation to improve biological nutrients removal in domestic wastewater. *Bioresour. Technol.* **2018**, *260*, 105–114. [CrossRef] [PubMed]

45. Khiewwijit, R.; Keesman, K.J.; Rijnaarts, H.; Temmink, H. Volatile fatty acids production from sewage organic matter by combined bioflocculation and anaerobic fermentation. *Bioresour. Technol.* **2015**, *193*, 150–155. [CrossRef] [PubMed]

46. Hao, J.; Wang, H. Volatile fatty acids productions by mesophilic and thermophilic sludge fermentation: Biological responses to fermentation temperature. *Bioresour. Technol.* **2015**, *175*, 367–373. [CrossRef] [PubMed]

47. Ma, H.; Chen, X.; Liu, H.; Liu, H.; Fu, B. Improved volatile fatty acids anaerobic production from waste activated sludge by pH regulation: Alkaline or neutral pH? *Waste Manag.* **2016**, *48*, 397–403. [CrossRef] [PubMed]

48. Xiao, K.; Zhou, Y.; Guo, C.; Maspolim, Y.; Ng, W.J. Impact of undissociated volatile fatty acids on acidogenesis in a two-phase anaerobic system. *J. Environ. Sci.* **2016**, *42*, 196–201. [CrossRef] [PubMed]

49. Infantes, D.; del Campo, A.G.; Villasenor, J.; Fernandez, F.J. Kinetic model and study of the influence of pH, temperature and undissociated acids on acidogenic fermentation. *Biochem. Eng. J.* **2012**, *66*, 66–72. [CrossRef]

50. Sundberg, C.; Al-Soud, W.A.; Larsson, M.; Alm, E.; Yekta, S.S.; Svensson, B.H.; Sørensen, S.J.; Karlsson, A. 454 pyrosequencing analyses of bacterial and archaeal richness in 21 full-scale biogas digesters. *FEMS Microbiol. Ecol.* **2013**, *85*, 612–626. [CrossRef]

51. Sun, L.; Müller, B.; Westerholm, M.; Schnürer, A. Syntrophic acetate oxidation in industrial CSTR biogas digesters. *J. Biotechnol.* **2014**, *171*, 39–44. [CrossRef]

52. Westerholm, M.; Moestedt, J.; Schnürer, A. Biogas production through syntrophic acetate oxidation and deliberate operating strategies for improved digester performance. *Appl. Energy* **2016**, *179*, 124–135. [CrossRef]

53. Appels, L.; Baeyens, J.; Degréve, J.; Dewil, R. Principles and potential of the anaerobic digestion of waste-activated sludge. *Prog. Energy Combust.* **2008**, *34*, 755–781. [CrossRef]

54. Worm, P.; Müller, N.; Plugge, C.M.; Stams, A.J.M.; Schink, B. Syntrophy in Methanogenic Degradation. In *(Endo)symbiotic Methanogenic Archaea*; Springer: Berlin/Heidelberg, Germany, 2010.

55. Bohak, I.; Back, W.; Richter, L.; Ehrmann, M.; Ludwig, W.; Schleifer, K.H. *Lactobacillus amylolyticus* sp. nov., isolated from beer malt and beer wort. *Syst. Appl. Microbiol.* **1998**, *21*, 360–364. [PubMed]

56. Reddy, G.; Altaf, M.; Naveena, B.J.; Venkateshwar, M.; Kumar, E.V. Amylolytic bacterial lactic acid fermentation—A review. *Biotechnol. Adv.* **2008**, *6*, 22–34. [CrossRef] [PubMed]

57. Ma, J.; Xie, S.; Yu, L.; Zhen, Y.; Zhao, Q.; Frear, C.; Chen, S.; Wang, Z.W.; Shi, Z. pH shaped kinetic characteristics and microbial community of food waste hydrolysis and acidification. *Biochem. Eng. J.* **2019**, *146*, 52–59. [CrossRef]

58. Lindgren, S.E.; Axelsson, L.T.; McFeeters, R.F. Anaerobic ʟ-lactate degradation by *Lactobacillus plantarum*. *FEMS Microbiol. Lett.* **1990**, *66*, 209–213. [CrossRef]

59. Grause, G.; Igarashi, M.; Kameda, T.; Yoshioka, T. Lactic acid as a substrate for fermentative hydrogen production. *Int. J. Hydrog. Energy* **2012**, *37*, 16967–16973. [CrossRef]

60. Cselovszky, J.; Wolf, G.; Hammes, W.P. Production of formate, acetate, and succinate by anaerobic fermentation of *Lactobacillus pentosus* in the presence of citrate. *Appl. Microbiol. Biotechnol.* **1992**, *37*, 94–97. [CrossRef]

61. STANTON, T.B.; SAVAGE, D.C. Roseburia cecicola gen. nov., sp. nov., a Motile, Obligately Anaerobic Bacterium from a Mouse Cecum. *Int. J. Syst. Evol. Microbiol.* **1983**, *33*, 618–627. [CrossRef]

62. Duncan, S.H.; Hold, G.L.; Barcenilla, A.; Stewart, S.S.; Flint, H.J. Roseburia intestinalis sp. nov., a novel saccharolytic, butyrate-producing bacterium from human faeces. *Int. J. Syst. Evol. Microbiol.* **2002**, *52*, 1615–1620.

63. Duncan, S.H.; Aminov, R.I.; Scott, K.P.; Louis, P.; Stanton, T.B.; Flint, H.J. Proposal of Roseburia faecis sp. nov., Roseburia hominis sp. nov. and Roseburia inulinivorans sp. nov., based on isolates from human faeces. *Int. J. Syst. Evol. Microbiol.* **2006**, *56*, 2437–2441. [CrossRef]

64. Falsen, E.; Collins, M.D.; Welinder-Olsson, C.; Song, Y.; Finegold, S.M.; Lawson, P.A. Fastidiosipila sanguinis gen. nov., sp. nov., a new Gram-positive, coccus-shaped organism from human blood. *Int. J. Syst. Evol. Microbiol.* **2005**, *55*, 853–858. [CrossRef] [PubMed]

65. Xie, Z.; Wang, Z.; Wang, Q.; Zhu, C.; Wu, Z. An anaerobic dynamic membrane bioreactor (AnDMBR) for landfill leachate treatment: Performance and microbial community identification. *Bioresour. Technol.* **2014**, *161*, 29–39. [CrossRef]

66. He, Q.; Li, L.; Zhao, X.; Qu, L.; Wu, D.; Peng, X. Investigation of foaming causes in three mesophilic food waste digesters: Reactor performance and microbial analysis. *Sci. Rep.* **2017**, *7*, 13701. [CrossRef] [PubMed]

67. Kim, E.; Lee, J.; Han, G.; Hwang, S. Comprehensive analysis of microbial communities in full-scale mesophilic and thermophilic anaerobic digesters treating food waste-recycling wastewater. *Bioresour. Technol.* **2018**, *259*, 442–450. [CrossRef] [PubMed]

68. Cardinali-Rezende, J.; Rojas-Ojeda, P.; Nascimento, A.M.; Sanz, J.L. Proteolytic bacterial dominance in a full-scale municipal solid waste anaerobic reactor assessed by 454 pyrosequencing technology. *Chemosphere* **2016**, *146*, 519–525. [CrossRef] [PubMed]

69. Chen, S.; Dong, B.; Dai, X.; Wang, H.; Li, N.; Yang, D. Effects of thermal hydrolysis on the metabolism of amino acids in sewage sludge in anaerobic digestion. *Waste Manag.* **2019**, *88*, 309–318. [CrossRef] [PubMed]

70. Slobodkin, A. The Family Peptostreptococcaceae. In *The Prokaryotes*; Rosenberg, E., Ed.; Springer: Berlin/Heidelberg, Germany, 2014.

71. Stackebrandt, E. The Family Eubacteriaceae. In *The Prokaryotes-Firmicutes and Tenericutes*; Rosenberg, E., Ed.; Springer: Berlin/Heidelberg, Germany, 2014.

Article

Degradation of Veterinary Antibiotics in Swine Manure via Anaerobic Digestion

Ali Hosseini Taleghani [1], Teng-Teeh Lim [1,*], Chung-Ho Lin [2], Aaron C. Ericsson [3] and Phuc H. Vo [2]

[1] Division of Food Systems and Bioengineering, University of Missouri, Columbia, MO 65211, USA; ahbq6@mail.missouri.edu

[2] School of Natural Resources, University of Missouri, Columbia, MO 65211, USA; LinChu@missouri.edu (C.-H.L.); phucvh2410@gmail.com (P.H.V.)

[3] Department of Veterinary Pathobiology, University of Missouri, Columbia, MO 65201, USA; ericssona@missouri.edu

* Correspondence: limt@missouri.edu; Tel.: +1-573-882-9519

Received: 10 August 2020; Accepted: 2 October 2020; Published: 9 October 2020

Abstract: Antibiotic-resistant microorganisms are drawing a lot of attention due to their severe and irreversible consequences on human health. The animal industry is considered responsible in part because of the enormous volume of antibiotics used annually. In the current research, veterinary antibiotic (VA) degradation, finding the threshold of removal and recognizing the joint effects of chlortetracycline (CTC) and Tylosin combination on the digestion process were studied. Laboratory scale anaerobic digesters were utilized to investigate potential mitigation of VA in swine manure. The digesters had a working volume of 1.38 L (in 1.89-L glass jar), with a hydraulic retention time (HRT) of 21 days and a loading rate of 1.0 g-VS L^{-1} d^{-1}. Digesters were kept at $39 \pm 2\,^{\circ}C$ in incubators and loaded every two days, produced biogas every 4 days and digester pH were measured weekly. The anaerobic digestion (AD) process was allowed 1.5 to 2 HRT to stabilize before adding the VAs. Tests were conducted to compare the effects of VAs onto manure nutrients, volatile solid removal, VA degradation, and biogas production. Concentrations of VA added to the manure samples were 263 to 298 mg/L of CTC, and 88 to 263 mg/L of Tylosin, respectively. Analysis of VA concentrations before and after the AD process was conducted to determine the VA degradation. Additional tests were also conducted to confirm the degradation of both VAs dissolved in water under room temperature and digester temperature. Some fluctuations of biogas production and operating variables were observed because of the VA addition. All CTC was found degraded even only after 6 days of storage in water solution; thus, there was no baseline to estimate the effects of AD. As for Tylosin, 100% degradation was observed due to the AD (removal was 100%, compared with 24–40% degradation observed in the 12-day water solution storage). Besides, complete Tylosin degradation was also observed in the digestate samples treated with a mixture of the two VAs. Lastly, amplicon sequencing was performed on each group by using the 50 most variable operational taxonomic units (OTUs)s and perfect discriminations were detected between groups. The effect of administration period and dosage of VAs on Phyla *Firmicutes* Proteobacteria, Synergistetes and Phylum Bacteroides was investigated. These biomarkers' abundance can be employed to predict the sample's treatment group.

Keywords: anaerobic digester; antibiotics removal; antimicrobial; biogas; chlortetracycline; Tylosin

1. Introduction

Nowadays, a wide variety of antibiotics are being used in animal farms to cure, prevent and also to improve the growth of animals, accounting for more than 52% of total antibiotics consumption in the world [1–3]. Due to the rapid effect of antibiotics and low cost of them, daily use of it rocketed during the last two decades) [4]. Although in 2017, attempts to restrict using antibiotics took place, from 2009 to 2016 use of several veterinary antibiotics (VAs) was raised by 36.8% on average [4]. Many of these compounds have weak absorption within the animal gut and intestine during digestion, resulting in the excretion of potent parent and daughter products [5]. A high percentage of the antibiotics (60–90%) is excreted without metabolism in urine and feces, leading to potential human and ecological health risks for soil and water [6–9]. In addition, based on a study of Alexy et al. [10], most of the antibiotics being used are not biodegradable, with degradation extents varying between 4 and 27%.

Since a vast volume of manure is being produced each year due to concentrated animal feeding operations (CAFOs) and is mostly being applied to solid materials as a fertilizer, the long-term presence of such antibiotics in manure with even trace concentrations (i.e., ng/L) could lead to the formation of antibiotic resistance genes (ARGs) [11]. The microbacterial resistance will result in higher medical costs, longer treatment periods, and increased mortality [12,13]. Animal farms typically utilize simple treatment systems, which are mainly AD, stockpiling, composting, wetlands, or lagoons ([3,14]). These simple treatments might not be sufficient to prevent the appearance of ARGs [15]. There are new approaches to remove or reduce antibiotics, including activated carbon adsorption, membrane filtration, advanced oxidation processes ([14,16–19]), however, they require advanced technical supervision as well as extreme expenses. Furthermore, the VAs could pollute the soil and water, then the human food chain through crops and animal-derived foods [20–24]. Moreover, the residue of VAs in the AD process could sustain microbes under the minimum inhibitory concentration (MIC), fostering the selection for ARGs by microbes [25,26].

Anaerobic digesters can produce biogas as well as removing VAs and ARGs [27–29]. In contrast, composting [30] requires less monitoring, has a more stable working system and a generally higher removal rate for certain VAs, but consumes energy and occasionally has less capability to remove ARGs than AD. Xie et al. [31] concluded that thermophilic composting of cow manure would result in ARG mitigation, lowering 16S rRNA with tetracycline, sulfonamide and fluoroquinolone resistance genes, however, not effective with aadA, aadA2, qacED1, tetL, cintI1, intI1, and tnpA04. A similar study of dairy manure composts showed satisfactory treatment of antibiotic-resistant E. coli and Salmonella, yet some antibiotic-resistant Enterobacter spp. and multidrug-resistant Pseudomonas spp. population raised after application of these composts to rangeland soils in Texas [29]. Prado et al. [32] aimed to use an aerobic reactor with activated sludge to track the fate of tetracycline (TC) and Tylosin as antibiotics. Both TC and Tylosin were not biodegradable in this type of reactor. Research also determined that the biosorption of both antibiotics appeared to be most favorable for TC.

Joy et al. [33] investigated the behavior of three antibiotics (bacitracin, chlortetracycline, and tylosin) and two classes of ARGs (Tet and Erm), which were monitored in swine manure slurry under anaerobic conditions. First-order decay rates were determined for each antibiotic with half-lives ranging from 1 day (chlortetracycline) to 10 days (tylosin). Angenent and Wrenn [34] examined the effects of an anaerobic sequencing batch reactor (ASBR) on the removal of antibiotic tylosin. They observed no inhibitory effect on biogas production, but some macrolide–lincosamide– streptogramin B (MLSB)-resistant bacteria appeared. Shi et al. [35], discovered that a certain dosage of tetracycline (TC) and sulfamethoxydiazine (SMD) could reduce biogas production. They also noticed the rapid disappearance of antibiotics (more than 50%) in the first 12 h. However, they were not sure about whether it was being degraded or just absorbed into solid materials. A similar study was conducted by Beneragama et al. [36], who confirmed the efficiency of AD of antibiotics in dairy manure. They also utilized thermophilic microorganisms (working in 55 °C or 131 °F). Results showed no inhibition in gas production and the efficiency of the reactor.

Approximately 80% of the 16,000 metric tons of antibiotics sold annually in the U.S. are used in animal husbandry [37]. These antibiotics can be transported to run-off water, groundwater, soil, and finally, plants [38–43]. In 2030, antibiotic resistance would cost USD 3.4 trillion due to subsequent mortality and substitute treatment [44]. Therefore, according to this extensive use of antibiotics and their stability in the environment, seeking an efficient and financially feasible method is vital.

Besides, limited findings are available about the threshold concentrations of different classes of antibiotics in manure that can be removed during the AD process and the interaction between anaerobic digesters and antibiotics.

The objectives of this study were to evaluate (i) anaerobic digestion efficiency on the removal of chlortetracycline (CTC) and Tylosin, (ii) inhibitory behavior of VAs on the reactors, and (iii) the effect of these antibiotics on microbial dynamics in anaerobic digesters with swine manure. The novelty of this research is the imitation of on-farm mesophilic anaerobic digesters that were loaded frequently with manure from a commercial pig farm, while operating over several months. The research focused on widely used antibiotics and emphasized proper concentration and duration at which the antibiotics were administered (following manufacturer's recommendation and average pig weights), and amount of antibiotics excreted by the animals based on a literature review. Critical operating variables of the digesters, including pH and biogas productions were monitored closely, similar to what most on-farm AD technicians are employing to monitor AD performance, which require no sophisticated analytical expertise. The bench-scale anaerobic digesters were relatively larger and semi-continuously loaded for over several months, while many of the previous studies only focused on inhibition effect and usually using batch reactors, which are different from actual on-farm AD conditions.

2. Methods

The current study focuses on utilizing an AD process to assist in removing VAs and finding the efficiency and practicality of the reactors. Antibiotics are chosen to be spiked meticulously, based on their importance, usage in feedstock and their danger to the environment. Besides, the dosage of antibiotics was close to concentrations administrated for animals, absorbed, and then excreted, to imitate the real condition. Our reactors are fed with swine manure, which has been tested for background concentration of VAs, to diminish the chance of interference. CTC was injected into anaerobic sequencing batch reactors (ASBR) with doses of 263, 280, and 298 mg/L for each spike and a total of three injections every two days while Tylosin doses were 88, 175, and 263 mg/L for each injection and a total of 5 injections every two days.

Several factors are contributing to reactor performance and biogas production. Temperature is one of them. Different types of bacteria work on various temperature ranges and some of them are highly susceptible to temperature fluctuation. Besides, the pH and alkalinity of the environment in which bacteria are growing should be near neutral and consistent. Therefore, the temperature was kept at around 39 °C (102 °F). In addition, we were recording incubator temperature and humidity for tracking the performance of our incubator. The methane forming bacteria are very sensitive to slight changes in organic loading, pH, and temperature (a temperature change greater than 2 degrees of Fahrenheit per day will affect the methane formers).

2.1. Feedstock for Anaerobic Digestion

Manure samples were collected from a mid-central Missouri commercial swine farm. The farm was VA free for the finishing pigs, located in Versailles, Missouri, USA. Furthermore, to make sure that no antibiotics existed in the solid manure used, it was analyzed to eradicate any interference or error. Because the farm has shallow pits, the manure would be less than one month old. After collecting manure, buckets full of manure were kept frozen at −20 °C (−4 °F) until they were used as feedstock for the reactors. Once manure was needed, one of these big buckets was thawed down and separated into a small bucket (usually 4 L (L) in volume). Just one of these small buckets was in the refrigerator for feeding; the rest were kept in a freezer to keep it unchanged as much as possible. Total and volatile solids (TS and VS) of each big bucket were tested to evaluate the proper feeding ratio. The total solid (TS) of the solid manure was 25.89%, and the volatile solid (VS) was 82.02% of the TS. There was no test conducted to verify the potential effect of the freezing, although there were little observed changes in biogas production between refrigerated and frozen manure in the last year of AD tests.

The inoculum was collected from semi-continuous AD jars of previous tests (Wang et al. [45]), which were steadily producing biogas for over three months, and the feedstock was swine manure with organic loading rate (OLR) of 1 g-VS L^{-1} d^{-1} only. The total solid (TS) of inoculum was 2.20%, and the volatile solid (VS) was 64.92% of TS.

2.2. Experimental Design

Tests were carried out with laboratory size jars as reactors (adjusted for AD). Antibiotics were added to reactors with different concentrations of CTC and Tylosin to monitor antibiotic removal and gas production variations in those mesophilic reactors. Nine laboratory-scale jars as anaerobic bioreactors with the working volume of 1.375 L were kept at 39 ± 2 °C (102 °F to 105 °F) in the incubator. The jars are being fed with VA-free swine manure at 1g-VS per L-day, with 21 days hydraulic retention time (HRT). The volume of the feed given every two days is measured based on HRT and our reactor volume. Because our HRT is 21 days and the reactor volume is 1.375 L, so 0.131 L of our reactor liquid were removed and replace by feedstock (the digesters were fed every two days) [46].

Each jar was connected to 10-L Tedlar bags to collect produced biogas, and the volume was measured every four days [47]. A custom-built device was used to help distribute the biogas evenly in the bag, so the height of the bag could be measured more accurately. By utilizing a predetermined model, the volume of each bag was then estimated by bag height. Besides, to prevent any leakage of the Tedlar bags, each time two of the bags were randomly tested for possible leakage before emptying. Additionally, tubes, caps, and any connective parts were tested for leakage. After biogas measurements, bags were emptied safely and burned.

The experiment consisted of nine jars; three of them were spiked with CTC, three with Tylosin and the last three with both CTC and Tylosin, to observe the combined effect or any interaction between two types of antibiotics (Figure 1). Furthermore, to investigate the efficiency of the AD process, we added six more jars, filled with distilled water and the headspace with N$_2$. Three of these jars were being kept in the incubator at the same temperature of the digester jars (39 °C), the rest were being kept in the room temperature to monitor the effect of the temperature. The same pattern of antibiotics concentration was conducted for control jars, two groups of three jars. Retention time and sampling procedures were identical.

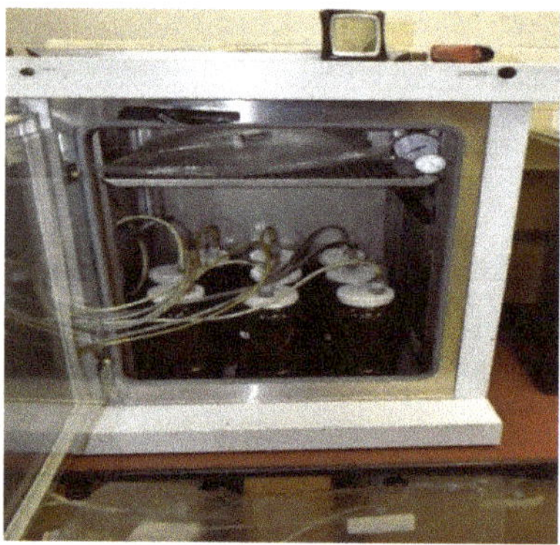

Figure 1. The scheme of the reactors, tubes transferring biogas and the incubator.

2.3. Antibiotics

The two most widely used antibiotics were selected based on consumption rate and market share of different antibiotic classes in the United States [4]. Antibiotics used in this experiment were CTC as chlortetracycline HCl and tylosin as tylosin tartrate. Commercial grade CTC was bought from "PharmGate Animal Health; Omaha, Nebraska" with the brand of "Pennchlor 64". Commercial grade tylosin used was from "Elanco Animal Health; Indianapolis, Indiana" with the brand of "Elanco". Moreover, to prepare standard samples for LCMS/MS, both antibiotics were ordered as the analytical grade from "Sigmaaldrich", St. Louis, Missouri. Chlortetracycline hydrochloride, VETRANAL™, analytical standard, with CAS number of 64-72-2 and tylosin, United States Pharmacopeia (USP) Reference Standard, with CAS number of 1401-69-0 were used for standard solutions.

We followed the prescription on the labels to imitate the real condition in a barn. Consequently, the recommended dosage for tylosin was 66 ppm in drinking water. For Swine Dysentery, adding tylosin to drinking water should be continued for 3 to 10 days, depending upon the severity of the infection. For CTC, the recommended dosage was 22 mg/kg body weights per day. The duration of treatment is 3 to 5 days depending on the infection. Pigs are generating 4.28 L of manure per day on average. Additionally, it is assumed that the average body weight of a pig is around 68 kg (finishing pigs weigh around 45–113 kg). These assumptions would help us estimate the concentration we should inject in our reactors, by considering the excretion rate and metabolism percentage.

The stabilization time for the reactors and the microbial community was expected to be one to three months, until the biogas production, digestate pH, and alkalinity trend became flat. For the current research, the digester was fed for eight weeks or 2.5 times the retention time. Important operating variables, including organic loading rate based on total volatile solids (TVS), solid content, temperature, mixing (swirling the jar daily), and foaming (if any) were recorded. Digester alkalinity and pH were monitored weekly by measuring the digestate.

For Lower range concentration, the lowest factor in each section was used. For instance, to calculate the lower band of tylosin, 11.35 L per day as pig's drinking volume, 50% excretion level applied and 4.28-L excretion per day was selected. For upper range concentration, the highest factor in each section was used. For instance, to calculate the upper band of tylosin, 18.93 L per day as pig's drinking volume, 90% excretion level and 4.28-L excretion per day was selected. The average concentration is the average of the lower and upper concentration. Recalling that jars were loaded every two days with a mixture of solid manure and water, VAs added with feed had a concentration of day 1 plus day 2.

Since antibiotics are being added to the water part of the feeding (not to solid part), therefore solubility of the VAs should be checked. Table 1 is a summary of the solubility of CTC and tylosin in the water at 20 °C:

Table 1. Chlortetracycline (CTC) and tylosin water solubility.

Reference	CTC	Tylosin
Manufacture Info.	264 mg/L	528 mg/L
Merck Index	500 mg/L	6000 mg/L
Sigma	8.6 mg/mL	50 mg/mL

Considering the solubility of CTC and tylosin in the water at 20 °C, there was no problem with CTC and tylosin solving limit individually. However, one set of three jars was used, which we decided to use to test the combined effect of antibiotics, so we had to mix two antibiotics in the same volume of water (0.103 L). There is always a chance of interference between two types of chemicals, especially when they are being added near their solubility limit. Thus, the decision was made to add CTC directly to the water, transfer it to the reactor and then add tylosin powder separately to the reactor. Other solvents such as methanol or ACN were dismissed because of their adverse effect and interference with the reactor's performance (an independent test was conducted to evaluate the impact of adding methanol onto AD performance; details are not included in this paper). Table 2 summarizes the recommended VA concentrations based on the manufacturer's recommendation and corresponding dosages considering the ranges of dosage, water consumption, and excretion rate.

Table 2. Summary of antibiotics' prescription and concentrations added every two days.

Antibiotic	Dosage	Manure Per Day (L/day)	Consumption	Treatment Duration	Excretion Level	Concentration (Lower Band) (mg/L)	Concentration (Upper Band) (mg/L)	Avg. (mg/L)
Tylosin	66 mg/liter	4.28	Drinking 11.35–18.93 L/day	3–10 days	50 to 90%	87.67	263	175.34
CTC	22 mg/kg.-body weight, day	4.28	Average Pig weight = 68 kg	3–5 days	75%	263	298	280.54

As previously mentioned, we recorded biogas production for at least two HRTs, before and after introducing the antibiotics. Table 3 illustrates the added VAs concentration in each reactor. As shown below, the first group is being administrated only with CTC, the second group with both CTC and tylosin and the last, with only tylosin.

Table 3. Veterinary antibiotics (VAs) concentration spiked in each reactor.

Jar #	CTC Concentration (ppm)	Tylosin Concentration (ppm)
1	263	0
2	280.54	0
3	298	0
4	263	87.67
5	280.54	175.34
6	298	263
7	0	87.67
8	0	175.34
9	0	263

The pH of the digestate was measured every two days while adding antibiotic, with pH meter (PINPOINT, American Marine Inc., Ridgefield, CT, USA). Using pH data, microbial activity of the digester and the reactor performance is projected. However, pH can also be affected by alkalinity. For quality assurance, alkalinity tests were also carried out.

The CO_2 concentration of the biogas was measured with a standard combustion analyzer (Bacharach Fyrite Classic Combustion Analyzer, New Kensington, PA, USA) every eight days. The concentration of CO_2 was measured every four days during the antibiotic addition period. Comparative tests using a gas spectrometry were used to check how accurate our measurements were. Below is a comparative table that illustrates accuracy control values (Table 4). The gas chromatograph device was (GC-2014, Shimadzu, US) with a thermal conductivity detector (TCD) using a ShinCarbon ST 80/100 Column (Restek, US) [48].

Table 4. Comparative table of methane content between data collected by Bacharach Fyrite Classic Combustion Analyzers and gas chromatography.

Sample #	Retention Time (ms)			Detected Volume		Bacharach Fyrite Classic Results	GC Results	
				CH_4 (µL)	CO_2 (µL)	CO_2 (%)	CH_4/100%	CO_2/100%
1	120,043	384,994	283,623	117.17	57.50	28	56.51	27.73
2	67,869.1	441,945	300,395	134.25	60.85	28	62.86	28.49
3	68,710.1	455,328	311,024	138.27	62.98	28	62.86	28.63
4	131,615	411,986	302,588	125.26	61.29	26	56.33	27.56
5	131,197	421,019	309,690	127.97	62.71	26	56.52	27.70

Because the administration times for CTC and tylosin were different (6 days for CTC and 10 days for tylosin), jars with CTC spikes were sampled at the end of day 6, while tylosin-spiked jars were sampled at day 10. For jars with the combined CTC and tylosin, samples were taken at both day 6 and day 10. Since VAs were added every two days with feed, the sampling would occur two days after the last spike. Samples were frozen at −20 °C immediately. Gas production, pH and CO_2 level were considered the vital data, which were recorded before, during and past spikes.

3. Extraction and Chemical Analysis

3.1. Sample Preparation

Two grams of the sample were transferred to 50-mL "Corning™ PP Centrifuge Tubes (polypropylene) and 5 mL of phosphate buffer (0.14M) was added. Following the pH adjustment, 200 µL of internal standard (Sulfamethazine phenyl-13C6) was fortified and the antibiotics were extracted with 25 mL of acetonitrile (ACN) with sonication for an hour. Following the sonication, the samples were centrifuged for 15 min in 4000rpm, at 4 °C (39.2 °F) with a Sorvall LYNX 6000 Superspeed Centrifuge (Thermo Scientific™, Waltham, MA, USA), and the supernatant was collected. The same extraction process was repeated with 15 mL ACN; both supernatants were combined. Twenty-milliliters of the extract was transferred to the test tube, and the solvent was further evaporated to 2 mL under a stream of nitrogen gas. The extract was diluted with 18 mL of DI water before solid-phase extraction (SPE).

3.2. Solid-Phase Extraction

The antibiotics were extracted by a Waters Oasis-HLB SPE cartridge (Oasis HLB 12 cc Vac Cartridge, 500 mg Sorbent per Cartridge, 60 µm Particle Size). The solid-phase extraction cartridges were preconditioned in an order with 10 mL ACN, 10 mL DI water all with the rate of (2 mL/min). The sample was subsequently introduced to the cartridge at a flow rate of 2 mL/min. The impurity in the cartridge was washed by using 10 mL DI water for 5 min of vacuum drying. The antibiotics retained on the cartridges were eluted with 8 mL of methanol followed by 8 mL of ACN with a flow rate of 2 mL/min. The eluate was evaporated by a gentle stream of nitrogen at 15 L/min in a water bath at 35 °C and concentrated to 10–20 µL. The extract was further filtered via a 0.22-µm Anotop inorganic filter (Sigma Aldrich) and was ready for antibiotic analysis [49].

3.3. LC-MS/MS Analysis

The concentrations of antibiotics were determined by a Waters Alliance 2695 High-Performance Liquid Chromatography (LC-MS/MS) system coupled with Waters Acquity TQ triple quadrupole mass spectrometer (MS/MS). The analytes were separated by a Phenomenex (Torrance, CA) Kinetex C18 (100 mm × 4.6 mm; 2.6 µm particle size) reverse-phase column. The mobile phase consisted of 10 mM ammonium acetate and 0.1% formic acid in water (A) and 100% acetonitrile (B). The gradient conditions were 0–0.5 min, 2% B; 0.5–7 min, 2–80% B; 7.0–9.0 min, 80–98% B; 9.0–10.0 min, 2% B; 10.0–15.0 min, 2% B at a flow rate of 0.5 mL/min. The ion source in the MS/MS system was electrospray ionization (EI) operated in either positive (ES+) mode with a capillary voltage of 1.5 kV. The ionization sources were programmed at 150 °C and the desolvation temperature was programmed at 450 °C. The MS/MS system was in the multi-reaction monitoring (MRM) mode with the optimized collision energy. The ionization energy, MRM transition ions (precursor and product ions; Table 5), capillary and cone voltage, desolvation gas flow and collision energy were optimized by the Waters IntelliStart™ optimization software package [50]. The retention time, calibration equations, and limits of the detection for the analyses of metabolites are summarized in Table 5.

Table 5. The ionization mode, retention times and optimized precursor/product ions for analysis of the VAs by the developed LC-MS/MS method.

	Chemical	Ionization Mode	Retention Time	Precursor/Product Ions
1	Ceftiofur (Excenel)	ESI+	7.04	523.8/210
2	Penicillin G Potassium salt	ESI+	7.32	335/160
3	Carbodox	ESI+	6.27	263/90
4	Chlortetracycline hydrochloride	ESI+	7.57	479/444
5	Tiamulin (Denaguard)	ESI+	8.64	494.3/192.1
6	Tylosin	ESI+	8.63	917/174
		ESI+	8.63	917/772
7	Enrofloxacin-d5	ESI+	6.98	365/321

3.4. Statistical Analysis

The statistical analyses were carried out using a two-sample t-test with unequal variances from the statistical analysis (R Core Team, 2013) to compare biogas inhibition between groups and between different VAs concentrations. Significance was accepted at probabilities $p \leq 0.05$ for all analysis. In addition, for amplicon sequencing, Bray–Curtis similarities and Jaccard similarities methods are used for this comparison. The Bray–Curtis dissimilarity is a method used to measure the structural variation between two different groups, based on counts at each group. Mathematically, the index of dissimilarity is:

$$\mathrm{BCij} = \frac{1 - 2\mathrm{Cij}}{\mathrm{Si} + \mathrm{Sj}} \tag{1}$$

where Cij is the sum of the lesser values for only those species on the intersection of two sets. Si and Sj are the total numbers of specimens at both sites. The range is between 0 and 1 [51].

The Jaccard similarity index compares members for two sets of data to quantify the resemblance between them, with a range from 0 to 1. The closer the number is to 1, the more similar the two populations are.

$$J(A, B) = \frac{(A \cap B)}{(A \cup B)} \tag{2}$$

3.5. Sampling and DNA Isolation

Raw and digested manure samples have been analyzed by the MU Metagenomics Center for the microbial/taxonomy analysis using the 16S rRNA library sequencing methodology. The results show that over 60k sequences were identified, confirming that the taxonomy analysis of manure samples can be analyzed using the specific method.

In total, twelve samples were collected into 50-mL sterile centrifuge plastic tubes. The first three samples were taken from CTC-added jars, with low, medium, and high concentrations, sampled 6 days after the first addition of VAs. The next three were sampled from jars with the addition of a mixture of CTC and tylosin on day 6 and day 10. The last group, including samples 9 to 12 were taken from jars administrated only with tylosin and were sampled on day 10 of VAs addition. Prior to sampling, each jar was mixed thoroughly with a hand mixer for 1 min. During the time after sampling and before starting the amplicon sequencing, samples were frozen to prevent any interference with oxygen. According to the TissueLyser II (Qiagen, Venlo, The Netherlands), samples were incubated at 70 °C for 20 min with intervallic vortexing. Then, samples were centrifuged at 5000× *g* for five minutes at room temperature, and the supernatant conveyed to a 1.5-mL Eppendorf tube. Next, ammonium acetate was added, mixed, incubated on ice and centrifuged. The supernatant was then blended completely with a unit volume of chilled isopropanol and for 30 min incubated on ice. Products were then centrifuged at 16,000× *g* for 15 min at 4 °C. The supernatant was evaporated and removed; the DNA pellet was cleaned several times with 70% ethanol and resolved in 150 µL of Tris-EDTA. The rest of the method was performed, according to Ericsson et al. [52,53].

3.6. 16S rRNA Library Preparation and Sequencing

The DNA of extracted samples was tested at the University of Missouri DNA Core Facility. Bacterial 16S rDNA amplicons were created with a magnification of the V4 hypervariable region of the 16S rDNA gene with universal primers (U515F/806R) formerly established against the V4 region, edged by Illumina standard adapter sequences [54]. A single forward primer and reverse primers with a unique 12-base index were used in all reactions. PCR amplification was completed as follows: 98 °C(3:00) + (98 °C(0:15) + 50 °C(0:30) + 72 °C(0:30)) × 25 cycles + 72 °C(7:00) [52,53]. The amplified product from each reaction was mixed entirely; then purified and incubated at room temperature for 15 min. Products were washed with 80% ethanol several times and the dried pellet was resuspended in Qiagen EB Buffer (32.5 µL), incubated at room temperature for 2 min, and then placed on the magnetic stand for 5 min. The final amplicon pool was assessed using the Advanced Analytical Fragment Analyzer automated electrophoresis system, quantified with the Qubit fluorometer using the Quant-iT HS dsDNA reagent kit (Invitrogen), and diluted according to Illumina's standard protocol for sequencing on the MiSeq [52].

3.7. Informatics Analysis

Constructing, data binning, and descriptive analysis of DNA sequences was performed at the MU Informatics Research Core Facility. FLASH software [55] was employed to group the contiguous sequences of DNA, and contigs were discarded if they turned out to be less than 31 after trimming for a base quality. Qiime v1.7 [56] software was used to carry out de novo and reference-based chimera detection and exclusion, and other contigs were allocated to operational taxonomic units (OTUs) with a significance of 97% nucleotide identity. Taxonomy was appointed to selected OTUs using BLAST [57] in comparison to the Greengenes database [58] of 16S rRNA sequences and taxonomy.

4. Results and Discussion

The presence of VAs in anaerobic digesters could have an inhibitory effect on biogas production [36,59,60], because VAs could disrupt microorganisms' dynamics, especially when the concentration is high. Because AD is not efficient in degrading VAs completely, in the long term, AD reactors can also become a fostering environment for VAs that would help the development of new ARGs [61]. By scrutinizing the figures derived, some abnormalities were visible one week after the last spike, recalling that October 18 was the start date of the spiking antibiotics and final day was October 28 (Figure 2a). This biogas fluctuation started with a decline in samples spiked with tylosin and also a mixture of tylosin and CTC, immediately after the first spike. For CTC samples, this drop was delayed until early November. On November 11, it grew again and then reached its lowest point on November 23.

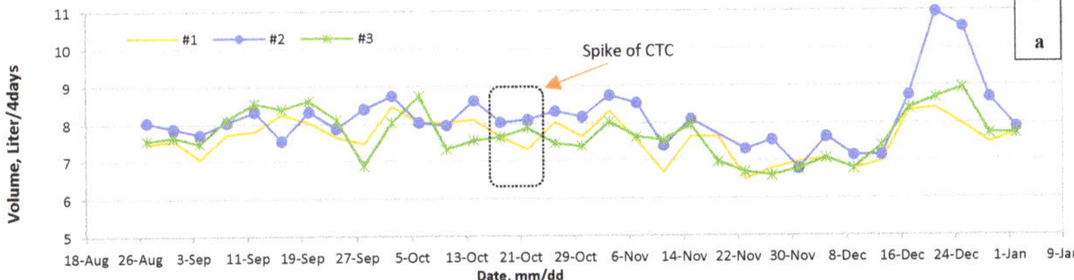

Figure 2. *Cont.*

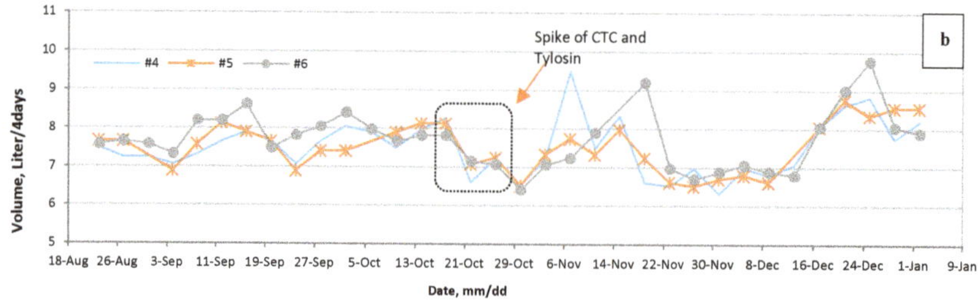

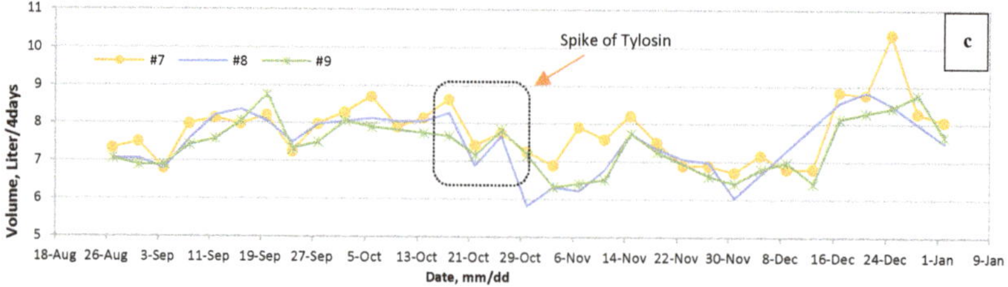

Figure 2. Biogas production in 2017–2018, (**a**) #1 low, #2 medium and #3 high concentration, before and after adding CTC, (**b**) #4 low, #5 medium and #6 high concentration, before and after adding both CTC and tylosin, and (**c**) #7 w, #8 medium and #9 high concentration, before and after adding tylosin and the dotted box shows the administration period of the VAs.

For the CTC plus tylosin, after a drop on October 22, and again on October 30, we witness a surge after that. Reactor #4 peaks on November 7 and reactor #6 peaks on November 19. Reactor #5 climbs steadily during this period. They start to drop in mid-November and reach their low at the end of November. Similarly, Figure 2b shows the same behavior, declining after the first spike until the end of October (last spike), followed by an upward trajectory. Likewise, this trend hits its bottom in early December.

Running a T-Test on biogas data implies that AD bacterial activity was immediately inhibited for samples that have tylosin in them (Figure 2b,c) (p-value = 0.005). Still, the bacteria either adapted or the inhibiting compound was removed from the system after a few weeks. Biogas production was untouched for CTC samples, yet for the mixture of CTC and tylosin, and tylosin alone, it was significantly lower, immediately after VA addition. The tylosin concentration in this experiment was 92 mg/L and less, complying with the findings of Mitchel et al. [62]. They concluded that the bioreactor containing 92 mg/L tylosin had less biogas for nearly 30 d until the system recovered. The biogas reduction for samples with tylosin and CTC was close to 14%, and for tylosin, samples was between 8 and 19%, with no dose-dependent relationship. On the other hand, Chelliapan et al. [63] found no biogas inhibition in an up-flow anaerobic stage reactor (UASR) containing 100–800 mg/L tylosin.

Erythromycin, another macrolide antibiotic caused 6–24% biogas reduction with 6–100 mg/L, and no dose-dependent relationship [64].

But CTC did not disturb the bacterial activity, substantiate the evidence that CTC antibiotic may present minimal AD biogas inhibition at concentrations less than approximately 70 mg/L occurring in the current study. Yin et al. [65] observed similar results; for a mesophilic anaerobic digester with the manure and CTC concentrations of 0, 20, 40, and 60 mg/kg. TS, no significant inhibition in biogas production occurred. Dreher et al. [66] showed that no inhibition of biogas production happened in anaerobic sequencing batch reactor with 28 mg/L CTC but that the volumetric composition of methane decreased by about 13–15%. Mixed results of the inhibition in the literature could be due to various reactor types, inoculum/manure ratio, inoculum and manure age and source, reactor size, and batch or continuous operation [67]. In this experiment, CTC concentration was probably lower than its required inhibitory level.

4.1. pH and CO$_2$

pH value can demonstrate how well Acetogenesis and Methanogenesis bacteria are working. At the beginning of AD performance, Acetogenesis bacteria start to produce volatile acids that cause the pH to decrease. Subsequently, Methanogenesis bacteria convert the volatile acids to methane and CO$_2$, and cause pH to increase. At HRTs with more than five days, the methane-forming bacteria begin to consume the volatile acids.

By comparing before and after the addition of VAs, it is evident that reactors are experiencing a fluctuating pH status (Figure 3a–c). The graph shows that variations immediately after antibiotic spike have increased intensively, with a rising trend. Following up, in the first week of November, almost all reactors reach their plateau. From then on, the gradual decline continued until November 17. Subsequently, reactors seemed to recover themselves with an increase in pH. By the end of November, pH returns to its average level of around 7.8.

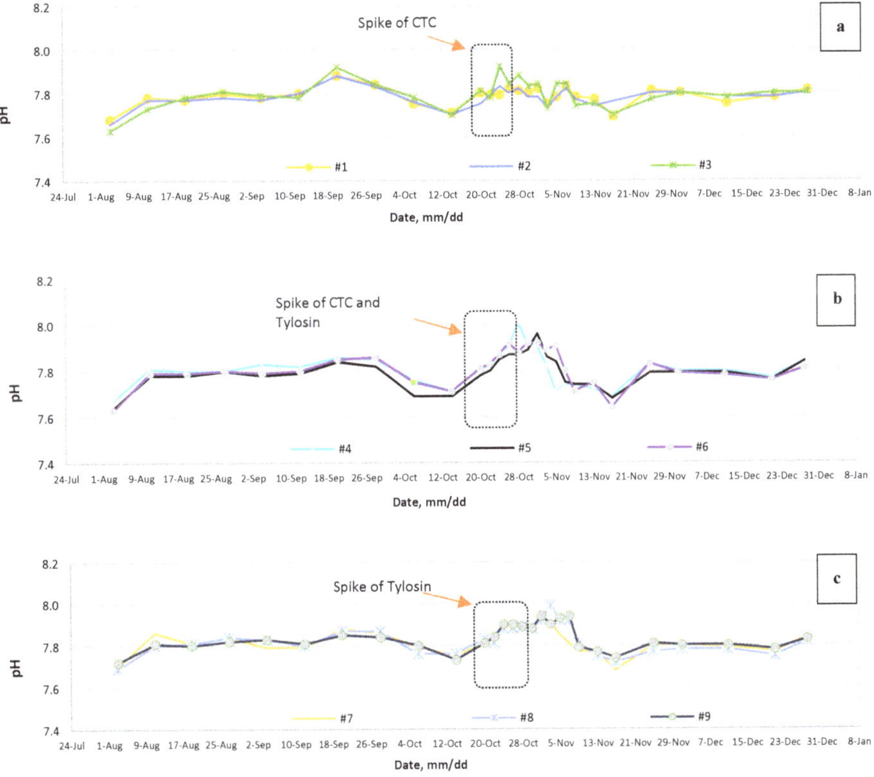

Figure 3. Digestate pH levels for (**a**) before and after CTC spike, (**b**) before and after CTC plus tylosin spike, and (**c**) before and after tylosin spike and the dotted box shows the administration period of the VAs.

The T-test on the pH data shows that pH values were significantly lower for the samples with tylosin in them (*p*-value = 0.05). However, CTC did not affect the pH significantly. Since fluctuations in pH level are not sharp, this indicates that VAs did not disturb the bacterial community substantially.

Nevertheless, none of the reactors became upset or affected intensively by the VA addition. pH fluctuation was ±0.16 maximum and it never dropped under 7.60. Similarly, the CO_2 level has detectable alteration around the antibiotic spike date (Figure 4a–c).

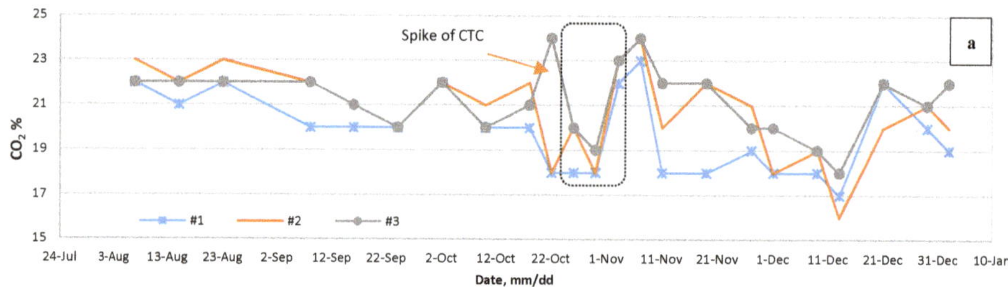

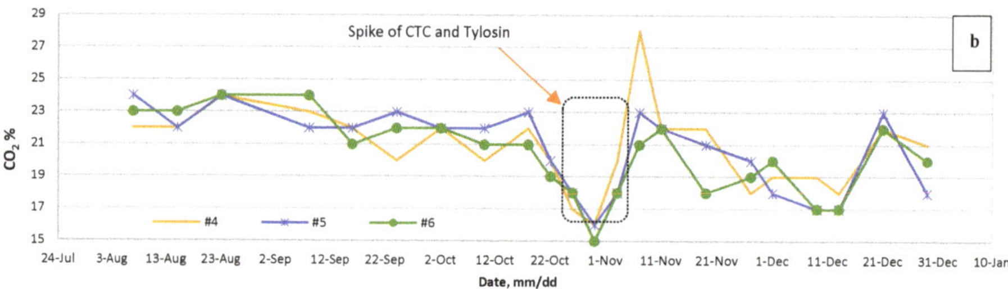

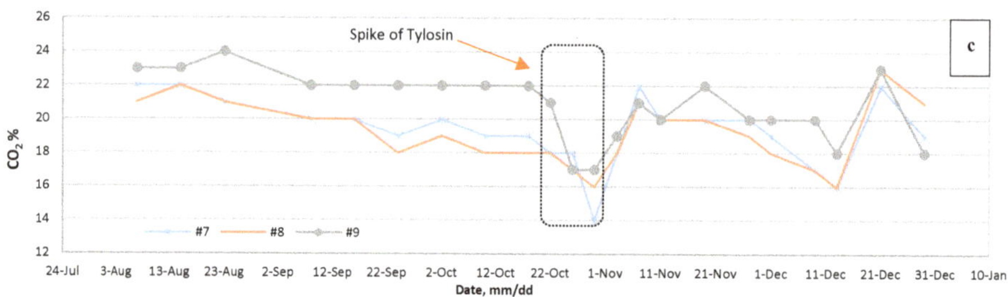

Figure 4. The concentrations of CO_2 (**a**) before and after CTC spike, (**b**) before and after CTC plus tylosin spike, and (**c**) before and after tylosin spike and the dotted box shows the administration period of the VAs.

Biogas produced is consisting of almost 50–75% of methane, 25–40% of carbon dioxide and other gases, depending on organic material [68]. By comparing CO_2 data and performing a T-test, results imply that CTC had a significant effect on the biogas methane content (*p*-value = 0.05). At the same time, samples with tylosin only were not affected considerably. The reason could simply be that the CTC is active primarily against Gram-negative organisms by blending with the A location of the 30S subunit of bacterial ribosomes. So, they prevent peptide growth and the protein synthesis effect, which finally leads to bacteria death [65,69]. Methanogen bacteria are Gram-negative bacteria [70]. Thus, at a certain level of CTC, significant biogas inhibition should be imposed to the bioreactor.

Values of VS in the digestate before and after VA addition are shown in Figure 5. In general, every treatment sample except for the medium tylosin concentration, showed an increase in the VS percentage after VA addition. The VS values agree with the slight fluctuations observed in Figures 2–4, that the microbial communities were slightly affected by the VA addition and the biogas production was not halted. To recall, the initial manure VS loading was 2 g-VS/L/day and the sampling for VS were conducted between 6 and 10 days after the first injection and two days after the last injection of VA. Thus, it complies with reduction in methane production which was around 13–15% studied by Dreher et al. [66]. Angenent et al. [34] also reported a temporary decrease in VS removal which recovered quickly. The average VS level before and after VA addition is 0.56% and 0.64%; and VS removal is 1.44% and 1.36%, respectively.

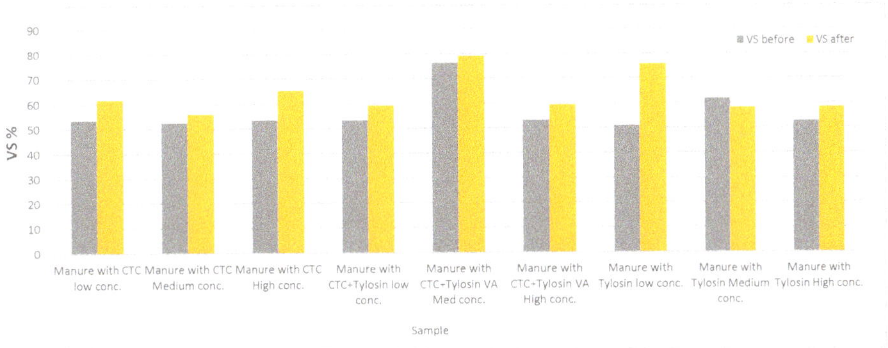

Figure 5. Digestate volatile solid (VS) concentrations before and after VA additions.

4.2. LC-MS/MS Results and Adjustments

The plan was to try using the measured Enrofloxacin concentrations to calculate adjustment factors for the other VAs. With these factors, sample concentrations after the dilution and short-term loading in the lab digesters were recalculated, assuming there was no degradation or absorption. Should there be significant differences, these would then be caused by sampling error, degradation due to AD, or the error of the LC-MS/MS measurement, including the SPE. Table 6 shows all of the samples, their added VAs and a comparison between spiked concentration, detected concentration by LC-MS/MS and recalculated concentration using adjustment factors.

Table 6. Samples, their content and comparison between their calculated, detected by LC-MS/MS and adjusted concentrations.

Sample #	Sample Type	Calculated CTC Conc. (ppm)	CTC Detected Conc. with LC-MS/MS (ppm)	Calculated Tylosin Conc. (ppm)	Tylosin Detected Conc. with LC-MS/MS (ppm)	CTC Recovery and Purity Adj. (ppm)	Tylosin Recovery and Purity Adj. (ppm)	Enrofloxacin Conc. (ppb)
1	Digestate with CTC low conc.	61	0	0	0	−1	0	226
2	Digestate with CTC Medium conc.	65	0	0	0	−2	0	523
3	Digestate with CTC High conc.	69	0	0	0	−1	0	1377
4	Digestate with Mixture VA low conc.	61	0	20	1	0	−6	1069
5	Digestate with Mixture VA Med conc.	65	0	40	1	−2	−5	940
6	Digestate with Mixture VA High conc.	69	0	61	1	−2	−4	568
7	Digestate with Mixture VA low conc.	50	0	31	1	−1	−5	1434
8	Digestate with Mixture VA Med conc.	53	0	61	1	−1	−4	600
9	Digestate with Mixture VA High conc.	56	0	92	2	−2	−2	834
10	Digestate with tylosin low conc.	0	0	31	1	−2	−5	881
11	Digestate with tylosin Medium conc.	0	0	61	1	−2	−4	1186
12	Digestate with tylosin High conc.	0	0	92	2	−2	−3	891
13	VA in water, Heat treated	69	0	0	1	−1	−1	236
14	VA in water, Heat treated	56	0	92	106	−1	64	3197
15	VA in water, Heat treated	0	0	92	116	−1	70	1668
16	VA in water, Room Temp.	69	0	0	2	−1	−1	1775
17	VA in water, Room Temp.	56	0	92	95	−1	57	2809
18	VA in water, Room Temp.	0	0	92	93	−1	55	2957
19	Diluted Manure without VA	0	0	0	1	−2	−4	839
20	Diluted Manure without VA	0	0	0	1	−2	−4	777
21	Diluted Manure without VA	0	0	0	1	−2	−5	613
22	Diluted Manure with CTC + tylosin	234	15	78	40	131	74	675
23	Diluted Manure with CTC + tylosin	234	36	78	40	319	75	874
24	Diluted Manure with CTC + tylosin	234	29	78	44	251	83	552

4.3. Relatively High Recovery of Enrofloxacin in the Water-Only Samples

Although the spiked Enrofloxacin in digestate and manure samples had a very low recovery rate (226 ppb to 1433 ppb vs. 4444 ppb spiked values) (Table 7), all but one water sample detected relatively higher Enrofloxacin concentrations (1667 ppb to 3197 ppb, Table 1). The Enrofloxacin concentration in the first water sample (236 ppb, samples #1) was only a fraction of water samples. The water samples were made with distilled water and VAs, no solid manure. Relatively higher recovery rates suggest that there is a systematic bias in measuring the Enrofloxacin in the samples that have solids (manure and digestate).

Table 7. Concentrations of VAs were detected in the water samples.

Sample #	Sample Type	Compound	Detected CTC Conc. (ppb)	Detected Tylosin Conc. (ppb)	Enrofloxacin Conc. (ppb)
13	VA in water, Heat treated	CTC	24	859	236
14	VA in water, Heat treated	CTC + tylosin	15	106,243	3197
15	VA in water, Heat treated	Tylosin	4	116,324	1668
16	VA in water, Room Temp.	CTC	17	1787	1775
17	VA in water, Room Temp.	CTC + tylosin	9	94,837	2809
18	VA in water, Room Temp.	Tylosin	17	93,002	2957

Therefore, when sample 13 was excluded, the average of the water sample group was 2481 ppb, while the digestate samples averaged 825 ppb. On the other hand, if we disregard the Enrofloxacin concentrations as an adjustment factor and just compare the LC-MS/MS values with our calculated concentrations (assuming no degradation), provides a better outcome. In this way, external standards are utilized to evaluate samples with only water and VAs, to monitor whether the removal of VAs is due to AD or not.

4.4. Very High Recovery of Tylosin in the Water-Only Samples

As shown in Table 8, concentrations for tylosin are very close to and sometimes higher than what we were expecting (LC-MS/MS measured 106 ppm, we expected 92 ppm for sample 2). Furthermore, the water samples that were not spiked with tylosin did yield very low tylosin concentrations (0.86 ppm and 1.79 ppm, samples 13 and 16).

Table 8. Comparison of calculated VAs concentrations and LC-MS/MS detected levels.

Sample #	Sample Type	Compound	Calculated CTC Conc. (ppm)	CTC Detected Conc. with LC-MS/MS (ppm)	Calculated Tylosin Conc. (ppm)	Tylosin Detected Conc. with LC-MS/MS (ppm)
13	VA in water, Heat treated	CTC	68.74	0.02	0.00	0.86
14	VA in water, Heat treated	CTC + tylosin	56.27	0.01	92.07	106.24
15	VA in water, Heat treated	Tylosin	0.00	0.00	92.07	116.32
16	VA in water, Room Temp.	CTC	68.74	0.02	0.00	1.79
17	VA in water, Room Temp.	CTC + tylosin	56.27	0.01	92.07	94.84
18	VA in water, Room Temp.	Tylosin	0.00	0.02	92.07	93.00

It is a different case for CTC; yet the reasons for low CTC detection are still unknown. The trend for CTC concentration shows they are disappearing so fast, which may be due to its half-life degradation or anaerobic reactor removal; alternatively, this may simply be because the CTC we used was already degraded, see the discussion below.

4.5. Consistent and Proportional LC-MS/MS Tylosin Results in the Digestate Samples

Based on the tylosin results being more consistent than CTC and Enrofloxacin results, the LC-MS/MS results of the digestate samples were meticulously scrutinized. Even though the LC-MS/MS detected concentration values of tylosin that were lower than expected, they were consistently proportional to the concentrations. For example, the expected concentration of tylosin was 20, 40, and 60 ppm for samples 4, 5, and 6, with zero degradation assumption, the LC-MS/MS values were 0.5, 0.9, and 1.2 ppm. No significant correlation for CTC was found. Figure 6 shows the LC-MS/MS results (*Y*-axis, ppm) vs. spiked values (*X*-axis, ppm). The consistently lower measured concentrations in the digestate samples and the high recovery rates in the water samples suggest that there was significant tylosin degradation due to the AD process.

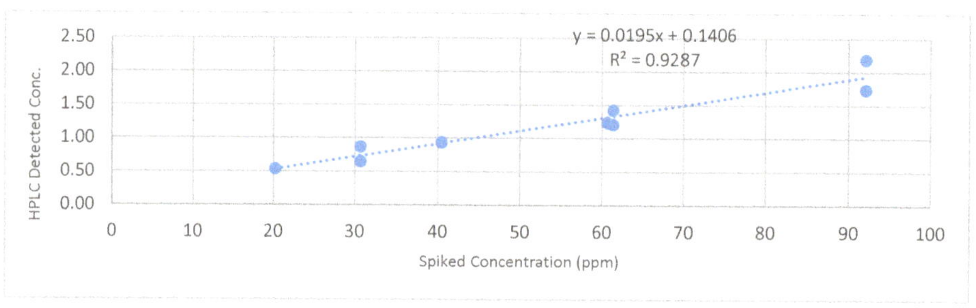

Figure 6. Correlations of LC-MS/MS measured and spiked tylosin concentrations.

4.6. Consistent LC-MS/MS Tylosin Measurements in the Manure External Standards

The detected tylosin concentrations of the three external standard samples were similar and had low deviation, Table 9. For tylosin, the detected levels ranged from 39.8 to 44.3 ppm and averaged 41.5 ppm, while the expected concentration was 77.9 ppm. For CTC, the measured concentrations were again a small fraction of the expected level.

Table 9. Concentrations of VAs were detected in the manure external standard samples.

Sample #	Sample Type	Compound	Calculated CTC Conc. (ppm)	CTC Detected Conc. with LC-MS/MS (ppm)	Calculated Tylosin Conc. (ppm)	Tylosin Detected Conc. with LC-MS/MS (ppm)
1	Diluted Manure with VAs	CTC + tylosin low Conc.	233.78	14.96	77.93	39.84
2	Diluted Manure with VAs	CTC + tylosin low Conc.	233.78	36.18	77.93	40.41
3	Diluted Manure with VAs	CTC + tylosin low Conc.	233.78	28.52	77.93	44.32

4.7. Relatively Higher Recovery of CTC in the External Standard Samples, and Two Additional CTC Standards

Studying the results of the three external standard samples showed that there was a relatively higher recovery rate for CTC. As an instance, compared with the concentration of 233.78 ppm, LC-MS/MS detected 14.96, 36.16 and 28.52 ppm. Compared with previous CTC samples, this group had a much higher recovery rate, which was fresh samples made with diluted manure and VAs. Besides, the external standards were prepared with commercial-grade antibiotics instead of analytical grade.

Table 10 shows results for freshly prepared samples with diluted manure and CTC antibiotic, at concentrations of 4ppm and 40 ppm, and the LC-MS/MS measured concentrations were 1.63 ppm and 11.7 ppm, respectively. The recovery rates of CTC were 29% and 41%. Since the samples were freshly prepared, the probability of degradation due to AD or half-life degradation was eliminated. Other possibilities are absorption to organic matter, and that inconsistent purity or degradation had already happened before application. For tylosin, LC-MS/MS detected higher concentrations (12 ppm and 67.6 ppm detected for 4 ppm and 40 ppm samples, respectively), which gives a detection rate of 300% and 169%.

Table 10. Concentrations of measured calibration standards.

Spiked Conc. (ppm)	LC-MS/MS Detected Conc. (ppm)	
	CTC	Tylosin
0	0.03	1.26
4	1.63	12.17
40	11.69	67.62

4.8. Applying External Standard Correction Factors

Because the internal standards (Enrofloxacin) did not yield consistent measurement, the correction was made based on external standards instead. By applying the external standard adjustment factor, the VA concentrations were corrected accordingly. The adjustment factor was obtained from samples with manure and spiked antibiotics, without retention time for AD. In other words, we just spiked different concentrations of antibiotics in samples made with manure, then prepared those for LC-MS/MS, immediately. In this way, we may be able to track other important factors contributing to our results, such as absorption, ion suppression or enhancement and recovery rate. Figure 7 presents the measured and corrected VA concentrations.

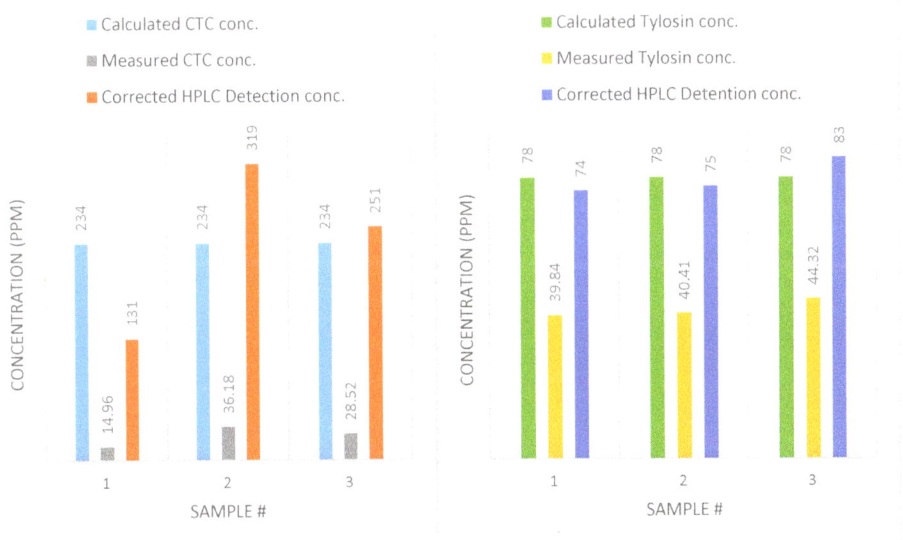

Figure 7. Left: concentrations of CTC, comparing calculated, measured, and corrected concentrations. **Right**: concentrations of tylosin, comparing calculated, measured, and corrected concentrations.

By using the external standard correction factor instead of the Enrofloxacin correction factor, data are more consistent, especially for tylosin (less than 6% error). It suggests that Enrofloxacin failed to act as an ideal internal standard. The inconsistency could be due to Enrofloxacin binding to the abundant organic materials or the presence of Ca^{2+}, Mg^{2+} Ions.

4.9. Degradation of CTC

For CTC, results showed a high degradation rate for both the samples in water and AD (Figure 8). For instance, almost all CTC injections with various concentrations have close to zero concentration. The low concentrations were measured for AD-treated samples, and also for CTC dissolved in water stored at room temperature and 40 °C, the temperature of the AD. In addition, the concentrations of the external standards were 234 ppm. The results suggest that the CTC degrade much faster than the tylosin, which might be due to the shorter half-life (8 days) as reported by the manufacturer. It is also possible that the CTC powder we used had already degraded. CTC concentration in external standard samples was reduced to 131, 320 and 251 ppm from its original 234 ppm. Because the purity correction was already applied and recovery rate adjustment was made, also only low CTC concentration was detected for the water and digestate samples, the CTC probably just degraded itself over a short time. A study by Winckler and Grafe [71] showed that tetracycline in liquid manure was degraded by 50% in 82 days. Arikan [72] reported a 75% reduction in CTC concentration with AD after 33 days, with a half-life of 18 days. Cheng et al. [73] reported a high affinity between tetracycline and solid manure during AD. For future research, additional testing to examine the possibility of the adsorption by the glass jar used in this study as the AD reactors should be conducted, since there are very few investigations on this subject.

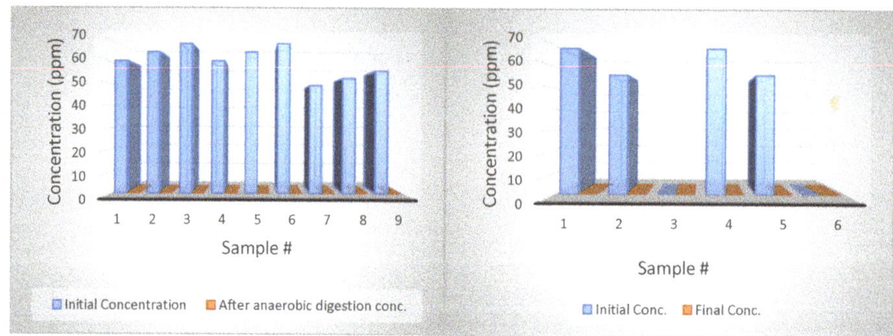

Figure 8. Left: CTC concentration change with anaerobic digestion. **Right**: CTC concentration change with reactors filled with diluted water.

4.10. Degradation of Tylosin

Tylosin degraded very well with the ASBR reactor working at 39 °C and loading with swine manure every two days. By comparing the degradation rate of tylosin in ASBRs with jars filled only with water, it shows that AD is effective in reducing tylosin (Figure 9). The degradation rate of tylosin in water averaged 33.5; however, the degradation was 100 percent with AD. A study by Kolz et al. [74] concluded no effective degradation for tylosin B and D in anaerobic conditions up to eight months. tylosin A was degraded under the aerobic conditions with a half-life of 2 to 40 days [5,75]. Stone et al. [59] also reported no significant degradation for tylosin.

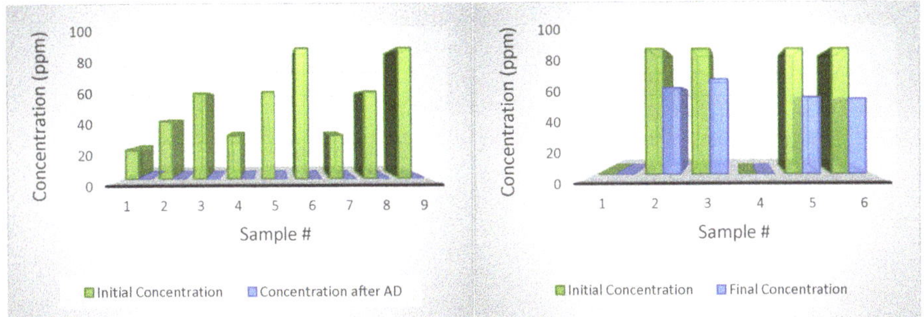

Figure 9. Left: Tylosin concentration change with anaerobic digestion. **Right**: Tylosin concentration change with reactors filled with diluted water.

4.11. Effects of Having Two Types of VA in the Digestion and Water

In these experiments, we planned to compare the effect of having two different antibiotics on the reduction efficiency of antibiotics. Identical injections and concentrations were applied in the mixture treatment, except for a series of samples, both antibiotics were spiked.

For tylosin, the results showed that there was little difference between samples. The tylosin removal was similar with or without CTC mixture, suggesting that the CTC addition had no adverse effect on tylosin degradation.

However, because CTC degradation was much faster than tylosin, and because the samples collected for CTC concentration measurements were not resolute enough (shorter time than the five-day sampling), the speed of the CTC degradation and effects of the tylosin addition could not be determined based on this dataset (Table 9).

4.12. Contrast Tylosin Reduction in Water and AD Reactors

Tylosin tartrate showed a relatively higher removal in AD treatment when compared to samples that were dissolved in water. Figure 10 depicts that the tylosin samples treated by AD for twelve days were removed entirely (100% removal). However, the removal of tylosin tartrate in water (dotted line) was 40% or less during the same period. The lower degradation in the water samples suggests that AD is effective in enhancing tylosin degradation in the animal manure, and that this could be the essential effect of the AD.

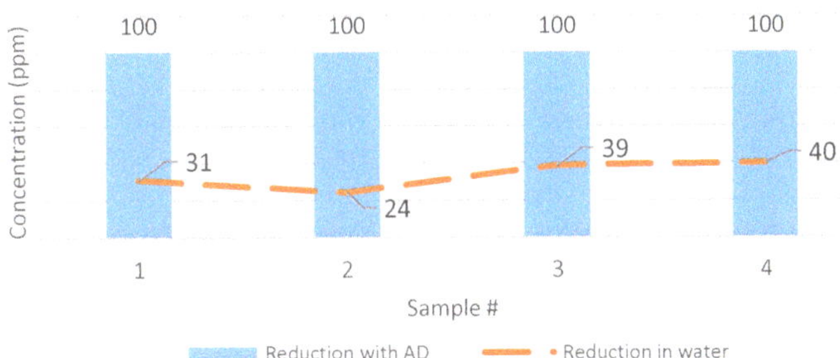

Figure 10. The removal of tylosin in water and the anaerobic digestion (AD) reactor.

4.13. Bacterial Community Dynamics

Phyla *Firmicutes, Bacteroidetes, Proteobacteria,* and *Synergistetes* were dominant or co-dominant in bacteria. Different types of *Clostridia* consisted mostly of the *Firmicutes. Methanomicrobia* was the dominant *Archaea* among our samples.

The first group was treated with CTC and showed a slight fluctuation in *Archaea* abundance, 4.37%, 5.19%, and 5.06% related to low, medium, and high concentrations of CTC, respectively. As shown in the Table 11, *Firmicutes* almost remained constant, and Bacteriodetes increased from 14.03 to 15.30% and then decreased to 12.50% for low, medium and high concentrations of CTC, respectively. The same pattern occurred for Synergistetes, going up from 1.67 to 1.75% and then dropping to 1.44%, with the mentioned level of CTC concentrations. However, the reverse happened for Proteobacteria: the abundance level reduced from 2.46 to 1.35% and then rose to 1.87%.

Table 11. Percentages of different microorganisms with various treatment plans.

	Total	CTC Low	CTC Med	CTC High	CTC + Tyl Low	CTC + Tyl Med	CTC + Tyl Med	CTC + Tyl High	Tyl Low	Tyl Med	Tyl High
D_0__Archaea; D_1__Euryarchaeota	5.90%	4.40%	5.20%	5.10%	7.80%	7.40%	5.20%	6.30%	4.80%	3.70%	8.70%
D_0__Bacteria; D_1__Actinobacteria	0.10%	0.10%	0.00%	0.10%	0.10%	0.10%	0.10%	0.10%	0.10%	0.10%	0.10%
D_0__Bacteria; D_1__Atribacteria	1.00%	1.00%	1.10%	1.50%	0.90%	1.10%	0.90%	1.30%	0.90%	0.50%	0.70%
D_0__Bacteria; D_1__Bacteroidetes	7.70%	14.00%	15.30%	12.50%	4.00%	5.90%	10.20%	5.50%	2.20%	5.30%	2.50%
D_0__Bacteria; D_1__Chloroflexi	0.40%	0.50%	.40%	0.30%	0.40%	0.30%	0.30%	0.70%	0.80%	0.30%	0.50%
D_0__Bacteria; D_1__Cloacimonetes	0.10%	0.10%	0.30%	0.10%	0.00%	0.10%	0.20%	0.00%	0.00%	0.10%	0.00%
D_0__Bacteria; D_1__Firmicutes	74.70%	72.80%	71.90%	75.30%	74.90%	79.70%	77.90%	68.10%	76.50%	82.70%	66.70%
D_0__Bacteria; D_1__Kiritimatiellaeota	1.00%	0.80%	0.80%	0.50%	1.60%	0.90%	0.70%	0.90%	1.50%	1.00%	1.60%
D_0__Bacteria; D_1__Planctomycetes	0.30%	0.40%	0.30%	0.30%	0.30%	0.10%	0.20%	0.10%	0.30%	0.20%	0.40%
D_0__Bacteria; D_1__Proteobacteria	6.00%	2.50%	1.40%	1.90%	7.00%	1.80%	1.90%	13.40%	10.20%	4.00%	16.00%
D_0__Bacteria; D_1__Synergistetes	2.00%	1.70%	1.70%	1.40%	2.60%	1.80%	1.30%	3.40%	2.10%	1.50%	2.60%
No blast hit; Other	0.60%	1.50%	1.50%	0.80%	0.20%	0.50%	0.90%	0.10%	0.40%	0.40%	0.10%

The second group, consisting of four samples, all are added with the mixture of CTC and tylosin with low, medium and high concentrations. Two samples were taken from medium range concentration because the administration duration of CTC was 6 days and tylosin was 10 days. Therefore, samples were taken at the end of the administration of each VA, at day 6 and day 10. There is a reverse relation between *Archaea* abundance and VAs concentration as well as administration duration. *Archaea* level is dropping with a higher concentration of VAs and longer retention time. The effect of the administration period is stronger than the dosage on the *Archaea* population. *Firmicutes*, on the other hand, have increased from 74.9 to 79.70% by increasing the dosage of VAs from low to medium. The trend is not consistent with shifting from medium to high concentration of VA; it would decrease the abundance of *Firmicutes*. The administration period has the same effect, but not as much as dosage. Phylum Bacteroides population increased with the rise of VAs concentration and doubled with an increase in retention time, from 6 to 10 days with a medium level of VAs. Proteobacteria and Synergistetes abundance both have dropped by increasing dosage for low to medium, but the fall is drastic for Proteobacteria, changing from 7 to 1.8%. By the end of CTC administration, when adding tylosin, their abundance recovered slightly. Surprisingly, in high concentrations of CTC and tylosin, both of these bacteria showed growth in their population. The proteobacteria population is almost doubled by having high dosage if VAs instead of low dosage.

The last group, which is being medicated with tylosin only, *Archaea*, decreased slightly and then almost doubled when moving from low concentration to high. For Bacteroides and *Firmicutes*, it is exactly the reverse, with maximum abundance around medium range concentration and the nearly same number for low and high concentrations. For proteobacteria, results showed a sharp drop with shifting from low to medium concentration, 10.2% changed to 4.03%. However, by raising the dosage, the abundance of proteobacteria returned to 15.97%.

Bacteroidia are the major classes found within the phylum of *Bacteroidetes;* and are abundant in digesters that use cow manure as feedstock [76]. *Firmicutes* phylum is mostly syntrophic bacteria that can decompose a variety of fatty acids, and exists in both activated sludge systems and anaerobic digesters [77]. Within the species of *Firmicutes*, *Clostridia* is the dominant class. The predominance of *Clostridia* in the AD sludge was related to the comparably fast hydrolysis and VFA (volatile fatty acids) fermentation happening in the digesters [78].

Fatty acid-oxidizing bacteria, including Synergistales group which have syntrophs are connecting bonds of the chain between the primary fermenters and methanogens [79], are abundant in thermophilic digesters [80,81]. The presence of *Synergistetes* (syntrophic acetate oxidizers) might be an indicator of a decent acetotrophic activity in the bioreactor [82].

There are two major categories of methanogens; *acetoclastic* which consumes acetate to produce methane or *hydrogenotrophic* that are converting CO_2 and H_2 to methane.

The *acetoclastic methanogenesis* is linked with the *Methanosarcinales* and the *hydrogenotrophic methanogenesis* is linked with the *Methanomicrobiaceae* family. In the current study, the *hydrogenotrophic* pathway with *Methanosarcinaceae* was dominant. Kim et al. [42], Nogueira et al. [78] and Padmasiri et al. [83] also detected a dominant *Methanomicrobiales* order on AD.

4.14. Statistical Analysis of the Effect of Different Treatments on Samples

There is a notable difference between the treated and control group, which shows antibiotics had a significant influence on altering the bacterial community in our digesters. Figure 11 shows the samples arranged using the same two similarity measures used to generate the PCoA plots but in the form of a dendrogram. Bray–Curtis similarities and Jaccard similarities method are used for this comparison.

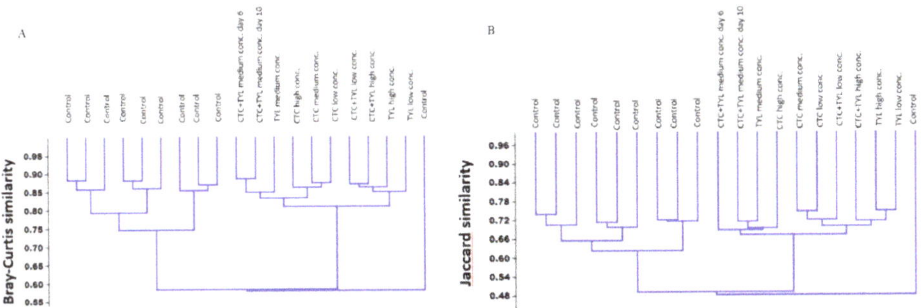

Figure 11. Dendrogram of bacterial community based on (**A**) Bray–Curtis similarities and (**B**) Jaccard similarities.

In both of the methods, the differences between all three treatment groups are modest and likely obscured by the variability introduced by the control samples.

Figure 12 shows a stacked bar chart at the best taxonomic resolution afforded by our primers. Again, the differences between the two datasets are stark, while the differences between treatment groups are more subtle (but present).

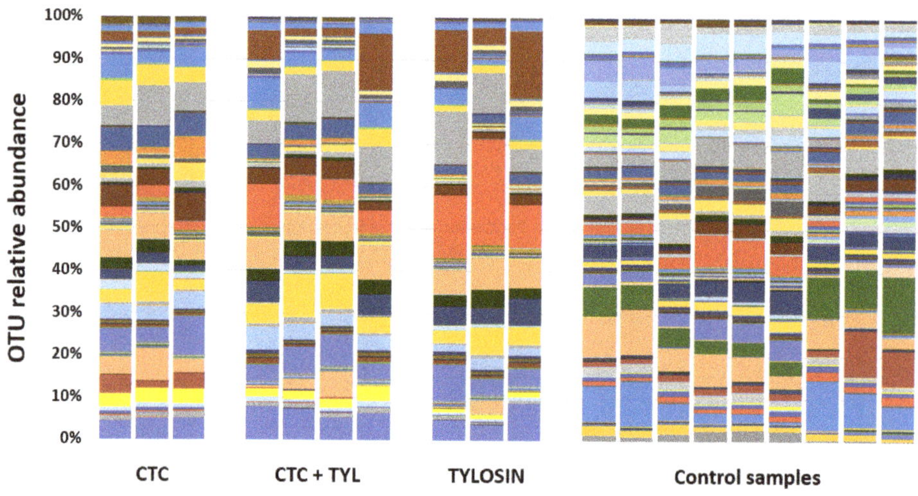

Figure 12. The stacked bar chart at the best taxonomic resolution.

Figure 13 shows a heat map in which samples (columns) are ordered according to similarity using a hierarchical method (UPGMA) based on the 50 OTUs (operational taxonomic unit) (rows) with the lowest p values following serial ANOVA testing of all 629 OTUs. In short, it shows perfect discrimination between groups when the samples are clustered using only the 50 most variable OTUs. Taxonomic identity of the microbes is listed on the right-hand side of the heat map.

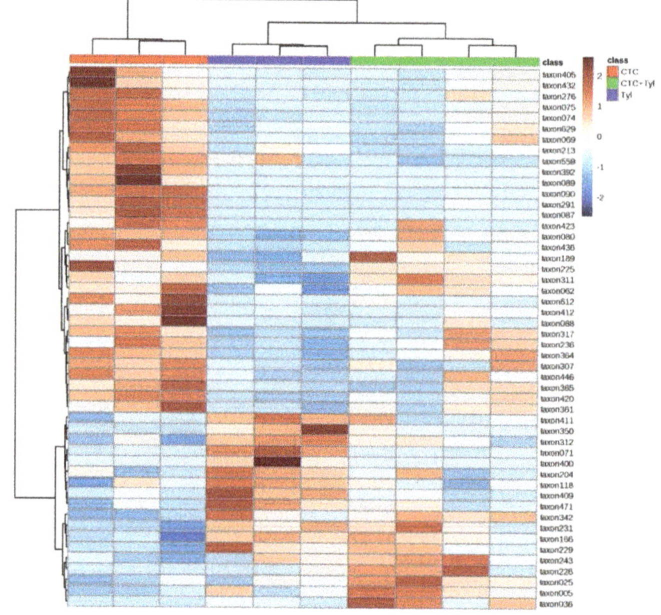

Figure 13. Heat map of treated samples using a hierarchical method (UPGMA) based on the 50 operational taxonomic units (OTUs) (rows) with the lowest p values.

Figure 14 shows box plots representing the relative abundance of eight of the OTUs with the lowest p values. One can easily note the very clear pattern of microbes with sensitivity to one or the other drugs, with Proteiniphilum (lower left) being the anomaly. Some other microbes with low p values mostly had these types of patterns, either down in CTC and CTC + tylosin, or down in tylosin and CTC + tylosin.

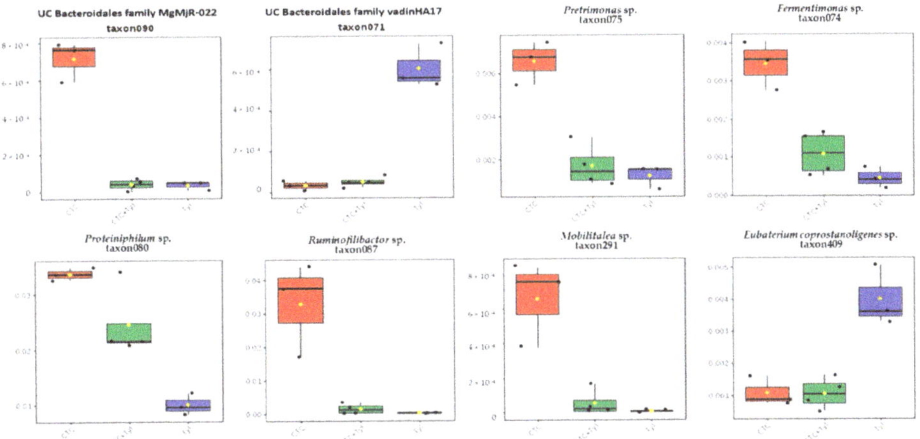

Figure 14. Box plots representing the relative abundance of 8 of the OTUs with the lowest *p* values.

Finally, Figure 15 is a random forest analysis looking for "biomarkers" of each treatment group. The greater the "MeanDecreaseAccuracy", the better that OTU is as a biomarker of the rankings shown to the right of the Figure. For example, Methanoculleus is apparently an excellent predictor of these groups by having a high relative abundance in the CTC samples and low abundance in the CTC + tylosin samples. Likewise, the Anaerorhabdus furcosa group and Flexilinea sp. can be found more on samples with high tylosin concentration and low CTC. Ruminiclostridium sp. can be abundant in conditions with high CTC levels and low tylosin.

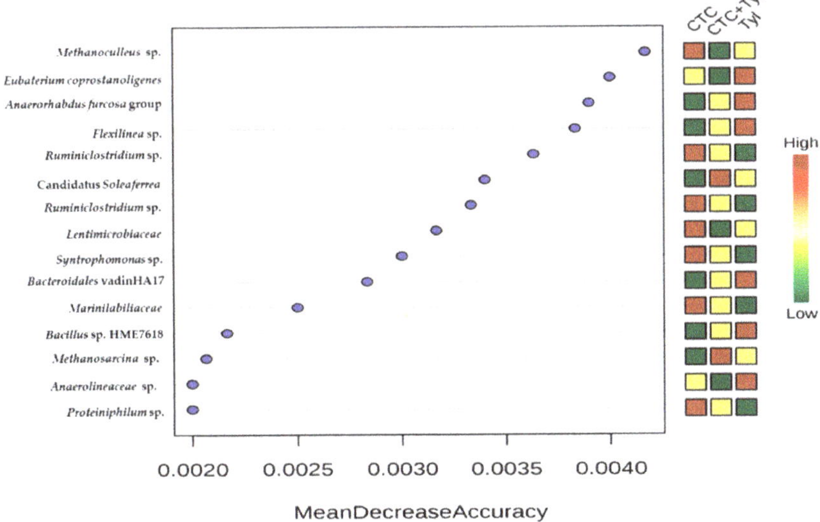

Figure 15. Random forest analysis of biomarkers based on MeanDecreaseAccuracy.

5. Conclusions

The results show that for both CTC and tylosin with maximum concentration added of 298 and 263 ppm, respectively, a negligible inhibitory effect on ASBR performance was observed. No harmful effect on the microbial community, pH or alkalinity was observed; however, microbial diversity was decreased. Efficient tylosin removal with AD occurred (removal was 100%, while removal in distilled water-filled reactors was around 40% or less), though, it cannot be proven for CTC. In addition, no difference was detected for using a mixture of tylosin and CTC, compared to the solo use of each. More research must be carried out on testing different VAs to discover the efficiency of AD reactors for VA removal. Besides, amplicon sequencing performed on each group by using the 50 most variable operational taxonomic units (OTUs)s and perfect discriminations were detected between groups. The effect of administration period and dosage of VAs on Phyla *Firmicutes* Proteobacteria, Synergistetes and Phylum Bacteroides was investigated. OTU's alteration is used to detect biomarkers. These biomarkers' abundance can be employed to predict the sample's contamination with these antibiotics.

Author Contributions: Conceptualization, T.-T.L.; methodology, T.-T.L., A.H.T., C.-H.L., A.C.E.; software, A.H.T., P.H.V., A.C.E.; formal analysis, A.H.T., P.H.V., A.C.E.; investigation, T.-T.L., A.H.T., C.-H.L., A.C.E.; data curation, A.H.T., T.-T.L.; writing—original draft preparation, A.H.T., T.-T.L.; writing—review and editing, T.-T.L., A.C.E.; supervision, T.-T.L., C.-H.L. All authors have read and agreed to the published version of the manuscript.

Funding: This research was funded by the University of Missouri Research Board and University of Missouri Extension Program.

Acknowledgments: Support of the University of Missouri Extension, manager of the finishing farm, Mel Gerber and research farm for the feedstock collection, and graduate students Haipeng Wang, Joshua Brown, Cuong Duong and Vu Danh were acknowledged.

Conflicts of Interest: The authors declare no conflict of interest.

References

1. Barton, M.D. Antibiotic use in animal feed and its impact on human healt. *Nutr. Res. Rev.* **2000**, *13*, 279–299. [CrossRef] [PubMed]

2. Zhang, Q.-Q.; Ying, G.-G.; Pan, C.-G.; Liu, Y.-S.; Zhao, J.-L. Comprehensive evaluation of antibiotics emission and fate in the river basins of China: Source analysis, multimedia modeling, and linkage to bacterial resistance. *Environ. Sci. Technol.* **2015**, *49*, 6772–6782. [CrossRef]

3. Wang, J.; Wang, S. Preparation, modification and environmental application of biochar: A review. *J. Clean. Prod.* **2019**, *227*, 1002–1022. [CrossRef]

4. 2016 Summary Report on Antimicrobials Sold or Distributed for Use in Food-Producing Animals. Available online: https://www.fda.gov/files/about%20fda/published/2016-Summary-Report-on-Antimicrobials-Sold-or-Distributed-for-Use-in-Food-Producing-Animals.pdf (accessed on 1 June 2018).

5. Teeter, J.S.; Meyerhoff, R.D. Aerobic degradation of tylosin in cattle, chicken, and swine excreta. *Environ. Res.* **2003**, *93*, 45–51. [CrossRef]

6. Kuppusamy, S.; Kakarla, D.; Venkateswarlu, K.; Megharaj, M.; Yoon, Y.-E.; Lee, Y.B. Veterinary antibiotics (VAs) contamination as a global agro-ecological issue: A critical view. *Agric. Ecosyst. Environ.* **2018**, *257*, 47–59. [CrossRef]

7. Chen, J.; Yu, Z.; Michel, F.; Wittum, T.; Morrison, M. Development and application of real-time PCR assays for quantification of erm genes conferring resistance to macrolides-lincosamides-streptogramin B in livestock manure and manure management systems. *Appl. Environ. Microbiol.* **2007**, *73*, 4407–4416. [CrossRef]

8. Lienert, J.; Bürki, T.; Escher, B. Reducing micropollutants with source control: Substance flow analysis of 212 pharmaceuticals in faeces and urine. *Water Sci. Technol.* **2007**, *56*, 87–96. [CrossRef]

9. Kumar, K.; Gupta, S.C.; Baidoo, S.K.; Chander, Y.; Rosen, C.J. Antibiotic uptake by plants from soil fertilized with animal manure. *J. Environ. Qual.* **2005**, *34*, 2082–2085. [CrossRef] [PubMed]

10. Alexy, R.; Kümpel, T.; Kümmerer, K. Assessment of degradation of 18 antibiotics in the Closed Bottle Test. *Chemosphere* **2004**, *57*, 505–512. [CrossRef] [PubMed]

11. Adeel, M.; Song, X.; Wang, Y.; Francis, D.; Yang, Y. Environmental impact of estrogens on human, animal and plant life: A critical review. *Environ. Int.* **2017**, *99*, 107–119. [CrossRef]

12. Brown, E.D.; Wright, G.D. Antibacterial drug discovery in the resistance era. *Nat. Cell Biol.* **2016**, *529*, 336–343. [CrossRef] [PubMed]

13. Liu, Y.; Liu, Y.; Li, H.; Fu, X.; Guo, H.; Meng, R.; Lu, W.; Zhao, M.; Wang, H. Health risk impacts analysis of fugitive aromatic compounds emissions from the working face of a municipal solid waste landfill in China. *Environ. Int.* **2016**, *97*, 15–27. [CrossRef] [PubMed]

14. Liu, P.; Zhang, H.; Feng, Y.; Yang, F.; Zhang, J. Removal of trace antibiotics from wastewater: A systematic study of nanofiltration combined with ozone-based advanced oxidation processes. *Chem. Eng. J.* **2014**, *240*, 211–220. [CrossRef]

15. Qiao, M.; Ying, G.-G.; Singer, A.C.; Zhu, Y.-G. Review of antibiotic resistance in China and its environment. *Environ. Int.* **2018**, *110*, 160–172. [CrossRef]

16. Zhao, W.; Sui, Q.; Mei, X.; Cheng, X. Efficient elimination of sulfonamides by an anaerobic/anoxic/oxic-membrane bioreactor process: Performance and influence of redox condition. *Sci. Total. Environ.* **2018**, *633*, 668–676. [CrossRef]

17. Yu, F.; Li, Y.; Han, S.; Ma, J.; Han, S. Adsorptive removal of antibiotics from aqueous solution using carbon materials. *Chemosphere* **2016**, *153*, 365–385. [CrossRef]

18. Wang, J.; Zhuan, R.; Chu, L. The occurrence, distribution and degradation of antibiotics by ionizing radiation: An overview. *Sci. Total. Environ.* **2019**, *646*, 1385–1397. [CrossRef]

19. Homem, V.; Santos, L. Degradation and removal methods of antibiotics from aqueous matrices—A review. *J. Environ. Manag.* **2011**, *92*, 2304–2347. [CrossRef]

20. Menkem, Z.E.; Ngangom, B.L.; Tamunjoh, S.S.A.; Boyom, F.F. Antibiotic residues in food animals: Public health concern. *Acta Ecol. Sin.* **2019**, *39*, 411–415. [CrossRef]

21. Phillips, I.; Casewell, M.; Cox, T.; De Groot, B.; Friis, C.; Jones, R.; Nightingale, C.; Preston, R.; Waddell, J. Does the use of antibiotics in food animals pose a risk to human health? A critical review of published data. *J. Antimicrob. Chemother.* **2003**, *53*, 28–52. [CrossRef]

22. Sivagami, K.; Vignesh, V.J.; Srinivasan, R.; Divyapriya, G.; Nambi, I. Antibiotic usage, residues and resistance genes from food animals to human and environment: An Indian scenario. *J. Environ. Chem. Eng.* **2020**, *8*, 102221. [CrossRef]

23. Tasho, R.P.; Cho, J.Y. Veterinary antibiotics in animal waste, its distribution in soil and uptake by plants: A review. *Sci. Total. Environ.* **2016**, *563*, 366–376. [CrossRef] [PubMed]

24. Chung, H.S.; Lee, Y.-J.; Rahman, M.; El-Aty, A.A.; Lee, H.S.; Kabir, H.; Kim, S.W.; Park, B.-J.; Kim, J.-E.; Hacımüftüoğlu, F.; et al. Uptake of the veterinary antibiotics chlortetracycline, enrofloxacin, and sulphathiazole from soil by radish. *Sci. Total. Environ.* **2017**, *605*, 322–331. [CrossRef] [PubMed]

25. Yang, L.; Zhang, S.; Chen, Z.; Wen, Q.; Wang, Y. Maturity and security assessment of pilot-scale aerobic co-composting of penicillin fermentation dregs (PFDs) with sewage sludge. *Bioresour. Technol.* **2016**, *204*, 185–191. [CrossRef]

26. Bengtsson-Palme, J.; Larsson, D.G.J. Concentrations of antibiotics predicted to select for resistant bacteria: Proposed limits for environmental regulation. *Environ. Int.* **2016**, *86*, 140–149. [CrossRef]

27. Taleghani, A.H.; Lim, T.T.; Lin, C.-H. Degradation of veterinary antibiotics in swine manure via anaerobic digestion. In Proceedings of the 2018 ASABE International Meeting, Michigan, MI, USA, 29 July–1 August 2018.

28. Miller, C.; Heringa, S.; Kim, J.; Jiang, X. Analyzing indicator microorganisms, antibiotic resistant *Escherichia coli*, and regrowth potential of foodborne pathogens in various organic fertilizers. *Foodborne Pathog. Dis.* **2013**, *10*, 520–527. [CrossRef]

29. Edrington, T.S.; Fox, W.E.; Callaway, T.; Anderson, R.C.; Hoffman, D.W.; Nisbet, D.J. Pathogen prevalence and influence of composted dairy manure application on antimicrobial resistance profiles of commensal soil bacteria. *Foodborne Pathog. Dis.* **2009**, *6*, 217–224. [CrossRef]

30. Dolliver, H.; Gupta, S.C.; Noll, S. Antibiotic degradation during manure composting. *J. Environ. Qual.* **2008**, *37*, 1245–1253. [CrossRef]

31. Xie, W.-Y.; Yang, X.-P.; Li, Q.; Wu, L.-H.; Shen, Q.-R.; Zhao, F.J. Changes in antibiotic concentrations and antibiotic resistome during commercial composting of animal manures. *Environ. Pollut.* **2016**, *219*, 182–190. [CrossRef]

32. Prado, N.; Ochoa, J.; Amrane, A. Biodegradation and biosorption of tetracycline and tylosin antibiotics in activated sludge system. *Process. Biochem.* **2009**, *44*, 1302–1306. [CrossRef]

33. Joy, S.R.; Li, X.; Snow, D.; Gilley, J.E.; Woodbury, B.; Bartelt-Hunt, S. Fate of antimicrobials and antimicrobial resistance genes in simulated swine manure storage. *Sci. Total. Environ.* **2014**, *481*, 69–74. [CrossRef] [PubMed]

34. Angenent, L.T.; Mau, M.; George, U.; Zahn, J.A.; Raskin, L. Effect of the presence of the antimicrobial tylosin in swine waste on anaerobic treatment. *Water Res.* **2008**, *42*, 2377–2384. [CrossRef]

35. Shi, J.; Liao, X.; Wu, Y.; Liang, J. Effect of antibiotics on methane arising from anaerobic digestion of pig manure. *Anim. Feed. Sci. Technol.* **2011**, 457–463. [CrossRef]

36. Beneragama, N.; Lateef, S.A.; Iwasaki, M.; Yamashiro, T.; Umetsu, K. The combined effect of cefazolin and oxytetracycline on biogas production from thermophilic anaerobic digestion of dairy manure. *Bioresour. Technol.* **2013**, *133*, 23–30. [CrossRef] [PubMed]

37. 2012 Summary Report on Antimicrobials Sold or Distributed for Use in Food-Producing Animals. Available online: https://www.fda.gov/media/89630/download (accessed on 1 July 2019).

38. Furtula, V.; Farrell, E.G.; Diarrassouba, F.; Rempel, H.; Pritchard, J.; Diarra, M.S. Veterinary pharmaceuticals and antibiotic resistance of Escherichia coli isolates in poultry litter from commercial farms and controlled feeding trials. *Poult. Sci.* **2010**, *89*, 180–188. [CrossRef]

39. Hu, D.; Coats, J. Aerobic degradation and photolysis of tylosin in water and soil. *Environ. Toxicol. Chem.* **2007**, *26*, 884–889. [CrossRef]

40. Jia, A.; Xiao, Y.; Hu, J.; Asami, M.; Kunikane, S. Simultaneous determination of tetracyclines and their degradation products in environmental waters by liquid chromatography-electrospray tandem mass spectrometry. *J. Chromatogr. A* **2009**, *1216*, 4655–4662. [CrossRef]

41. Kemper, N. Veterinary antibiotics in the aquatic and terrestrial environment. *Ecol. Indic.* **2008**, *8*, 1–13. [CrossRef]

42. Kim, M.; Ahn, Y.-H.; Speece, R.E. Comparative process stability and efficiency of anaerobic digestion; mesophilic vs. thermophilic. *Water Res.* **2002**, *36*, 4369–4385. [CrossRef]

43. Kümmerer, K. Antibiotics in the aquatic environment-A review-Part I. *Chemosphere* **2009**, *75*, 417–434. [CrossRef]

44. World Bank. Drug-Resistant Infections: A Threat to Our Economic Future. *World Bank Rep.* **2016**, *2*, 1–132. [CrossRef]

45. Wang, H.; Lim, T.T.; Duong, C.; Zhang, W.; Xu, C.; Yan, L.; Mei, Z.; Wang, W. Long-term mesophilic anaerobic co-digestion of swine manure with corn stover and microbial community analysis. *Microorganisms* **2020**, *8*, 188. [CrossRef] [PubMed]

46. Lim, T.-T.; Harvey, B.C.; Zulovich, J.M. Implementation of a Pilot Scale Anaerobic Digester System for Swine Finishing Barn. In Proceedings of the ASABE International Meeting, Kansas City, MO, USA, 21–24 July 2013.

47. Lim, T.-T. Start-up of a pilot scale anaerobic digestion system for swine finishing barn. In Proceedings of the 18th World Congress of CIGR, International Commission of Agricultural and Biosystems Engineering, Beijing, China, 16–18 September 2013.

48. Lin, C.-H.; Lerch, R.N.; Thurman, E.M.; Garrett, H.E.; George, M.F. Determination of Isoxaflutole (balance) and its metabolites in water using solid phase extraction followed by high-performance liquid chromatography with ultraviolet or mass spectrometry. *J. Agric. Food Chem.* **2002**, *50*, 5816–5824. [CrossRef] [PubMed]

49. Lin, C.-H.; Lerch, R.N.; Garrett, H.E.; George, M.F. Improved GC-MS/MS Method for determination of atrazine and its chlorinated metabolites in forage plants-Laboratory and field experiments. *Commun. Soil Sci. Plant. Anal.* **2007**, *38*, 1753–1773. [CrossRef]

50. Yang, S.F.; Lin, C.F.; Wu, C.J.; Ng, K.K.; Lin, A.Y.-C.; Andy Hong, P.K. Fate of Sulfonamide Antibiotics in Contact with Activated Sludge—Sorption and Biodegradation. *Water Res.* **2012**, *46*, 1301–1308. [CrossRef]

51. Bray, J.R.; Curtis, J.T. An ordination of the upland forest communities of Southern Wisconsin. *Ecol. Monogr.* **1957**, *27*, 325–349. [CrossRef]

52. Ericsson, A.C.; Turner, G.; Montoya, L.; Wolfe, A.; Meeker, S.; Hsu, C.; Maggio-Price, L.; Franklin, C.L. Isolation of segmented filamentous bacteria from complex gut microbiota. *BioTechniques* **2015**, *59*, 94–98. [CrossRef]

53. Ericsson, A.C.; Davis, D.J.; Franklin, C.L.; Hagan, C.E. Exoelectrogenic capacity of host microbiota predicts lymphocyte recruitment to the gut. *Physiol. Genom.* **2015**, *47*, 243–252. [CrossRef]

54. Walters, W.A.; Caporaso, J.G.; Lauber, C.L.; Berg-Lyons, D.; Fierer, N.; Knight, R. Primer Prospector: De novo design and taxonomic analysis of barcoded polymerase chain reaction primers. *Bioinformatics* **2011**, *27*, 1159–1161. [CrossRef]

55. Magoč, T.; Salzberg, S.L. FLASH: Fast length adjustment of short reads to improve genome assemblies. *Bioinformatics* **2011**, *27*, 2957–2963. [CrossRef]

56. Kuczynski, J.; Stombaugh, J.; Walters, W.A.; González, A.; Caporaso, J.G.; Knight, R. Using QIIME to analyze 16S rRNA gene sequences from microbial communities. *Curr. Protoc. Bioinform.* **2011**, *36*, 10.7.1–10.7.20. [CrossRef] [PubMed]

57. Altschul, S.F.; Madden, T.L.; Schäffer, A.A.; Zhang, J.; Zhang, Z.; Miller, W.; Lipman, D.J. Gapped BLAST and PSI-BLAST: A new generation of protein database search programs. *Nucleic Acids Res.* **1997**, *25*, 3389–3402. [CrossRef] [PubMed]

58. DeSantis, T.Z.; Hugenholtz, P.; Larsen, N.; Rojas, M.; Brodie, E.L.; Keller, K.; Huber, T.; Dalevi, D.; Hu, P.; Andersen, G.L. Greengenes, a chimera-checked 16S rRNA gene database and workbench compatible with ARB. *Appl. Environ. Microbiol.* **2006**, *72*, 5069–5072. [CrossRef] [PubMed]

59. Stone, J.J.; Clay, S.A.; Zhu, Z.; Wong, K.L.; Porath, L.R.; Spellman, G.M. Effect of antimicrobial compounds tylosin and chlortetracycline during batch anaerobic swine manure digestion. *Water Res.* **2009**, *43*, 4740–4750. [CrossRef] [PubMed]

60. Huang, B.; Wang, H.; Cui, D.; Zhang, B.; Chen, Z.-B.; Wang, A.-J. Treatment of pharmaceutical wastewater containing β-lactams antibiotics by a pilot-scale anaerobic membrane bioreactor (AnMBR). *Chem. Eng. J.* **2018**, *341*, 238–247. [CrossRef]

61. Spielmeyer, A.; Breier, B.; Groißmeier, K.; Hamscher, G. Elimination patterns of worldwide used sulfonamides and tetracyclines during anaerobic fermentation. *Bioresour. Technol.* **2015**, *193*, 307–314. [CrossRef]

62. Mitchell, S.M.; Ullman, J.L.; Teel, A.L.; Watts, R.J.; Frear, C. The effects of the antibiotics ampicillin, florfenicol, sulfamethazine, and tylosin on biogas production and their degradation efficiency during anaerobic digestion. *Bioresour. Technol.* **2013**, *149*, 244–252. [CrossRef]

63. Chelliapan, S.; Wilby, T.; Sallis, P.; Yuzir, A. Tolerance of the antibiotic Tylosin on treatment performance of an Up-flow Anaerobic Stage Reactor (UASR). *Water Sci. Technol.* **2011**, *63*, 1599–1606. [CrossRef]

64. Gartiser, S.; Urich, E.; Alexy, R.; Kümmerer, K. Anaerobic inhibition and biodegradation of antibiotics in ISO test schemes. *Chemosphere* **2007**, *66*, 1839–1848. [CrossRef]

65. Yin, F.; Dong, H.; Ji, C.; Tao, X.; Chen, Y. Effects of anaerobic digestion on chlortetracycline and oxytetracycline degradation efficiency for swine manure. *Waste Manag.* **2016**, *56*, 540–546. [CrossRef]

66. Dreher, T.M.; Mott, H.V.; Lupo, C.D.; Oswald, A.S.; Clay, S.A.; Stone, J.J. Effects of chlortetracycline amended feed on anaerobic sequencing batch reactor performance of swine manure digestion. *Bioresour. Technol.* **2012**, *125*, 65–74. [CrossRef] [PubMed]

67. Álvarez, J.; Otero, L.; Lema, J.; Omil, F. The effect and fate of antibiotics during the anaerobic digestion of pig manure. *Bioresour. Technol.* **2010**, *101*, 8581–8586. [CrossRef] [PubMed]

68. Graaf, D.; Fendler, R. *Biogas Production in Germany*; Federal Environment Agency: Dessau-Rosslau, Germany, 2010.

69. Stone, J.J.; Oswald, A.S.; Lupo, C.D.; Clay, S.A.; Mott, H.V. Impact of chlortetracycline on sequencing batch reactor performance for swine manure treatment. *Bioresour. Technol.* **2011**, *102*, 7807–7814. [CrossRef]

70. Thauer, R.K. Special Biochemistry of methanogenesis: A tribute to Marjory Stephenson. In Proceedings of the 140th Ordinary Meeting of the Society for General Microbiology, Nottingham, UK, 30 March–2 April 1998; pp. 2377–2406.

71. Winckler, C.; Grafe, A. Use of veterinary drugs in intensive animal production. *J. Soils Sediments* **2001**, *1*, 66–70. [CrossRef]

72. Arikan, O.A. Degradation and Metabolization of Chlortetracycline during the Anaerobic Digestion of Manure from Medicated Calves. *J. Hazard. Mater.* **2008**, *158*, 485–490. [CrossRef]

73. Cheng, D.; Ngo, H.H.; Guo, W.; Chang, S.W.; Nguyen, D.D.; Liu, Y.; Shan, X.; Nghiem, L.D.; Nguyen, L.N. Removal process of antibiotics during anaerobic treatment of swine wastewater. *Bioresour. Technol.* **2020**, *300*, 122707. [CrossRef]

74. Kolz, A.C.; Moorman, T.B.; Ong, S.K.; Scoggin, K.D.; Douglass, E.A. Degradation and metabolite production of tylosin in anaerobic and aerobic swine-manure lagoons. *Water Environ. Res.* **2005**, *77*, 49–56. [CrossRef]

75. Ingerslev, F.; Toräng, L.; Loke, M.-L.; Halling-Sørensen, B.; Nyholm, N. Primary biodegradation of veterinary antibiotics in aerobic and anaerobic surface water simulation systems. *Chemosphere* **2001**, *44*, 865–872. [CrossRef]

76. Wang, H.; Vuorela, M.; Keränen, A.-L.; Lehtinen, T.M.; Lensu, A.; Lehtomäki, A.; Rintala, J. Development of microbial populations in the anaerobic hydrolysis of grass silage for methane production. *FEMS Microbiol. Ecol.* **2010**, *72*, 496–506. [CrossRef]

77. García-Peña, E.I.; Parameswaran, P.; Kang, D.; Canul-Chan, M.; Krajmalnik-Brown, R. Anaerobic digestion and co-digestion processes of vegetable and fruit residues: Process and microbial ecology. *Bioresour. Technol.* **2011**, *102*, 9447–9455. [CrossRef]

78. Nogueira, R.G.S.; Lim, T.T.; Wang, H.; Rodrigues, P.H.M. Performance, microbial community analysis and fertilizer value of anaerobic co-digestion of cattle manure with waste kitchen oil. *Appl. Eng. Agric.* **2019**, *35*, 239–248. [CrossRef]

79. Schink, B. Syntrophic associations in methanogenic degradation. *Mol. Basis Symbiosis* **2006**, *41*, 1–19. [CrossRef]

80. Hatamoto, M.; Imachi, H.; Ohashi, A.; Harada, H. Identification and cultivation of anaerobic, syntrophic long-chain fatty acid-degrading microbes from mesophilic and thermophilic methanogenic sludges. *Appl. Environ. Microbiol.* **2006**, *73*, 1332–1340. [CrossRef] [PubMed]

81. Hattori, S. Syntrophic acetate-oxidizing microbes in methanogenic environments. *Microbes Environ.* **2008**, *23*, 118–127. [CrossRef] [PubMed]

82. Ito, T.; Yoshiguchi, K.; Ariesyady, H.D.; Okabe, S. Identification and quantification of key microbial trophic groups of methanogenic glucose degradation in an anaerobic digester sludge. *Bioresour. Technol.* **2012**, *123*, 599–607. [CrossRef]

83. Padmasiri, S.I.; Zhang, J.; Fitch, M.; Norddahl, B.; Morgenroth, E.; Raskin, L. Methanogenic population dynamics and performance of an anaerobic membrane bioreactor (AnMBR) treating swine manure under high shear conditions. *Water Res.* **2007**, *41*, 134–144. [CrossRef]

Article

A Small Study of Bacterial Contamination of Anaerobic Digestion Materials and Survival in Different Feed Stocks

Lauren Russell [1,2], Paul Whyte [2], Annetta Zintl [2], Stephen Gordon [2], Bryan Markey [2], Theo de Waal [2], Enda Cummins [3], Stephen Nolan [4], Vincent O'Flaherty [4], Florence Abram [4], Karl Richards [5], Owen Fenton [5] and Declan Bolton [1,*]

[1]   Teagasc Food Research Centre, Ashtown, Dublin 15, Ireland; lauren.russell@teagasc.ie
[2]   School of Veterinary Medicine, University College Dublin, Belfield, Dublin 4, Ireland;
      paul.whyte@ucd.ie (P.W.); annetta.zintl@ucd.ie (A.Z.); stephen.gordon@ucd.ie (S.G.);
      bryan.markey@ucd.ie (B.M.); theo.dewall@ucd.ie (T.d.W.)
[3]   School of Biosystems and Food Engineering, University College Dublin, Belfield, Dublin 4, Ireland;
      enda.cummins@ucd.ie
[4]   School of Natural Sciences, National University of Ireland, Galway, Ireland; stiofnolan@gmail.com (S.N.);
      vincent.oflaherty@nuig.ie (V.O.); florence.abram@nuigalway.ie (F.A.)
[5]   Teagasc Environmental Research Centres, Johnstown Castle, Wexford, Ireland;
      karl.richards@teagasc.ie (K.R.); owen.fenton@teagasc.ie (O.F.)
*   Correspondence: Declan.Bolton@teagasc.ie

Received: 20 July 2020; Accepted: 20 September 2020; Published: 22 September 2020

Abstract: If pathogens are present in feedstock materials and survive in anaerobic digestion (AD) formulations at 37 °C, they may also survive the AD process to be disseminated in digestate spread on farmland as a fertilizer. The aim of this study was to investigate the prevalence of *Salmonella* spp., *Escherichia coli* O157, *Listeria monocytogenes*, *Enterococcus faecalis* and *Clostridium* spp. in AD feed and output materials and survival/growth in four formulations based on food waste, bovine slurry and/or grease-trap waste using International Organization for Standardization (ISO) or equivalent methods. The latter was undertaken in 100 mL Ramboldi tubes, incubated at 37 °C for 10 d with surviving cells enumerated periodically and the T_{90} values (time to achieve a 1 log reduction) calculated. The prevalence rates for *Salmonella* spp., *Escherichia coli* O157, *Listeria monocytogenes*, *Enterococcus faecalis* and *Clostridium* spp. were 3, 0, 5, 11 and 10/13 in food waste, 0, 0, 2, 3 and 2/3 in bovine slurry, 1, 0, 8, 7 and 8/8 in the mixing tank, 5, 1, 17, 18 and 17/19 in raw digestate and 0, 0, 0, 2 and 2/2 in dried digestate, respectively. Depending on the formulation, T_{90} values ranged from 1.5 to 2.8 d, 1.6 to 2.8 d, 3.1 to 23.5 d, 2.2 to 6.6 d and 2.4 to 9.1 d for *Salmonella* Newport, *Escherichia coli* O157, *Listeria monocytogenes*, *Enterococcus faecalis* and *Clostridium sporogenes*, respectively. It was concluded that AD feed materials may be contaminated with a range of bacterial pathogens and *L. monocytogenes* may survive for extended periods in the test formulations incubated at 37 °C.

Keywords: *Salmonella* spp.; *Escherichia coli* O157; *Listeria monocytogenes*; *Enterococcus faecalis*; *Clostridium* spp.; anaerobic digestion; digestate; pathogens; sustainable farming

---

1. Introduction

Anaerobic digestion (AD) is a cheap and efficient method for processing the large amounts of organic waste produced by farming (manures and slurries), food processing and sewage treatments (sludge) while contributing to international renewable energy targets. Co-digestion of combined wastes produces biogas (methane and carbon dioxide) and digestate, a nutrient rich fertilizer [1] while recycling nutrients from biowaste back into food production (an important activity in sustainable farming) [2]. In its most basic form, AD involves mechanical pretreatment of the feed waste materials to reduce particle size and mix the formulations, followed by anaerobic digestion, which produces biogas and digestate, the latter of which is usually subject to a treatment (pasteurization or drying) before use as a soil fertilizer (Figure 1).

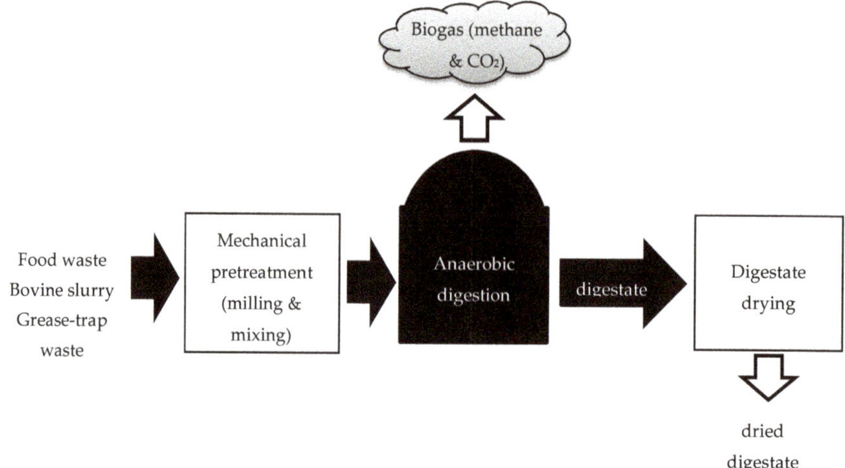

Figure 1. The basic steps in the anaerobic digestion process.

There are four stages in anaerobic digestion; hydrolysis, acidogenesis, acetogenesis and methanogenesis [3]. During hydrolysis the lipids, carbohydrates and protein present in the feed materials are broken down into fatty acids, sugars and amino acids, respectively. This is followed by acidogenesis, during which fermentative bacteria produce volatile fatty acids (VFAs), including propionic acid, butyric acid, acetic acid as well as ethanol, ammonia, carbon dioxide and hydrogen sulphide (H_2S). In the third stage (acetogenesis), the products of acidogenesis are converted into acetic acid, carbon dioxide and hydrogen while during methanogenesis (fourth stage), the products of the preceding stages are converted into methane, carbon dioxide and water [4]. The byproduct, digestate, is a nutrient rich fertilizer.

However, feedstocks may be contaminated with a range of bacterial, viral and parasitic pathogens of veterinary and public health concern [5], which may survive the process, depending on a combination of factors including initial load, feedstock, microbial competition, pH, temperature and ammonia production [6], to be disseminated on farms in contaminated digestate [2,7,8]. Thus, EC Regulations 1069/2009 and 142/2011 require that AD raw materials or digestate must be heat treated at 70 °C or 90 °C for a minimum of 60 min or equivalent. Regardless, it is generally agreed that such treatments are only sufficient to kill vegetative bacteria like *Salmonella*, *Listeria* and *Escherichia coli*, while spore-forming organisms such as *Clostridium* spp. will survive. The application of digestate as a fertiliser is therefore banned in some countries [9].

Farm based AD plants in Ireland currently operate at mesophilic temperatures and typically co-digest animal slurry with food waste [10]. Data on bacterial contamination and survival during the different stages of the AD process is limited. Although the process parameters such as temperature are set to optimise biogas production, other factors such as the composition of feedstock and retention time could be manipulated, if necessary, to promote the destruction of target pathogenic bacteria without negatively impacting on the efficiency of the process [11]. The aims of this study were to test a range of AD input and output materials for the presence of *Salmonella* spp., *E. coli* O157, *L. monocytogenes*, *Enterococcus faecalis* and *Clostridium* spp. and to investigate the survival of representative strains of these bacteria in four AD feedstock materials/formulations, stored at 37 °C in a laboratory-scale batch system previously used in similar studies [12].

2. Materials and Methods

2.1. Pathogen Evaluation/Survey

2.1.1. AD Samples

Food waste (a mixture of dairy and vegetable wastes; $n = 13$), bovine slurry ($n = 3$), mixing tank ($n = 8$), raw digestate ($n = 19$) and dried digestate ($n = 2$) samples were collected from 3 separate commercial AD facilities located in the east of Ireland. These materials were not preselected but were the feedstock materials being used on the day of each visit. Each plant was visited on one occasion and the samples aseptically removed using a sterile scoop (Sterileware, Fisher Scientific Ireland, Dublin, Ireland) and sterile containers (VWR, Dublin, Ireland). All samples were transported to the laboratory in a cool box at 2–4 °C within 3 h.

2.1.2. Microbiological Analysis

Exactly 25 g of each sample was diluted and/or enriched in 225 mL of diluent or broth before plating on selective agar and incubated at 37 °C for 24 h, unless otherwise indicated (Table 1). Presumptive colonies were confirmed using culture based and PCR methods (also Table 1). All media (except BBL Enterococcosel broth, which was supplied by Becton Dickinson (Limerick, Ireland)) were Oxoid products and purchased from Fannin Ltd., (Dublin, Ireland), as were the AnaeroGen sachets. Immunomagnetic separation (IMS) beads by Dynal® BeadRetriever were supplied by Thermo Fisher Scientific (Dublin, Ireland) while the Sifin anti-coli O157 sera test and defibrinated horse blood were provided by Cruinn Diagnostics Ltd., (Dublin, Ireland).

Table 1. The isolation and confirmation methods used to test the samples for the target bacteria.

Detection		Confirmation	
Treatment	Selective Agar	Culture Based	Molecular
Salmonella spp.			
buffered peptone water	modified semi-solid Rappaport Vassiliadis medium with novobiocin supplement (20 mg/L), incubated at 42 °C for 24 h	Xylose lysine deoxycholate (XLD) agar	Pathmanathan et al. [13]
E. coli O157			
modified tryptone soya broth (mTSB) containing cefixime (50 µg/L) and vancomycin (6 mg/L)	Immunomagnetic separation with plating on sorbitol MacConkey agar supplemented with cefixime-tellurite (CT-SMAC)	Eosin methyl blue agar and plate count agar (PCA) followed by agglutination testing using the Sifin anti-coli O157 sera test	Paton and Paton [14].
L. monocytogenes			
half strength Fraser broth, incubated overnight at 30 °C followed by full strength Fraser broth incubated at 37 °C for 48 h	*Listeria* Selective Oxford agar and Brilliance *Listeria* agar (BLA), incubated at 37 °C for 48 h	PCA	Terzi et al. [15]
E. faecalis			
BBL Enterococcosel broth and plated on Slanetz and Bartley agar (SBA) incubated at 37 °C for 24 h, followed by 44 °C for an additional 24 h	Pink colonies were streaked on PCA and stabbed in rows into well-dried bile aesculin agar plates, incubated at 44 °C for 24 h.	PCA	Dutka-Malen et al. [16]
Clostridium spp.			
Maximum recovery diluent before plating on reinforced clostridial agar (RCA) incubated anaerobically (AnaeroGen sachets in BioMérieux GENbox jars (Hampshire, UK) at 37 °C for 48 h	Columbia blood agar supplemented with 5% defibrinated horse blood		Song et al. [17]

2.2. Survival Studies

2.2.1. Inoculum Preparation

Salmonella Newport, *E. coli* O157 (NCTC 12900), *L. monocytogenes* and *E. faecalis* (NCTC 12697) strains were obtained from the Teagasc culture collection. The *S.* Newport and *L. monocytogenes* strains had a streptomycin resistance (1000 µg/mL) marker to facilitate recovery. To prepare the inoculum, a culture bead from frozen storage was streaked on TSA and incubated at 37 °C for 24 h. A single colony was then selected and placed into 10 mL of tryptone soya broth (TSB; Oxoid, Fannin Ltd., Ireland) and incubated overnight at 37 °C. The culture obtained was centrifuged and washed 3 times with phosphate buffered saline (PBS; Oxoid, Fannin Ltd., Ireland), before resuspension in PBS and serially diluted to obtain a cell concentration of approximately 10^5 cfu/mL.

Freeze-dried *C. sporogenes* DSM 767 obtained from the Deutsche Sammlung von Mikroorganismen und Zellkulturen (DSMZ, Braunschweig, Germany) were rehydrated as per the instructions provided. Twenty tubes of cooked meat medium (CMM; Oxoid, Fannin Ltd., Ireland) broth (20 mL) were inoculated with 100 µL rehydrated *C. sporogenes*, and incubated in an anaerobic cabinet for 12–18 h at 37 °C. Clostridium sporulation agar was prepared as described by [18] and placed in a Whitley A35 anaerobic chamber (Don Whitley Scientific, West Yorkshire, UK) overnight using the ANO_2 gas mixture (10% H_2, 10% CO_2 and 80% N_2; Air Products Ireland, Dublin, Ireland) to exclude all oxygen. Aliquots (300 µL) of the overnight CMM broth were then spread onto 300 plates of CSA (inside the anaerobic chamber) before transfer to anaerobic boxes (GenBOX jars; BioMérieux UK Ltd., Basingstoke, UK; AnaeroGen sachets; Oxoid, Fannin Ltd., Ireland) and incubated at 37 °C for 12 d. The CSA plates were then inspected to ensure sufficient spore growth for harvesting. Spore harvesting took place in a laminar flow hood. Approximately 4–5 mL ice-cold sterile distilled water was placed onto the surface of the CSA plates, agitating the surface of the agar with a sterile spreader to release the spores. The suspension was then transferred to the next agar plate and the scraping process repeated. This method was repeated until spores had been harvested from all of the 300 CSA plates. The suspensions were pooled in 50 mL tubes, centrifuged at 7000 RPM at 4 °C for 10 min and washed with iced water, reducing the amount of liquid over the course of repeated cycles until a spore suspension of approximately 10^7 spores/mL (estimated by phase contrast microscope examination), which was then confirmed by plating out on Columbia blood agar (CBA; Oxoid, Fannin Ltd., Ireland) with 5% defibrinated horse blood (Cruinn diagnostics, Ireland). The spore preparations (1 mL aliquots) were stored at −80 °C. Prior to inoculation, spore preparations were thawed at room temperature, prior to heat treatment at 80 °C for 10 min to ensure the exclusion of vegetative cells.

2.2.2. AD Commercial Formulation Preparation

Four feedstock mixtures; [1] 100% food waste (primarily vegetable matter with small amounts of cooked meats and bakery product waste); [2] slurry (bovine) and food waste (1:3); [3] slurry and food waste (3:1) and [4] slurry and grease-trap waste (from restaurants) (2:1) were formulated on a volumetric basis as per the advice of our commercial AD stakeholders. Food waste was supplied by local restaurants, slurry by beef farms in counties Galway, Louth and Meath and grease-trap waste from the Bioenergy and Organic Fertilizer Services (BEOFS) AD plant in Camphill, County Kilkenny, Ireland. Before use all samples were tested to ensure the target bacteria were absent.

2.2.3. The Laboratory Model System

Exactly 70 model reactors were prepared for each of the four mixtures. Each contained 10 mL of fresh seed material (obtained from a commercial AD bioreactor) mixed with 20 mL of the feedstock material in a sterile 100 mL tube (Ramboldi tubes, VWR, Ireland). For each mixture, 14 tubes were randomly assigned to each of the bacteria being studied. The bacterial cells/spores, prepared as described above, were then added to 1 mL MRD to give a final concentration of approximately 10^4 cells or approximately 10^7 spores/mL. The tubes were then incubated anaerobically (GenBOX jars; bioMérieux UK Ltd., Basingstoke, UK; AnaeroGen sachets; Oxoid, Fannin Ltd., Ireland) at 37 °C. Duplicate tubes were removed periodically (0 (immediately after inoculation), 1, 2, 3, 4, 5 and 10 d), from the vortexed tubes, the pH recorded (Eutech pH 150 probe (Thermo Scientific, Waltham, MA, USA), which was calibrated using pH 4, 7 and 10 standards prior to use) and the surviving cells/spores enumerated.

2.2.4. Enumeration of Surviving Cells

The extracted samples (1 mL) were diluted in 9 mL MRD and serial dilutions prepared. Surviving cells/spores were enumerated as described in Table 2. All media and the AnaeroGen sachets were Oxoid products and purchased from Fannin Ltd., (Dublin, Ireland). Streptomycin sulphate was obtained from Sigma Aldrich Ireland Ltd., (Wicklow, Ireland). Agar plates were incubated at 37 °C for 24 h, unless otherwise indicated.

Table 2. Methods for enumerating surviving cells or spores.

	Enumeration	PCR Confirmation
S. Newport	XLD, supplemented with streptomycin sulphate (1000 µL/g)	Pathmanathan et al. [13]
E. coli O157	CT-SMAC	Paton and Paton [14].
L. monocytogenes	BLA, supplemented with streptomycin sulphate (1000 µL/g) incubated at 37 °C for 48 h	Terzi et al. [15].
E. faecalis	SBA incubated at 37 °C for 24 h, followed by 44 °C for a further 24 h	Dutka-Malen et al. [16].
C. sporogenes	RCA, incubated anaerobically (AnaeroGen sachets in BioMérieux GENbox jars (Hampshire, UK) at 37 °C for 48 h	Song et al. [17] and Morandi et al. [19].

2.3. Data Analysis

The survival study, as described above, was performed in duplicate and repeated on three separate occasions. Bacterial counts were converted into $\log_{10}$ cfu/mL and the T_{90}-values (the time required to achieve a 90% (1 log) reduction in the population) were determined by linear regression using GraphPad Prism 7 software (San Diego, CA, USA), considering each replicate Y-value as an individual point. Differences between slopes were examined using ANOVA and Tukey's multiple comparison tests (GraphPad Prism 7.02). Statistical significance was set at the 5% level ($p < 0.05$).

3. Results

The results of the survey of commercial AD inputs and outputs are shown in Table 3. *Salmonella* spp. were detected in the food waste (3 positive out of 13 samples tested (3/13)), mixing tank (1/8) and raw digestate (5/19) samples. *E. coli* O157 was only detected in one sample (raw digestate). In contrast *L. monocytogenes*, *E. faecalis* and *Clostridium* spp. were common in food waste (5, 11 and 10/13), slurry (2, 3 and 2/3), mixing tank (8, 7 and 8/8) and raw digestate (17, 18 and 17/19) samples. The latter two bacteria were also detected in the two dried digestate samples tested.

Table 3. Detection of the target pathogens in the different types of samples.

Pathogen / Type of samples	*Salmonella* spp.	*E. coli* O157	*L. monocytogenes*	*E. faecalis*	*Clostridium* spp.
Pre anaerobic digestion					
food waste (13) [1]	Positive (3) [2]	negative	positive (5)	positive (11)	positive (10)
bovine slurry (3)	negative	negative	positive (2)	positive (3)	positive (2)
mixing tank (8)	positive (1)	negative	positive (8)	positive (7)	positive (8)
Post anaerobic digestion					
raw digestate (19)	positive (5)	positive (1)	positive (17)	positive (18)	positive (17)
dried digestate (2)	negative	negative	negative	positive (2)	positive (2)

[1] total number of samples tested; [2] number of positive samples.

In the model 100 mL tubes, the pH of the food waste (100%) and slurry and food waste (1:3) formulations decreased from pH 7.1 to 5.8. and from pH 7.2 to 6.0, respectively (data not shown). In contrast the pH values in the slurry and food waste (3:1) increased from pH 7.5 to 8.0 while the pH was stable at pH 8.0 in the slurry and grease-trap waste (2:1) over the 10 d of the study.

The results of the regression analysis are provided in Figure 2 and Table 4. An initial period of growth (1–3 d) was observed in food waste (100%; *S.* Newport, *E. coli* O157 and *C. sporogenes*), slurry and food waste (1:3; *S.* Newport, *E. coli* O157 and *E. faecalis*), slurry and food waste (3:1; *S.* Newport and *E. faecalis*) and in slurry and grease-trap waste (2:1; *E. coli* O157). The time required toachieve a 1 log reduction in the *S.* Newport and *E. coli* O157 populations ranged from 1.5–2.8 d, with significantly ($p < 0.05$) higher T_{90}-values observed for slurry when combined with food (3:1) and grease-trap waste (2:1). In contrast, the T_{90}-values for *L. monocytogenes* were significantly lower in these two formulations (3.5 and 3.1 d, respectively) as compared to those obtained for the same bacteria in food waste (6.2 d) and slurry and food waste (1:3). The latter provided an environment where any reduction was minimal (slope = 0.04), resulting in an estimated 23.5 d required to achieve a 90% population reduction. T_{90}-values for *E. faecalis* ranged from 2.2 to 6.6 d with the latter obtained in slurry and food waste (3:1). *C. sporogenes* T_{90}-values ranged from 2.4 to 9.1 d, with significantly different values obtained in each of the formulations in the order of; slurry and grease-trap waste (2:1) > food waste > slurry and food waste (1:3) > slurry and food waste (3:1).

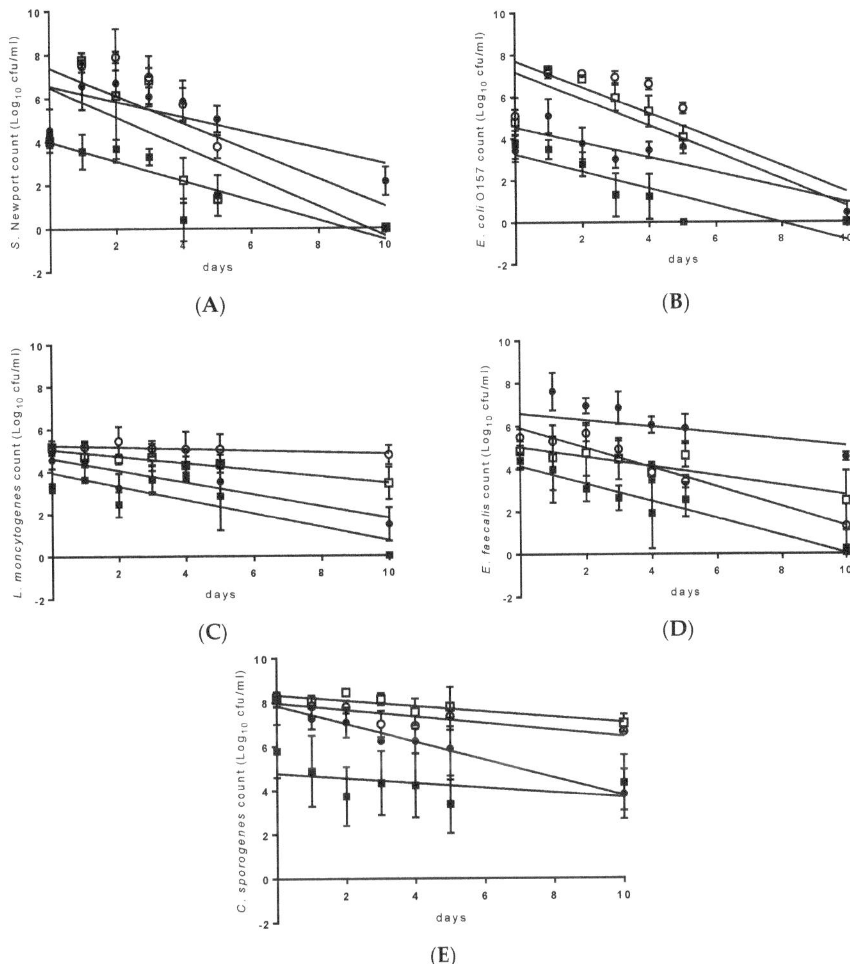

Figure 2. Linear regression graphs for *Salmonella* Newport (**A**), *Escherichia coli* O157 (**B**), *Listeria monocytogenes* (**C**), *Enterococcus faecalis* (**D**) and *Clostridium sporogenes* (**E**), in food waste (□), slurry and food waste (1:3) (○); slurry and food waste (3:1) (•) and slurry and grease-trap waste (2:1) (■). Each point is the mean of 6 data points (*n* = 6) and the error bar is the standard deviation.

Table 4. Observed growth and decay rate (T_{90}-values; the time for the bacterial concentration to decrease by 1 log unit) for the 5 pathogens (*Salmonella* spp., *Escherichia coli* O157, *Listeria monocytogenes*, *Enterococcus faecalis* and *C. sporogenes*) in the 4 different AD feedstock recipes.

Pathogen	Recipe	Growth			Decay Rate				
		Yes/No	Period	Maximum Concentration ($\log_{10}$ cfu/mL)	Slope	SE	R^2-Value	T_{90}-Value (d)	n
S. Newport	[1] FW	yes	1d	7.8	−0.69	0.110	0.49	1.5 [A]	42
	[2] SF1	yes	1d	7.3	−0.64	0.089	0.56	1.6 [A]	42
	[3] SF2	yes	1d	6.7	−0.36	0.029	0.45	2.8 [B]	42
	[4] SGW	no	[5] NA	[6] NA	−0.45	0.051	0.66	2.2 [B]	42
E. coli O157	FW	yes	1d	7.3	−0.64	0.062	0.77	1.6 [A]	42
	SF1	yes	1d	7.2	−0.63	0.073	0.64	1.6 [A]	42
	SF2	no	ND	NA	−0.36	0.044	0.62	2.8 [B]	42
	SGW	yes	1d	5.1	−0.41	0.049	0.64	2.5 [B]	42
L. monocytogenes	FW	no	ND	NA	−0.16	0.016	0.49	6.2 [B]	42
	SF1	no	ND	NA	[7] −0.04	0.027	0.05	23.5 [C]	42
	SF2	no	ND	NA	−0.28	0.039	0.77	3.5 [A]	42
	SGW	no	ND	NA	−0.32	0.050	0.51	3.1 [A]	42
E. faecalis	FW	no	ND	NA	−0.22	0.053	0.31	4.5 [B]	42
	SF1	yes	1d	7.6	−0.46	0.030	0.85	2.2 [A]	42
	SF2	yes	1d	7.6	−0.15	0.060	0.14	6.6 [C]	42
	SGW	no	ND	NA	−0.41	0.049	0.63	2.4 [A]	42
C. sporogenes	FW	yes	3d	7.1	−0.13	0.025	0.38	8.0 [C]	42
	SF1	no	ND	NA	−0.15	0.024	0.50	6.5 [B]	42
	SF2	no	ND	NA	−0.41	0.039	0.74	2.4 [A]	42
	SGW	no	ND	NA	−0.11	0.073	0.54	9.1 [D]	42

[1] FW = food waste; [2] SF1 = slurry and food waste (1:3); [3] SF2 = slurry and food waste (3:1); [4] SGW = slurry and grease-trap waste (2:1); [5] ND = not detected; [6] NA = not applicable; [7] slope is very close to zero (0.04) hence the R^2 value is almost zero. Statistical analysis: for a given bacteria a different capital letter (A, B, C or D) indicates significantly different T_{90}-values at the 5% level ($p < 0.05$).

4. Discussion

The commercial AD feedstock samples (food waste, bovine slurry and mixing tank materials) were contaminated with pathogens of public health significance including *Salmonella* spp., *L. monocytogenes*, *E. faecalis* and *Clostridium* spp. but not *E. coli* O157. Although there is little or no data for food waste or mixing tank materials, bovine faeces has been extensively tested and previous Irish studies have reported *Salmonella* spp. and *L. monocytogenes* contamination rates of 2–3% [20,21] and 5–12% [21,22], respectively, while 0.7–2.4% of samples are contaminated with *E. coli* O157 [23,24].

Salmonella and *Clostridium* spp. have also been detected in other AD feed materials [5,7,25]. To the best of our knowledge this is the first study reporting the presence of *L. monocytogenes* and *E. faecalis*, but this was not unexpected as these bacteria are widespread in the natural environment [26]. Of greater concern was the presence of all the target bacteria, including *Salmonella* spp. and *E. coli* O157, in raw digestate. *Salmonella* has been previously detected in digestate, suggesting these bacteria survive the AD process [5,7], although the possibility of post-reactor contamination cannot be ruled out. In contrast, only *E. faecalis* and *Clostridium* spp. were detected in the dried digestate, suggesting the drying process is sufficient to kill most but not all the bacteria of concern. This is an important finding, as several countries (including Ireland), have a standard requirement for the absence of *Salmonella* in 25 g before this material can be used as a fertiliser [7].

This study also investigated the survival of *Salmonella* spp., *E. coli* O157, *L. monocytogenes*, *E. faecalis* and *Clostridium* spp. in four AD feedstock formulations at 37 °C in a small scale laboratory system. Although previously shown to be a useful study tool [12], laboratory-scale batch systems may not be representative of full-scale continuous commercial bioreactors due to differences in inoculation methods, rheology and hydrodynamic factors [27]. Moreover, as our feedstock mixtures were formulated on a volumetric basis, it is possible that the organic load could have been different between the various formulations. This would affect the production of VFAs, ethanol, ammonia, hydrogen disulphide, etc., by the bacteria present, thereby influencing pathogen survival. Thus, while the survival data obtained provides a good indication of the relative resistance of each bacteria in the materials and under the conditions tested, further research would be required to obtain a more accurate representation of how these organisms behave in large scale commercial systems.

The T_{90}-values for *S.* Newport ranged from 1.5 to 2.8 d, regardless of the feed stock formulation. Interestingly, these values are similar to those previously reported for the decline of *Salmonella* spp. in the initial stages of the AD process, which typically range from 0.2 d in sewage sludge [28] to 7 d in a mixture of plant waste, cattle manure and cattle slurry [29–31]. The *E. coli* O157 T_{90} values (1.6–2.8 d) were similar to those of *S.* Newport and within the range of 0.5–6.5 d reported in previous AD pathogen survival studies [31–35]. Considering these bacteria survive for extended periods (at least 3 months) in bovine slurry [36,37] our data supports the hypothesis that AD is an effective process for *Salmonella* and *E. coli* O157 removal from animal waste.

In three of the four formulations the population of *L. monocytogenes* decreased by 1 $\log_{10}$ cfu/mL after approximately 3–6 d but in slurry and food waste (1:3) the population was almost stable resulting in a regression slope close to zero (−0.04). While previous studies have reported typical T_{90}-values of 1.5–2.2 d, in AD formulations [38–40], *L. monocytogenes* may also achieve a steady state during AD where the population is maintained for extended periods and the T_{90} values are as high as 12.3 d in batch slurry and 35.7 d in semi-continuous digestion. This is not unexpected as *L. monocytogenes* have a host of molecular mechanisms that facilitate survival in a range of different environments [41]. The T_{90}-value for *E. faecalis* ranged from 2.2 to 6.6 d, with significantly higher endurance in food waste and in slurry and food waste (3:1). These values compare to the 0.1–7 d previously reported for *Enterococcus* spp. in different feed-stocks (dairy waste, cattle slurry, swine manure and sewage sludge) [31–33,35,40,42,43] and is of particular significance as enterococci are considered to be good indicators of the fate of bacterial vegetative cells during AD [43]. *C. sporogenes* survival rates were lower than expected, with T_{90} values of 2.4–9.1 d. While comparable data for *C. sporogenes* is not available, Froschle et al. [25] found it required approximately 35 d to achieve a 1 log reduction in the population of *Clostridium botulinum* in laboratory scale digesters at 38 °C, while Chauret et al. [40] observed no change in the concentration of *C. perfringens* in the mesophilic digestion of sewage sludge after 20 d. Our observations are inconsistent with these findings and may be the result of the experimental design, for example elevated carbohydrate concentrations stimulating early VFA production, but further investigation is required.

When the different formulations were compared the results were mixed and there was no one mixture that consistently provided higher or lower T_{90} values for all of the bacteria tested. Food waste, alone and when combined with slurry, supported an initial growth phase (1 d) for *S.* Newport, *E. coli* O157 and/or *E. faecalis*, which are metabolically similar under anaerobic conditions, but also provided the lowest T_{90}-values for these bacteria. Interestingly, increasing the proportion of slurry in these mixtures resulted in significantly higher T_{90}-values for these bacteria but the opposite was observed with *L. monocytogenes* and *C. sporogenes*. Thus, while the bacteria tested decreased, the reduction rate was dependent on factors other than the formulation, as previously reported [44].

5. Conclusions

It was concluded that AD feed materials might be contaminated with a range of bacterial pathogens. However given the large volumes used in commercial bioreactors these would be diluted out and present at very low concentrations. In the laboratory-scale batch system used in our experiments, the survival rates of *S.* Newport, *E. coli* O157 and *E. faecalis* were similar to those previously reported while *C. sporogenes* declined more rapidly than expected. This requires further investigation as does the ability of *L. monocytogenes* to survive for extended periods during AD, perhaps necessitating mandatory pasteurisation of digestate.

Author Contributions: Conceptualization, D.B.; Methodology, D.B., L.R., P.W.; Formal Analysis, L.R., D.B.; Investigation, L.R.; Resources, D.B.; Data Curation, L.R., D.B.; Writing—Original Draft Preparation, D.B., L.R.; Writing—Review and Editing, D.B., L.R., P.W., A.Z., E.C., S.G., B.M., T.d.W., S.N., V.O., F.A., K.R., O.F.; Supervision, D.B., P.W.; Project Administration, D.B.; Funding Acquisition, D.B. All authors have read and agreed to the published version of the manuscript.

Funding: This study was funded by the Food Institute Research Measure administered by the Department of Agriculture, Food and Marine (Project 14/SF/487).

Acknowledgments: Lauren Russell was supported by the Teagasc Walsh Scholarship Programme (number 2014239).

Conflicts of Interest: The authors declare no conflict of interest.

References

1. Alkanok, G.; Demirel, B.; Onay, T.T. Determination of biogas generation potential as a renewable energy source from supermarket wastes. *Waste Manag.* **2014**, *34*, 134–140. [CrossRef]
2. Johansson, M.; Emmoth, E.; Salomonsson, A.C.; Albihn, A. Potential risks when spreading anaerobic digestion residues on grass silage crops—Survival of bacteria, moulds and viruses. *Grass Forage Sci.* **2005**, *60*, 175–185. [CrossRef]
3. Ramos-Suárez, J.; Arroyo, N.C.; González-Fernández, C. The role of anaerobic digestion in algal biorefineries: Clean energy production, organic waste treatment, and nutrient loop closure. In *Algae and Environmental Sustainability*; Singh, B., Kuldeep, B., Faizal, B., Eds.; Springer: New Delhi, India, 2015; pp. 53–76.
4. Anukam, A.; Mohammadi, A.; Naqvi, M.; Granström, K. A review of the chemistry of anaerobic digestion: Methods of accelerating and optimizing process efficiency. *Processes* **2019**, *7*, 504. [CrossRef]
5. Sidhu, J.P.S.; Toze, S.G. Human pathogens and their indicators in biosolids: A literature review. *Environ. Int.* **2009**, *35*, 187–201. [CrossRef]
6. Orzi, V.; Scaglia, B.; Lonati, S.; Riva, C.; Boccasile, G.; Alborali, G.; Adani, F. The role of biological processes in reducing both odor impact and pathogen content during mesophilic anaerobic digestion. *Sci. Total Environ.* **2015**, *526*, 116–126. [CrossRef] [PubMed]
7. Bonetta, S.; Bonetta, S.; Ferretti, E.; Fezia, G.; Gilli, G.; Carraro, E. Agricultural reuse of the digestate from anaerobic co-digestion of organic waste: Microbiological contamination, metal hazards and fertilizing performance. *Water Air Soil Pollut.* **2014**, *225*, 2046. [CrossRef]
8. Bonetta, S.; Ferretti, E.; Bonetta, S.; Fezia, G.; Carraro, E. Microbiological contamination of digested products from anaerobic co-digestion of bovine manure and agricultural by-products. *Lett. Appl. Microbiol.* **2011**, *53*, 552–557. [CrossRef] [PubMed]
9. Stutz, L.H. Risk Assessment of Input of Pathogens Residing in Co-Substrates into the River from Sewage Treatment Plant with Mesophilic Anaerobic Digestion. A Case Study of *Salmonella* and *Campylobacter* Evaluation in the Sewage Treatment Plant of Bern (ARA Bern). Bachelor's Thesis, Life Sciences and Facility Management, Swiss Federal Institute of Technology, Zurich, Switzerland, 2015.
10. Auer, A.; Burgt, N.H.V.; Abram, F.; Barry, G.; Fenton, O.; Markey, B.K.; Nolan, S.; Richards, K.; Bolton, D.; De Waal, T.; et al. Agricultural anaerobic digestion power plants in Ireland and Germany: Policy and practice. *J. Sci. Food Agric.* **2016**, *97*, 719–723. [CrossRef] [PubMed]
11. Avery, L.M.; Anchang, K.Y.; Tumwesige, V.; Strachan, N.; Goude, P.J. Potential for pathogen reduction in anaerobic digestion and biogas generation in sub-Saharan Africa. *Biomass Bioenergy* **2014**, *70*, 112–124. [CrossRef]
12. Nolan, S.; Waters, N.R.; Brennan, F.; Auer, A.; Fenton, O.; Richards, K.; Bolton, D.J.; Pritchard, L.; O'Flaherty, V.; Abram, F. Toward assessing farm-based anaerobic digestate public health risks: Comparative investigation with slurry, effect of pasteurization treatments, and use of miniature bioreactors as proxies for pathogen spiking trials. *Fun. Environ. Micro.* **2018**, *2*, 1–11. [CrossRef]
13. Pathmanathan, S.G.; Cardona-Castro, N.; Sanchez-Jimenez, M.M.; Correa-Ochoa, M.M.; Puthucheary, S.D.; Thong, K.L. Simple and rapid detection of Salmonella strains by direct PCR amplification of the *hilA* gene. *J. Med. Microbiol.* **2003**, *52*, 773–776. [CrossRef] [PubMed]

14. Paton, A.W.; Paton, J.C. Detection and characterization of Shiga toxigenic *Escherichia coli* by using multiplex PCR assays for stx1, stx2, eaeA, enterohemorrhagic *E. coli* hlyA, rfbO111, and rfbO157. *J. Clin. Microbiol.* **1998**, *36*, 598–602. [CrossRef] [PubMed]

15. Terzi, G.; Gücükoğlu, A.; Çadirci, Ö.; Uyanik, T.; Alişarli, M. Serotyping and antibiotic susceptibility of *Listeria monocytogenes* isolated from ready-to-eat foods in Samsun, Turkey. *Turkish J. Vet. Anim. Sci.* **2015**, *39*, 211–217. [CrossRef]

16. Dutka-Malen, S.; Evers, S.; Courvalin, P. Detection of glycopeptide resistance genotypes and identification to the species level of clinically relevant enterococci by PCR. *J. Clin. Microbiol.* **1995**, *33*, 24–27. [CrossRef] [PubMed]

17. Song, Y.; Liu, C.; Finegold, S.M. Real-time PCR quantitation of clostridia in feces of autistic children. *Appl. Environ. Microbiol.* **2004**, *70*, 6459–6465. [CrossRef] [PubMed]

18. Casadei, M.A.; Ingram, R.; Skinner, R.J.; Gaze, J.E. Heat resistance of *Paenibacillus polymyxa* in relation to pH and acidulants. *J Appl. Microbiol.* **2000**, *89*, 801–806. [CrossRef]

19. Morandi, S.; Cremonesi, P.; Silvetti, T.; Castiglioni, B.; Brasca, M. Development of a triplex real-time PCR assay for the simultaneous detection of *Clostridium beijerinckii*, *Clostridium sporogenes* and *Clostridium tyrobutyricum* in milk. *Anaerobe* **2015**, *34*, 44–49. [CrossRef]

20. McEvoy, J.M.; Doherty, A.M.; Sheridan, J.J.; Blair, I.S.; McDowell, D.A. The prevalence of Salmonella spp. in bovine faecal, rumen and carcass samples at a commercial abattoir. *J. Appl. Microbiol.* **2003**, *94*, 693–700. [CrossRef]

21. Madden, R.H.; Murray, K.A.; Gilmour, A. Carriage of four bacterial pathogens by beef cattle in Northern Ireland at time of slaughter. *Lett. Appl. Microbiol.* **2007**, *44*, 115–119. [CrossRef]

22. Fox, E.; O'Mahony, T.; Clancy, M.; Dempsey, R.; O'Brien, M.; Jordan, K. *Listeria monocytogenes* in the Irish dairy farm environment. *J. Food Prot.* **2009**, *72*, 1450–1456. [CrossRef]

23. McEvoy, J.M.; Doherty, A.M.; Sheridan, J.J.; Thomson-Carter, F.M.; Garvey, P.; McGuire, L.; Blair, I.S.; McDowell, D.A. The prevalence and spread of *Escherichia coli* O157:H7 at a commercial beef abattoir. *J. Appl. Microbiol.* **2003**, *95*, 256–266. [CrossRef] [PubMed]

24. Thomas, K.M.; McCann, M.S.; Collery, M.M.; Logan, A.; Whyte, P.; McDowell, D.A.; Duffy, G. Tracking verocytotoxigenic *Escherichia coli* O157, O26, O111, O103 and O145 in Irish cattle. *Int. J. Food Microbiol.* **2012**, *153*, 288–296. [CrossRef] [PubMed]

25. Froschle, B.; Messelhausser, U.; Holler, C.; Lebuhn, M. Fate of *Clostridium botulinum* and incidence of pathogenic clostridia in biogass processes. *J. Appl. Microbiol.* **2015**, *119*, 936–947. [CrossRef] [PubMed]

26. Colleran, E. Hygienic and sanitation requirements in biogas plants treating animal manures or mixtures of manures and other organic wastes. In *Anaerobic Digestion: Making Energy and Solving Modern Waste Problems*; Ørtenblad, H., Ed.; AD-NETT, Herning Municipal Authoritie: Herning, Denmark, 2000; pp. 77–86.

27. Hofmann, J.; Müller, L.; Weinrich, S.; Debeer, L.; Schumacher, B.; Velghe, F.; Liebetrau, J. Assessing the effects of substrate disintegration on methane yield. *Chem. Eng. Technol.* **2020**, *43*, 47–58. [CrossRef]

28. Riau, T.; De la Rubia, M.A.; Pe´rez, M. Temperature-phased anaerobic digestion (TPAD) to obtain class A biosolids: A semi-continuous study. *Bioresour. Technol.* **2010**, *101*, 2706–2712. [CrossRef] [PubMed]

29. Kunte, D.P.; Yeole, T.Y.; Ranade, D.R. Inactivation of Vibrio cholera during anaerobic digestion of human night soil. *Bioresour. Technol.* **2000**, *75*, 149–151. [CrossRef]

30. Termorshuizen, A.J.; Volker, D.; Blok, W.J.; ten Brummeler, E.; Hartog, B.J.; Janse, J.D.; Knol, W.; Wenneker, M. Survival of human and plant pathogens during mesophilic digestion of vegetable, fruit and garden waste. *Eur. J. Soil Biol.* **2003**, *39*, 156–171. [CrossRef]

31. Santha, H.; Sandino, J.; Shrimp, G.F.; Sung, S. Performance evaluation of a sequential-batch temperature-phased anaerobic digestion (TPAD) Scheme for producing class A biosolids. *Water Environ. Res.* **2006**, *78*, 221–226. [CrossRef]

32. Olsen, J.E.; Larsen, H.E. Bacterial decimation times in anaerobic digestions of animal slurries. *Biol. Wastes* **1987**, *21*, 153–168. [CrossRef]

33. Cote, C.; Masse, D.I.; Quessy, S. Reduction of indicator and pathogenic microorganisms by psychrophilic anaerobic digestion in swine slurries. *Bioresour. Technol.* **2006**, *97*, 686–691. [CrossRef]

34. Higgins, M.J.; Chen, Y.; Murthy, S.N.; Hendrickson, D.; Farrel, J.; Schafer, P. Reactivation and growth of non-culturable indicator bacteria in anaerobically digested biosolids after centrifuge dewatering. *Water Res.* **2007**, *41*, 665–673. [CrossRef] [PubMed]

Bioengineering **2020**, *7*, 116

35. Masse, D.; Gilbert, Y.; Topp, E. Pathogen removal in farm-scale psychrophillic anaerobic digesters processing swine manure. *Bioresour. Technol.* **2011**, *102*, 641–646. [CrossRef] [PubMed]
36. Nicholson, F.A.; Groves, S.J.; Chambers, B.J. Pathogen survival during livestock manure storage and following land application. *Bioresour. Technol.* **2005**, *96*, 135–143. [CrossRef] [PubMed]
37. McGee, P.; Bolton, D.J.; Sheridan, J.J.; Earley, B.; Leonard, N. The survival of *Escherichia coli* O157:H7 in slurry from cattle fed different diets. *Lett. Appl. Microbiol.* **2001**, *32*, 152–155. [CrossRef]
38. Kearney, T.E.; Larkin, M.J.; Frost, J.P.; Levett, P.N. Survival of pathogenic bacteria during mesophilic anaerobic digestion of animal waste. *J Appl. Microbiol.* **1993**, *75*, 215–219. [CrossRef]
39. Horan, N.J.; Fletcher, L.; Betmal, S.M.; Wilks, S.A.; Keevil, C.W. Die-off of enteric pathogens during mesophilic anaerobic digestion. *Water Res.* **2004**, *38*, 1113–1120. [CrossRef]
40. Chauret, C.; Springthorpe, S.; Sattar, S. Fate of Cryptosporidium oocysts, Giardia cysts and microbial indicators during wastewater treatment and anaerobic sludge digestion. *Can. J. Microbiol.* **1999**, *45*, 257–632. [CrossRef]
41. Gahan, C.; Hill, C. *Listeria monocytogenes*: Survival and adaptation in the gastrointestinal tract. *Cell. Infect. Microbiol.* **2014**, *4*, 1–7. [CrossRef]
42. Pepper, I.L.; Brooks, J.P.; Sinclair, R.G.; Gurian, P.L.; Gerba, C.P. Pathogens and indicators in United States Class B biosolids: National and historic distributions. *J. Environ. Qual.* **2010**, *39*, 2185–2190. [CrossRef]
43. Viau, E.; Peccia, J. Survey of wastewater indicatorsand human pathogen genomes in biosolids produced byclass A and class B stabilization treatments. *Appl. Environ. Microbiol.* **2009**, *75*, 164–174. [CrossRef]
44. Smith, S.R.; Lang, N.L.; Cheung, K.H.M.; Spanoudaki, K. Factors controlling pathogen destruction during anaerobic digestion of biowastes. *Waste Manag.* **2005**, *25*, 417–425. [CrossRef] [PubMed]

Article

Performances of Conventional and Hybrid Fixed Bed Anaerobic Reactors for the Treatment of Aquaculture Sludge

Alessandro Chiumenti *, Giulio Fait, Sonia Limina and Francesco da Borso

Department of Agricultural, Food, Animal and Environmental Sciences (DI4A), University of Udine, via Delle Scienze 206, 33100 Udine, Italy; giulio.fait@uniud.it (G.F.); sonia.limina@uniud.it (S.L.); francesco.daborso@uniud.it (F.d.B.)
* Correspondence: alessandro.chiumenti@uniud.it

Received: 30 May 2020; Accepted: 25 June 2020; Published: 27 June 2020

Abstract: Aquaculture fish production is experiencing an increasing trend worldwide and determines environmental concerns mainly related to the emission of pollutants. The present work is focused on the improvement of the sustainability of this sector by assessing the anaerobic digestion (AD) of slurry. Wastewater from experimental plants for the production of trout (Udine, Italy) was subject to screening by a drum filter, and then to thickening in a settling tank. The thickened sludge, representing the input of AD, was characterized by total and volatile solids contents of 3969.1–9705.3 and 2916.4–7154.9 mg/L, respectively. The AD was performed in a containerized unit with two digesters (D1 and D2), biogas meters and monitoring of the temperature, pH and redox potential. Both reactors are mixed by a recirculation of the digestate, and reactor D2 is equipped with a fixed bed. The tests were performed at 38 °C with diversified loading rates and hydraulic retention times (HRT). HRT varied from 28.9 to 20.3 days for D1 and from 18.3 to 9.3 days for D2. Methane yields resulted as highest for the hybrid digester with the longest HRT (779.8 NL of CH_4/kg VS, 18.3 days). The conventional digester presented its best performance, 648.8 NL of CH_4/kgVS, with an HRT of 20.3 days.

Keywords: biogas; anaerobic digestion; methane; aquaculture; trout; sludge; wastewater; drum sieve; microfiltration; settling

---

1. Introduction

The aquaculture fish production sector is experiencing an increasing trend worldwide, surpassing fishing in 2016 [1]. The diffusion of sea fish farms, mainly characterized by floating cages, and of onshore fresh water intensive facilities, such as tanks, raceways and recirculating aquaculture systems (RAS), positively contributes to fulfill the demand of fish proteins, but conversely determines environmental concerns mainly related to the emission of pollutants. Anaerobic digestion (AD) represents a reliable technology widely implemented in the fields of management of organic wastes, livestock manure and agricultural by-products [2], but it has not been implemented for the treatment of effluents originated from aquaculture. The main advantage presented by the AD process is the production of biogas, a combustible gas formed mainly by CH_4 and CO_2, fuel commonly used in combined heat and power (CHP) units for the production of electric and thermal energy, or in some cases, upgraded to biomethane [3]. Furthermore, the AD process provides an important contribution to the protection of the environment by preventing the emission of greenhouse gasses derived from the uncontrolled fermentation of organic matter of different origin, including effluents from farming [2]. The fermentation process determines a sufficient degradation of the organic matter of input feedstocks along with their sanitization, producing an effluent, the digestate, that could be reused as fertilizer as a result of its content of nutrients, with very limited emissions of odors and avoiding spreading fish pathogens [4]. However, the treatment of fish effluents by AD presents some challenges mainly in relation to their excessive dilution, requiring pretreatments for the concentration of the organic matter, to excessive concentrations of free ammonia [5] and relevant concentrations of compounds such as sodium and sulphur, which are present mainly in the case of sea water and can be toxic for methanogenic populations [4].

Kugelman and van Gorder [6] and Mirzoyan et al. [7] treated freshwater aquaculture sludge with a TS content of 4 to 6% in a batch-type digester at 35 °C with hydraulic retention times (HRT) from 30 to 10 days, however, the free ammonia levels found were inhibitory to AD.

Lanari and Franci [5], instead, successfully treated diluted sludge from a trout RAS (1.3–2.4% of total solids (TS), 0.16–0.24 g Tot-N/L) at ambient temperature (25 °C), in a 0.4 m^3 anaerobic filter filled with polyurethane foam cubes, obtaining biogas production from 49.8 to 144.2 L/day, with a CH_4 content > 80% and methane yield from 0.40 to 0.46 L/g of volatile solids (VS).

Other scientific papers concerning marine or brackish water effluents or more frequently waste originated from fish slaughtering or industry were recently published, but little is still known about conventional or innovative anaerobic treatment systems for intensive fish farm effluents.

The aim of the present work is to present the results of AD experimental tests conducted on effluents from trout tanks, after filtration and thickening, in a pilot-scale, conventional completely mixed digester and hybrid up-flow, fixed bed digester. In more detail, specific objectives were to study the effects of thickening the influent sludge by sedimentation, the effects sorted by the reduction in HRT and the effects of the increase in the organic load rates (OLRs), both in conventional and hybrid fixed bed reactors.

2. Materials and Methods

Tests were performed at the experimental fish production unit of the University of Udine, Italy, where rainbow trout (Oncorhynchus mykiss Walbaum) were bred in 4 circular tanks of 2.1 m^3, with a fish density changing from 48.9 (initial) to 88.1 kg l.w./m^3 (final). Wastewater was spilled from the central bottom discharge of each tank and was subject to initial screening by a drum micro-sieve with a 20 μm mesh, and then to thickening in a settling tank.

The anaerobic digestion tests were performed in a containerized experimental unit equipped with a loading tank, with hydraulic mixing, and two anaerobic digesters (Figure 1): the first (D1) is completely mixed by means of programmed recirculation of the digestate, with a useful volume of 280 L; the second reactor (D2) is equipped with a plastic filling media characterized by a relevant specific surface (140 m^2/m^3) to promote the development of a fixed layer of anaerobic bacteria, with a useful volume of 260 L.

The digesters are made of steel, equipped with a loading input located on the top and side overflow discharge; different sampling points are located at different heights, while a valve is installed at the bottom of each digester to allow the removal of eventual sediments. The heating system is represented by a coil of electric resistance externally wrapping the wall of the tanks. The resistances are controlled by Pt100 temperature probes and digital thermostats. The digesters are insulated by a layer of rockwool (40 mm) externally protected by an aluminum sheet.

Biogas exits the dome of the digesters and flows into a hydraulic valve, that sets the pressure threshold (80 mm H_2O) and determines the condensation of vapor, and reaches the flowmeters, drum-type (wet-test) Gas Meters (TG1, Ritter DE). Biogas composition is determined by infrared analyzer (Ultramat 23, Siemens, Germany) for (CO_2 e CH_4), and by gas suction pump (ProTec, Germany) with H_2S vials (Kitagawa, Japan).

Process temperature was detected by means of resistive type sensors (Pt100), while pH and Redox potential by electrochemical probes (Steiel, Italy). All the electronic sensors were connected to a PLC (Eukrasia, Italy), which recorded data and also managed all equipment, as loading and mixing pumps.

Input and digestate were weekly sampled to perform the following analyses: Total Solids, Volatile Solids (VS), Chemical Oxygen Demand (COD), Biological Oxygen Deman (BOD$_5$), Total Kjeldhal Nitrogen (TKN), P, total alkalinity [8].

The tests were performed in mesophilic conditions (38 °C) with different loading rates intended to test different hydraulic retention times (HRT).

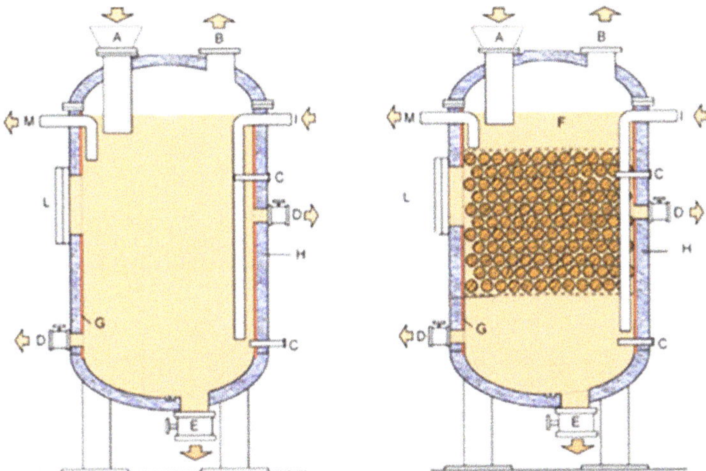

Figure 1. Schematic of the digesters D1 (left) and D2 (right): solid materials input (blocked in the present experimentation) (A), biogas output (B), temperature probe housings (C), sampling points (D), bottom discharge (blocked) (E), fixed bed (F), heating coil (G), insulation (H), sludge input and recirculation ports (I), maintenance openings (L) and digestate output (M).

2.1. Loading Rates

The tests lasted for a total of 71 days, divided in three periods with different and increasing loading rates (Table 1).

Table 1. Description of the different loading rates and hydraulic retention times (HRT) for D1 and D2.

	D1		D2	
Days	Load (L/day)	HRT (days)	Load (L/day)	HRT (days)
0–35	9.7	28.9	14.2	18.3
36–52	9.7	28.9	18.4	14.1
53–71	13.8	20.3	27.3	9.5

In particular, digester D1 was initially fed with 9.7 L/day of input until day n.52 and from that point on the daily load was increased to 13.8 L/day. Digester D2 was characterized by a higher load than D1 since the beginning, in detail 14.2 L/day up to day n.35, 18.4 L/day from day n.36 to day n.52and 27.3 L/day until the end of the tests. Consequently HRT varied from 28.9 to 20.3 for D1, and was reduced from 18.3 to 14.1 and 9.5 for D2.

2.2. Settling of the Input

The experimentation could be divided in two phases according to the variation in the characteristics of the input due to the introduction of a settling phase performed after the mechanical filtration from day 36.

The characteristics of the two different inputs will be described in the results.

3. Results and Discussion

3.1. Chemical Characteristics

Before day 35 (phase F), the effluent was not subjected to settling and hence resulted more diluted (Table 2). Instead, from day 36 to the end of the tests, the effluent presented higher concentrations of all parameters (phase FS, filtration and sedimentation).

Table 2. Characteristics of the input in the different phases of the tests (average and standard deviation).

	TS (mg/L)	VS (mg/L)	VS/TS (%)	COD (mg/L)	BOD_5 (mg/L)	TKN (mg/L)	TP (mg/L)	Alkalinity (mg/L)
Phase F Day 1–35	3969.1 ± 209.5	2916.4 ± 341.7	73.4 ± 7.2	6433.8 ± 587.7	3020.0 ± 408.7	205.2 ± 63.4	150.3 ± 21.8	1110.2 ± 71.9
Phase FS Day 36–71	9705.3 ± 1094.0	7154.9 ± 1060.8	74.1 ± 3.3	14511.0 ± 1484.1	6210.0 ± 690	521.3 ± 148.1	333.9 ± 23.9	1187.6 ± 10.8

Average TS content resulted in 3969.1 mg/L in phase 1 and 9705.3 mg/L in phase 2. The VS/TS ratio remained relatively constant, in the range between 73.4 and 74.1. COD (chemical oxygen demand) and BOD_5 (biological oxygen demand) respectively resulted in 6433.8 and 3020.0 mg/L in phase 1, and 3020.0 and 6210.0 in phase 2. Alkalinity resulted as quite stable in the two phases, with average values of 1110.2 and 1187.6 mg/L.

On the basis of these results (Table 2) and considering the volumetric loads (Table 1), it was possible to determine the organic load rates (OLR) that characterized the operation of the two digesters during the tests (Table 3). The OLR of reactor D1 was increased from an initial 0.100 to 0.352 kg VS/m^3 day, while for digester D2, this parameter was increased from 0.158 to 0.750 kg VS/m^3 day. These results are aligned with the experimentation of Lanari and Franci [5], where an OLR ranging between 0.227 and 0.751 kg VS/m^3 day was maintained in the AD of wastewater from trout fed at the regime of 1–2% of live weight/day. Alternately, these values can be considered as conservative in comparison with those indicated for swine or cattle manure, which normally are between 1.5 and 6.0 kg/VS day [2], with values below the range resulting in a possible lack of organic matter and low methanogenic bacteria, and values above the risk of acidification as an effect of the accumulation of volatile fatty acids (VFA) [2,3]. Previous tests conducted with the same digesters operating on swine manure presented values of 1.12 and 3.34 kg VS/m^3 day [2]. This is not to be considered negatively considering that optimal parameters for AD may vary from feedstock to feedstock.

In terms of COD, the organic load resulted between 0.222 and 0.715 kg COD/m^3 day for D1, and 0.350 and 1.522 kgCOD/m^3 day for D2. These values are lower than those maintained by Gebauer [4], who operated with an OLR up to 3.12 kgCOD/m^3 day on a lab scale with 15 L reactors treating Atlantic salmon sludge from a sieve and air-flushed ribbon strainer.

Table 3. Organic loading rates (OLR) in terms of VS and COD maintained in digester D1 and D2 during the experimental trial.

	OLR_{VS} (kg VS/m^3 day)		OLR_{COD} (kg COD/m^3 day)	
Days	D1	D2	D1	D2
0–35	0.100	0.158	0.222	0.350
36–52	0.248	0.506	0.502	1.026
53–71	0.352	0.750	0.715	1.522

3.2. Characteristics of Digestate and Organic Matter Removal Rate

The concentration of VS in the digestate resulted between 1495.6 and 4691.7 mg/L for D1, and between 1443.1 and 4044.8 mg/L for D2. The VS removal resulted as significant for the anaerobic filter reactor D2 from the beginning of the tests (52.4%) and remained steady during the tests, with an average value of 47.5%; the VS removal rate of D1 presented instead an increasing trend until day 36, reaching a plateau after with an average value of 34.7%, significantly lower than D2 (Figure 2).

The higher performance of anaerobic filters compared with traditional high-rate digesters is well known [2,9], and particularly during the first phase of the process, it can also be partially related to a mechanical retention of solids trapped in the packed bed filter, not only to increased solids degradation.

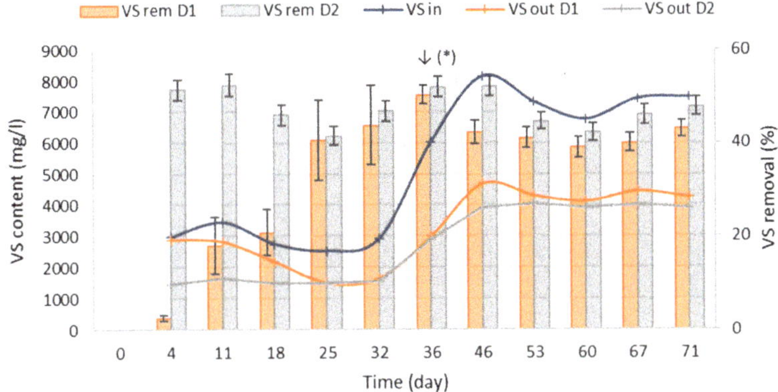

Figure 2. Evolution of VS concentration in the input and in digestates from D1 and D2, and of VS removal rates. ↓(*) = implementation of influent sedimentation.

3.3. Biogas Production

The evolution of biogas production presented an increasing trend, mainly in relation to the increase in the OLR of the different phases of the tests (Figure 3).

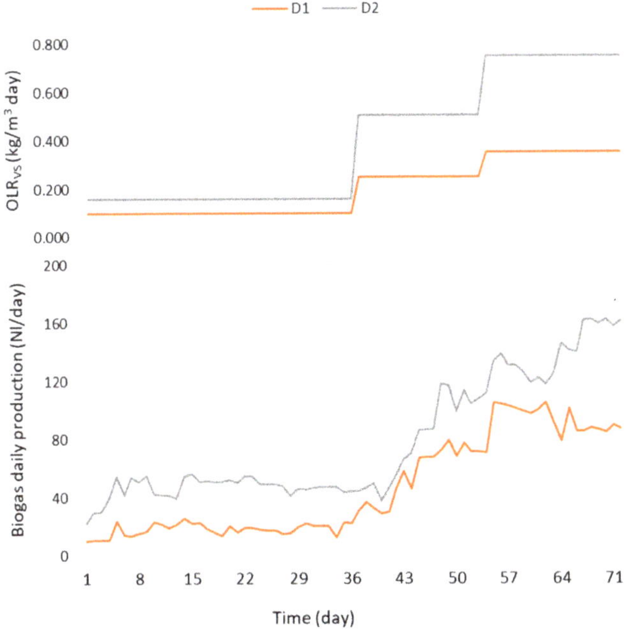

Figure 3. Evolution of OLR$_{VS}$ (above) and of daily biogas production (below).

Digester D2 presented, in general, a higher production than D1. In the first phase of the experimentation, with a higher dilution of the input and HRT (up to day 35), D1 showed a constant production with an average of 30.2 ± 4.7 NL/day, while D2 produced an average of 46.4 ± 7.6 NL/day of biogas. In the following phases, D1 presented at first an increasing trend, then a steady production till the end of the test: the average values resulted in 78.5 ± 28.5 and 139.4 ± 17.2 NL/day in the two periods until days 52 and 71, respectively.

The cumulative production of biogas resulted in 3366.8 and 5650.8 NL for digesters D1 and D2, corresponding to volumetric yields of 0.169 and 0.306 Nm^3/m^{3}day.

Other studies demonstrate comparable results, with values between 0.130 and 0.377 Nm^3/m^3 day, with similar effluents from trout, treated after filtration by the AD process in a pilot-scale system with a fixed bed [5]. Using the same pilot plant, previous tests conducted on filtered swine manure revealed 0.048 to 0.060 Nm^3/m^3 day for a traditional digester and 0.111 Nm^3/m^3 day for an anaerobic filter [2].

3.4. Biogas Quality

The concentration of methane in biogas resulted in the ranges 63.3–70.5% and 65.0–70.8% for D1 and D2, respectively (Figure 4), results comparable to those from the AD of swine manure [2] and higher than methane concentrations in biogas from bovine manure [3,10]. The literature on the AD of aquaculture effluents reports variable results in terms of methane concentration in biogas: Lanari and Franci [5], for example, reported 80% of CH$_4$ from the AD of trout effluents, while other authors presented values between 48.9 and 57.6% CH$_4$ in biogas from Atlantic salmon effluents [4].

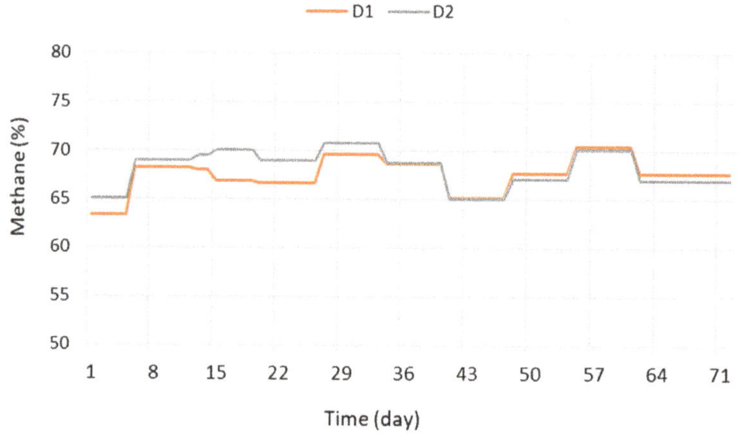

Figure 4. Evolution of methane concentration in biogas.

3.5. Methane Yield

Methane yield referred to VS resulted very high, between 396.8 (D2, days 36–52) and 779.8 NL/kg VS (D2, days 0–35) (Figure 5). Such values are not common in the treatment of effluents from animal farms or in the agricultural sector.

Comparable methane yields are generally obtained in the case of feedstocks with high energetic content, such as some residues from slaughter houses containing high concentrations of proteins and fat. Pitk and colleagues, for example, reported yields from 390 to 978 m^3 CH$_4$/t in the case of solid slaughterhouse waste rendering products [11].

The effluents used in the tests were not characterized in terms of the composition of the organic matter, but the trout feed was characterized by a high lipid content. Furthermore, the highest reported yield from the AD of aquaculture effluents was 460 L/kgVS [5], comparable to the lower values obtained in the current study.

In further detail, D1 demonstrated an increase in yield corresponding to the increase in the OLR, reaching a maximum of 648.8 NL/kgVS with an OLR of 0.352 kgVS/m^3 day and HRT of 20.3 days. Digester D2 showed a maximum yield of 779.8 NL/kgVS with an OLR of 0.158 kgVS/m^3 day and HRT of 18.3 days, lower than the HRT of D1. Furthermore, the D2 yield resulted as gradually descending with the increase in the organic load and decrease in HRT, resulting as particularly critical in the period between days 36 and 52, with an OLR of 1.026 kgVS/m^3 day and HRT of 14.1 days, when a yield of 396.8 NL/kgVS was achieved. Moreover, it must be noted that during the second and third phases, the methane production rates of the two digesters were not so different, as they were during the first phase of experimentation (Figure 5).

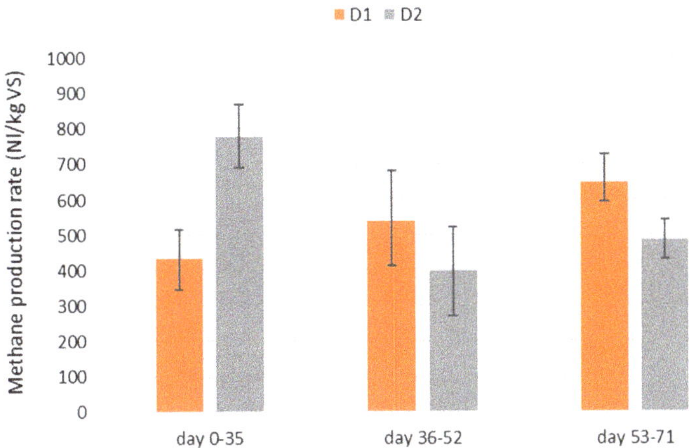

Figure 5. Methane production rates from the 2 digesters during the 3 trial phases.

4. Conclusions

Intensive trout production in tanks determines the production of large volumes of diluted effluents characterized by a low concentration of organic matter.

In the present study, microfiltration and settling were performed to concentrate this wastewater, obtaining an average TS content from 3969.1 (filtration) to 9705.3 mg/L (filtration and settling), with an average concentration of VS ranging between 73.4 and 74.1%TS.

The treated effluents were subjected to the AD process in mesophilic conditions, in pilot-scale conventional and fixed bed hybrid digesters. The evolution of the process was regular and without management problems.

The key results of this experimentation can be summarized as follows:

- The organic matter removal rate resulted as higher in the case of the hybrid digester and with the longest process time (average 52.4% VS removal with an HRT of 18.3 days);
- The CH$_4$ concentration in the biogas resulted as relevant and stable during the tests, presenting an average ranging from 63.3% to 70.8%;
- Methane yields resulted as very high, especially for the hybrid digester with the longest process time (779.8 NL of CH$_4$/kg VS with an HRT of 18.3 days). The conventional digester presented its best performance, 648.8 NL of CH$_4$/kgVS, with an HRT of 20.3 days;
- Operating with the hybrid digester and the lowest process time, corresponding to 9.3 days, it is still possible to achieve an optimal methane yield (486.7 NL CH$_4$/kg VS). Applied to a full-scale digester, this solution could determine a reduction in the digestion volume and consequently lower capital and operational costs (i.e., thermal energy for temperature control) while guaranteeing an optimal performance of the process.

Author Contributions: The contributions of Authors to conceptualization, methodology, validation, investigation, writing, review and editing of the present article must be considered equal. All authors have read and agreed to the published version of the manuscript.

Funding: This research was funded by REGIONE FRIULI VENEZIA GIULIA, Research Project "Nuove tecniche di abbattimento della sostanza organica e di sfruttamento ai fini energetici dei reflui in acquacoltura", art. 11 LR 11/2003.

Acknowledgments: The authors would like to thank Penny Lazo for the professional and kind contribution to the English language editing.

Conflicts of Interest: The authors declare no conflict of interest.

References

1. FAO Food and Agriculture Organization of the United Nations. The state of world fisheries and aquaculture. In *Meeting the Sustainable Development Goals*; FAO: Rome, Italy, 2018.
2. Chiumenti, R.; Chiumenti, A.; da Borso, F.; Limina, S.; Landa, A. Anaerobic Digestion of Swine Manure in Conventional and Hybrid Pilot Scale Plants: Performance and Gaseous Emissions Reduction. In Proceedings of the ASABE Annual Meeting, Reno, NV, USA, 21–24 June 2009; pp. 21–24.
3. Chiumenti, A.; da Borso, F.; Guercini, S.; Pezzuolo, A.; Zanotto, M.; Sgorlon, S.; Delle Vedove, G.; Miceli, F.; Stefanon, B. The Impact of the Dairy Cow Diet on Anaerobic Digestion of Manure. In Proceedings of the ASABE Annual International Meeting, Boston, MA, USA, 7–10 July 2019. [CrossRef]
4. Gebauer, R. Mesophilic anaerobic treatment of sludge from saline fish farm effluents with biogas production. *Bioresour. Technol.* **2004**, *93*, 155–167. [CrossRef] [PubMed]
5. Lanari, D.; Franci, C. Biogas production from solid wastes removed from fish farm effluents. *Aquat. Living Resour.* **1998**, *11*, 289–295. [CrossRef]
6. Kugelman, I.J.; van Gorder, S. Water and Energy Recycling in Closed Aquaculture Systems. In *Engineering Aspects of Intensive Aquaculture*; Northeast Regional Agricultural Engineering Service (NRAES)-49: Ithaca, NY, USA, 1991; pp. 80–87.
7. Mirzoyan, N.; Tal, Y.; Gross, A. Anaerobic digestion of sludge from intensive recirculating aquaculture systems: Review. *Aquaculture* **2010**, *306*, 1–6. [CrossRef]
8. American Public Health Association (APHA). *Standard Methods for the Examination of Water and Wastewater*, 21th ed.; APHA: Washington, DC, USA, 2005; pp. 5–41.
9. Ganesh, R.; Rajinikanth, R.; Thanikal, J.V.; Ramanujam, R.A.; Torrijos, M. Anaerobic treatment of winery wastewater in fixed bed reactors. *Bioprocess Biosyst. Eng.* **2010**, *33*, 619–628. [CrossRef] [PubMed]
10. Chiumenti, A.; Chiumenti, R.; da Borso, F.; Limina, S. Comparison between Dry and Wet Fermentation of Biomasses as Result of the Monitoring of Full Scale Plants. In Proceedings of the ASABE International Meeting, Dallas, TX, USA, 29 July–1 August 2012; p. 121340548. [CrossRef]
11. Pitk, P.; Kaparaju, P.; Vilu, R. Methane potential of sterilized solid slaughterhouse wastes. *Bioresour. Technol.* **2012**, *116*, 42–46. [CrossRef] [PubMed]

MDPI
St. Alban-Anlage 66
4052 Basel
Switzerland
Tel. +41 61 683 77 34
Fax +41 61 302 89 18
www.mdpi.com

Bioengineering Editorial Office
E-mail: bioengineering@mdpi.com
www.mdpi.com/journal/bioengineering